P9-CRE-833

TRANSISTOR CIRCUITS IN ELECTRONICS

TRANSISTOR CIRCUITS IN ELECTRONICS

Basic Principles for Amplifier, Oscillator and Switching Applications, including an introduction to Integrated Circuits

S. S. HAYKIN, B.Sc., Ph.D., D.Sc., M.I.E.E.
Professor of Electrical Engineering and Chairman of the Department, McMaster University, Hamilton, Canada

R. BARRETT, B.Sc., C.Eng., F.I.E.E.
Principal Lecturer in Electrical Engineering The Hatfield Polytechnic

LONDON
ILIFFE BOOKS

THE BUTTERWORTH GROUP
Butterworth & Co (Publishers) Ltd
London: 88 Kingsway, WC2B 6AB

AUSTRALIA
Butterworth & Co (Australia) Ltd
Sydney: 20 Loftus Street
Melbourne: 343 Little Collins Street
Brisbane: 240 Queen Street

CANADA
Butterworth & Co (Canada) Ltd
Toronto: 14 Curity Avenue, 374

NEW ZEALAND
Butterworth & Co (New Zealand) Ltd
Wellington: 49/51 Ballance Street
Auckland: 35 High Street

SOUTH AFRICA
Butterworth & Co (South Africa) (Pty) Ltd
Durban: 33/35 Beach Grove

First published in 1964 by
Iliffe Books, an imprint of the Butterworth Group
Second Edition . . . 1971

ISBN 0 592 00059 1 Standard
ISBN 0 592 00061 3 Limp

Printed and bound in England by
Chapel River Press
Andover, Hampshire

PREFACE TO FIRST EDITION

This book is intended as an introduction to the transistor and its various circuit applications; an attempt has been made to cover both the linear and non-linear aspects of transistor circuits to the same extent. Throughout the book worked examples have been introduced to illustrate the principles being discussed.

Chapter 1 deals with a brief introduction to transistor physics, and Chapter 2 is devoted to biasing and graphical analysis of transistor amplifier behaviour. Then the h-parameter and T-equivalent circuits for describing the small signal performance of a transistor are developed in Chapter 3 and they are used for evaluating the external circuit properties of the three basic configurations. This discussion ends with consideration of high-frequency effects. Next, d.c. amplifiers, wideband, tuned and power amplifiers are considered in Chapter 4. Treatment of linear aspects of transistors finishes with a chapter on the application of feedback to transistor circuits.

The remaining four chapters are devoted to switching applications of transistors. In Chapter 6 the transistor is introduced as a switch, and a number of specific non-regenerative circuits are considered. The multivibrator family of circuits and blocking oscillators are dealt with in Chapter 7. Boolean algebra is briefly introduced in Chapter 8 and is made use of in the development of logic circuits which are widely used in digital computers. The final chapter is concerned with the circuits required for the processes of modulation and demodulation; in this respect both frequency division and time division multiplexing are considered.

It is considered that the book will be useful to the student and the electronic engineer seeking a first introduction to the fundamentals of transistor circuits. No attempt has been made to make it a design book.

S. S. H.
R. B.

PREFACE TO SECOND EDITION

Since the publication of the first edition of this book in 1964 dramatic changes have occurred in the fabrication processes for semiconductor devices. Continual efforts have been made to miniaturise electronic equipment, urged on mainly by the space exploration programmes, resulting in the development of microelectronics. Monolithic integrated circuits, in which all circuit elements, both active and passive, are simultaneously formed in a single wafer of silicon by the diffused planar technique, have become of major importance and an introductory chapter covering this field has therefore been added. Apart from this addition the book is basically unchanged, since the fundamental principles on which it is based are as relevant as ever. Indeed with the accent now on silicon npn transistors, resulting in lower leakage current, the simplified circuits presented in the book are more practical The fact that the circuits are based on pnp transistors is unimportant, involving only a change in supply polarity. Complementary circuits, including both pnp and npn transistors in one circuit, are discussed in the chapter on integrated circuits.

At the time of publication of the first edition there was no comprehensive British Standard covering logic symbols, but an opportunity has been taken in this edition to bring the symbols used in Chapter 8 into line with BS3939, 1969, Section 21, in which logic elements are represented by semicircles. Since, in this chapter, the less positive potential is consistently selected as the 1-state, we are using a negative logic system and the logic convention indicator on the symbols is unnecessary.

Grateful acknowledgement is made to S.G.S. (United Kingdom) Ltd., Texas Instruments Inc., McGraw-Hill Book Company Inc., and the Orbit Publishing Company for permission to use certain material included in Chapter 10.

S. S. H.
R. B.

CONTENTS

CHAPTER I

TRANSISTOR CHARACTERISTICS

The first transistor was introduced in 1948 and was of the so-called point contact type invented by Bardeen and Brattain[1]. Then in 1949 Shockley[2] postulated the *junction transistor* on theoretical grounds. The first junction transistor was produced in 1951. In practice it is found difficult to manufacture point contact transistors with uniform characteristics. For this reason the transistors used nowadays are of the junction type, and the point contact transistor has become of theoretical interest only.

For its operation the transistor depends on the principle of *semiconduction* which was introduced by Wilson[3] as early as 1931. Semiconduction is exhibited by such materials as germanium and silicon. These materials, known as *semiconductors*, have a resistivity much greater than that of a conductor and yet much less than that of an insulator. Thus the resistivity of a conductor is of the order of 10^{-6} ohm-cm and that of an insulator is of the order of 10^{6} ohm-cm. The resistivity of many semiconductors is very sensitive to temperature variations and to minute additions of impurities. The resistivities of silicon and germanium decrease both with increasing temperature and with the addition of small amounts of such elements as antimony and indium.

In the next section we shall consider first the pure germanium and silicon atoms and then the consequences of adding small amounts of certain impurities to these two semiconductors. This discussion is followed by a consideration of the physical action in pn junctions and transistors.

[1] BARDEEN, J. and BRATTAIN, W. H.: 'The transistor, a semiconductor triode', *Phys. Rev.*, **74**, 230 (1948).

[2] SHOCKLEY, W., 'The theory of pn junctions in semiconductors and pn junction transistors', *Bell System Tech. J.*, **28**, 435 (1949).

[3] WILSON, A. H., 'Theory of electronic semiconductors', *Proc. Roy. Soc.*, **133**, 458 (1931).

1.1. SEMICONDUCTORS

When a semiconductor is free of all impurities it is said to be intrinsic. Pure germanium and pure silicon are examples of *intrinsic semiconductors.* On the other hand, when a semiconductor contains added impurities it is said to be *extrinsic.* The p-type and n-type semiconductors are examples of extrinsic semiconductors, and they will be considered later.

Silicon and germanium have atomic numbers of 14 and 32, respectively. Therefore, the silicon atom has 14 electrons and the germanium atom has 32 electrons. These electrons are located in shells around the nucleus and the shells have the following number of electrons.

SILICON

Shell No.	*Number of electrons*
1	2
2	8
3	4

GERMANIUM

Shell No.	*Number of electrons*
1	2
2	8
3	18
4	4

The innermost shells of 2 and 8 electrons of the silicon atom and 2, 8 and 18 electrons of the germanium atom are completely filled. They are, therefore, unaffected by external disturbances. The germanium and silicon atoms are referred to as *tetravalent atoms* because each contains 4 *valence* electrons in its outermost shell. A valence electron is one which enables an element to combine with another to form a compound.

In a germanium crystal, for example, each atom forms a *covalent bond* with each of 4 neighbouring germanium atoms in the manner illustrated in Fig. 1.1. Each covalent bond consists of 2 valence electrons shared between 2 neighbouring germanium atoms. These covalent bonds provide the forces that hold the respective atoms in their orderly positions inside the crystal.

At all temperatures above absolute zero, the atoms of the germanium crystal vibrate about their normal positions. The vibrations become more vigorous with increasing temperature. Occasionally an electron acquires sufficient energy to enable it to become free of its covalent bond, thereby leaving an electron vacancy, that is a *hole* behind it. The hole behaves like a positive charge carrier having an absolute charge value equal to that of an electron. However, the hole moves about inside the crystal more slowly than the electron.

At any temperature, therefore, there will be an equal number of holes and electrons inside an intrinsic semiconductor. This equality is necessary in order to ensure that the net electric charge neutrality of the intrinsic semiconductor is satisfied. The number of the

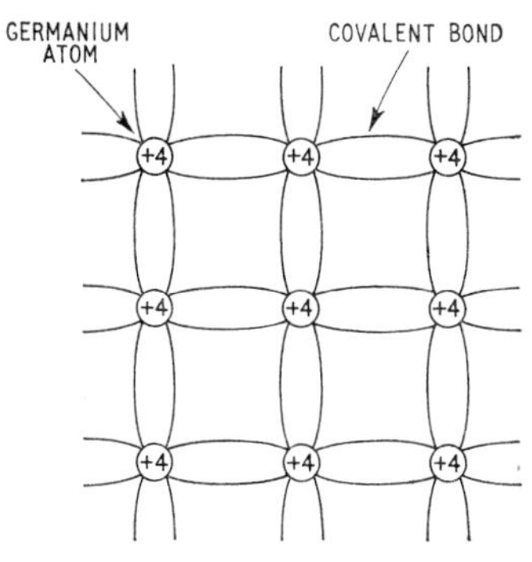

Fig. 1.1. Two-dimensional representation of germanium crystal

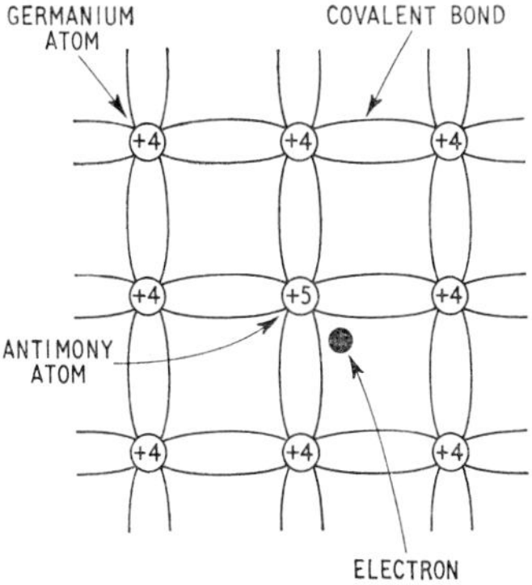

Fig. 1.2. Two-dimensional representation of n-type semiconductor

thermally generated holes and electrons of an intrinsic semiconductor increases with increasing temperature.

The properties of an intrinsic semiconductor can be modified by the addition of certain impurities. As an example, consider the addition of an antimony atom to replace one of the germanium atoms of the crystal. The antimony atom is *pentavalent* in that it has 5 valence electrons in its outermost shell. 4 of these 5 electrons of the antimony atom will form a covalent bond with each of 4 adjacent germanium atoms as illustrated in Fig. 1.2. The remaining electron is now so loosely held to the parent antimony atom that a relatively small amount of thermal energy is sufficient to release it. This leaves the antimony atom positively charged. Thus pentavalent atoms, such as antimony, will contribute free electrons. Such impurity atoms are, therefore, referred to as *donors*, and the extrinsic semiconductor containing these donors is said to be of the n-*type*.

In order to preserve the net charge neutrality of the n-type semiconductor it is essential that the density n of the electrons is equal to the sum of the density p of the holes and the density N_d of the positively charged donor atoms, that is,

$$n = p + N_d \tag{1.1}$$

Consequently, n is greater than p and so an n-type semiconductor contains a majority of electrons and a minority of holes. Further, if the density of the added donor atoms is large, then from

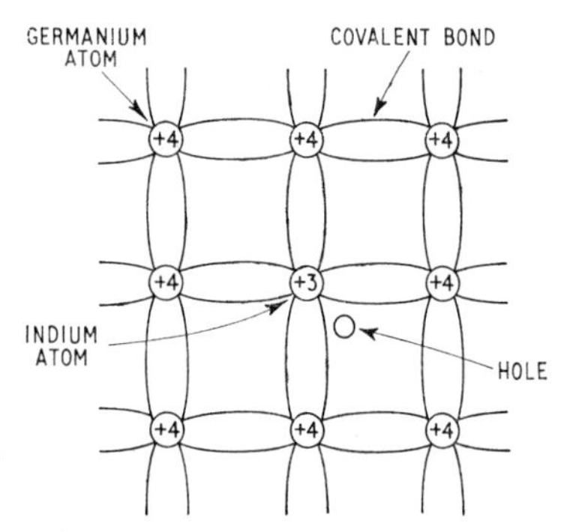

Fig. 1.3. Two-dimensional representation of p-type semiconductor

Equation 1.1 it follows that the number of electrons in an n-type semiconductor is nearly equal to the number of added donor atoms.

Next, consider the consequences of adding an indium atom to the germanium crystal. The indium atom is *trivalent* in that it has 3 valence electrons in its outer shell. These 3 electrons of the added indium atom will form a covalent bond with each of 3 neighbouring germanium atoms as illustrated in Fig. 1.3. In this case 1 electron vacancy, that is a hole, is created for every indium atom. The missing electron can be supplied by a neighbouring germanium atom. Then the added indium atom becomes a negatively charged immobile ion. Trivalent atoms such as indium, are referred to as *acceptors*, and the resulting extrinsic semiconductor is said to be of the p-*type*.

Again, in order to preserve the net charge neutrality of the p-type semiconductor it is necessary that the density p of the holes is equal to the sum of the density n of the electrons and the density N_a of the negatively charged acceptor atoms, that is,

$$p = n + N_a \tag{1.2}$$

Therefore, p is greater than n and so the p-type semiconductor contains a majority of holes and a minority of electrons. Further, if the number of the added acceptor atoms is large, then from Equation 1.2 we find that the number of holes is nearly equal to the number of added acceptor atoms.

1.2. THE JUNCTION DIODE

Consider the hypothetical case when a p-type and an n-type semiconductor are brought into contact with each other as in Fig. 1.4. The holes will tend to diffuse out of the p-type region, where their density is high, into the n-type region where their density is low. Similarly, the electrons tend to diffuse out of the n-type into the p-type region. Hence, in the immediate vicinity of the transition

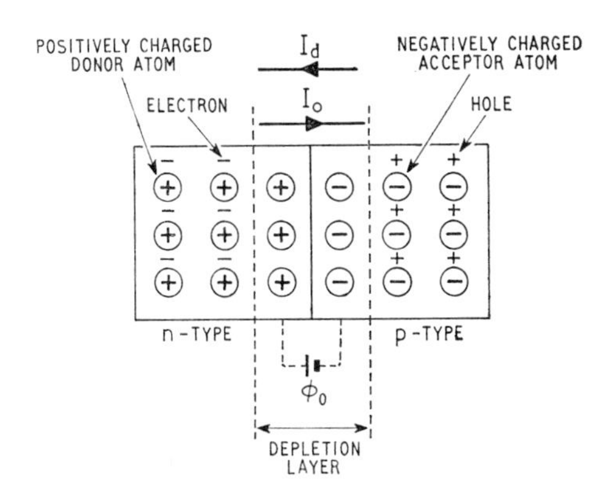

Fig. 1.4. pn junction

region the negatively charged immobile acceptor atoms of the p-type material and the positively charged immobile donor atoms of the n-type material are no longer electrically neutralised. Thereby an electrostatic potential difference ϕ_o is set up across the junction. This is usually referred to as the *barrier potential*, and its polarity is such as to confine the bulk of the holes and electrons to the p-type and n-type regions, respectively. The transition region containing the unneutralised and immobile ions is known as the *depletion layer* because it is depleted of mobile charge carriers, that is, holes and electrons.

The holes and electrons that succeed in diffusing into the n-type and p-type regions, respectively, contribute a *diffusion current* I_d flowing across the junction from the p-type into the n-type material.

However, in the absence of an externally applied voltage the total current across the junction must be zero. This can be achieved only if there is a current flow equal in magnitude and opposite in direction to I_d. Now, as pointed out earlier on page 2, the breaking up of covalent bonds due to thermal agitation results in a relatively small number of electrons and holes in the p-type and n-type regions respectively. The polarity of the barrier potential ϕ_o is such that it aids both the flow of the thermally generated electrons from the p-type into the n-type region, and the flow of the thermally generated

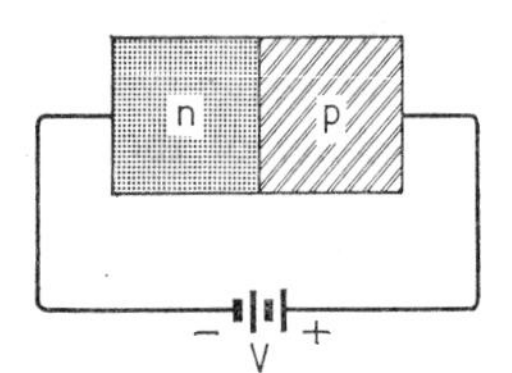

Fig. 1.5. Forward-biased pn junction

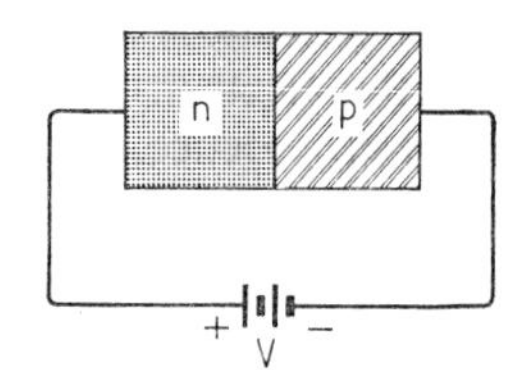

Fig. 1.6. Reverse-biased pn junction

holes in the opposite direction. This, therefore, produces the reverse current I_o that is required to balance the diffusion current I_d. The current I_o is termed the diode *saturation current* and it flows from the n-type into the p-type material. The magnitude of I_o is normally of the order of a few microamps in germanium pn junctions. It is virtually independent of the barrier potential ϕ_o, but its magnitude increases with increasing temperature.

If an external voltage V is applied across the pn junction, then the effective barrier potential becomes equal to $\phi_o - V$. The applied voltage V is positive when the p-type material is maintained at a positive potential with respect to the n-type material as shown in Fig. 1.5. This is the *forward-biased condition.* If, however, the externally applied voltage V is negative as in Fig. 1.6 then the pn junction is *reverse-biased.*

Consider first Fig. 1.5 where V is positive. In this case the effective barrier potential $\phi_o - V$ is lowered with the result that more holes are able to migrate from the p-type into the n-type material, and at the same time more electrons are able to migrate from the n-type into the p-type material. Therefore, applying a positive voltage V to the pn junction causes the diffusion current I_d to increase. In fact, I_d increases exponentially with V. However, as the saturation current is independent of the barrier potential, then it will remain equal to I_o. Consequently, when the pn junction is forward-biased the diffusion current will exceed the saturation current. Thus a net current will flow across the junction from the p-type into the n-type material, and its magnitude will increase with the applied voltage as illustrated in Fig. 1.7.

Next, consider Fig. 1.6 where V is negative. The effective barrier potential $\phi_o - V$ is now increased, thereby retarding the diffusion of the holes from the p-type into the n-type region, and electrons in the opposite direction. Thus I_d is reduced if the pn junction is reverse-biased. In fact, it is practically eliminated when the applied voltage V is only a few tenths of a volt negative. Hereafter, the

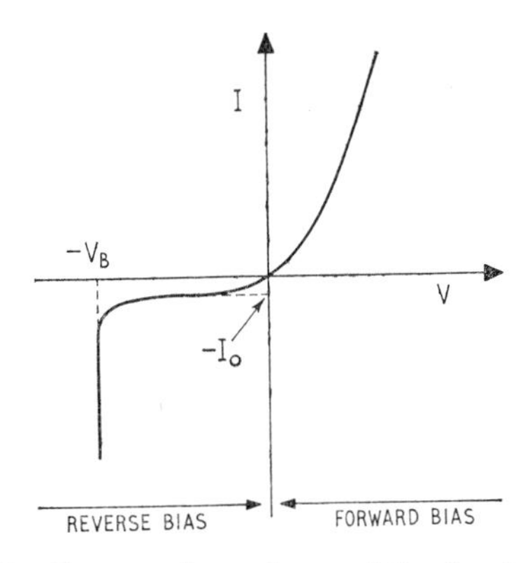

Fig. 1.7. Current-voltage characteristic of pn junction

current flowing across the junction reaches its limiting value of $-I_o$ as illustrated in Fig. 1.7.

In practice it is found that as the applied voltage is made more and more negative, then a region is reached where the reverse current begins to rise steadily with increasing reverse voltage. This process continues till the applied voltage reaches the critical value of $-V_B$, at which the current increases very rapidly and becomes self-maintaining. This is illustrated in Fig. 1.7. The critical voltage V_B is termed the *junction breakdown voltage.*

A widely accepted explanation of the breakdown phenomenon is the following, based on Townsend's theory of gaseous breakdown. As the voltage applied to the junction is made more negative, so the minority charge carriers which constitute the saturation current I_o will acquire sufficient kinetic energy to experience ionising collisions. Consequently 1 hole and 1 electron are produced for each collision. The original ionising charge carriers, together with the products of the collisions, will contribute to further carrier multiplication. The *secondary holes* and *electrons* thus produced will, therefore, account for the observed increase in reverse current.

A junction diode can be fabricated by growing a single crystal of semiconductor material in such a way that donor atoms predominate in one part of the crystal while acceptor atoms predominate in the other part. This type of diode is known as the *grown junction.* Alternatively, an n-type and a p-type semiconductor are actually fused together resulting in the *alloyed junction.*

1.3. THE JUNCTION TRANSISTOR

A junction transistor consists of an n-type region between 2 p-type regions as in Fig. 1.8. The n-type and p-type regions can be interchanged as in Fig. 1.9. The junction transistor of Fig. 1.8 is said

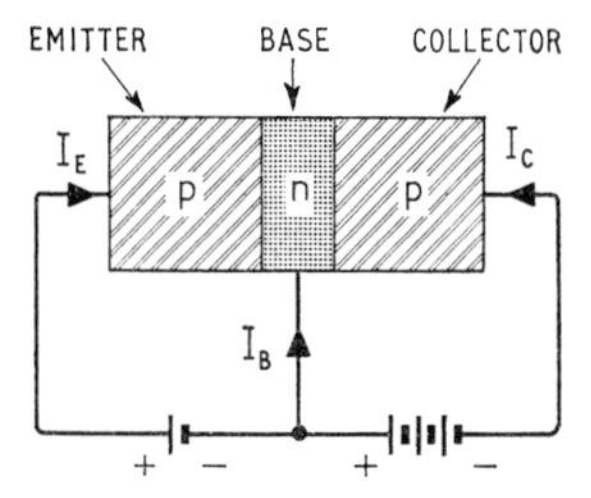

Fig. 1.8. pnp junction transistor

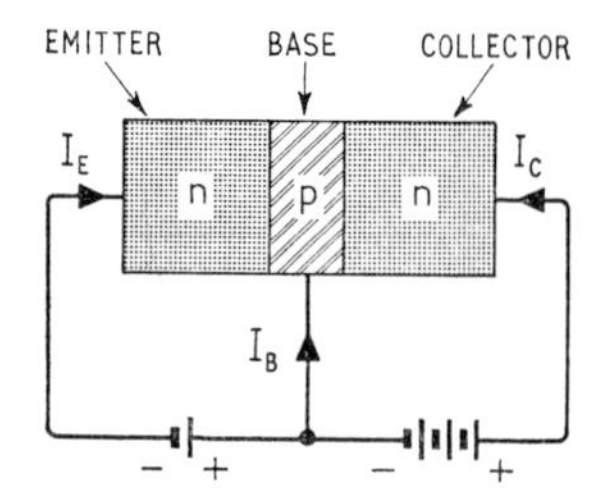

Fig. 1.9. npn junction transistor

to be of the pnp type and that of Fig. 1.9 is said to be of the npn type. In both diagrams the external biasing voltages have been included.

On examining Figs. 1.8 and 1.9 we see in both cases:

(1) The junction transistor consists basically of 2 pn junction diodes connected back to back.

(2) The 2 extrinsic semiconductors labelled the *emitter* and the *base* constitute a forward-biased pn junction diode.

(3) The base material and the extrinsic semiconductor labelled the *collector* form a reverse-biased pn junction diode.

In order to explain the transistor action consider a pnp junction transistor biased externally as in Fig. 1.8. As a result of forward-biasing the emitter–base junction, the emitter material injects holes into the base region while, simultaneously, the base material injects electrons into the emitter. Therefore, the emitter current consists of 2 components, namely, a hole current and an electron current.

However, only the former component contributes to the useful collector current. The ratio of the hole current to the total emitter current is termed the *emitter efficiency* γ. In an effort to make γ close to unity, as is normally desired, it is necessary to ensure that the density of holes injected into the base by the emitter is large compared to the density of electrons injected into the emitter by the base. This objective can be achieved by doping the emitter material more heavily than the base material, that is, by making the impurity concentration in the emitter greater than that in the base.

After the holes are injected into the base region, they propagate across it by a process of diffusion. During this transportation process some of the injected holes recombine with the electrons in the base. Therefore, the collector is deprived of some of its useful current and a base current flows so as to replace the loss of electrons in the base region. The ratio of the current reaching the collector–base junction to the hole current injected into the base by the emitter is termed the *transport factor* β. The transport factor can be made to approach unity by reducing the width of the base region. Consequently the holes will diffuse across the base region in a shorter time, affording less opportunity for the recombination process to take place.

As soon as the holes reach the reverse-biased collector–base junction, they are collected by the negatively biased collector terminal. However, if this junction is reverse-biased by a large voltage, then secondary holes and electrons are produced by the avalanche effect considered earlier on page 7. Consequently, the actual collector current can be greater than the incident hole current that leaves the base region. The ratio of the actual collector current to the incident hole current is termed the *collector efficiency* M. For a large range of collector–base voltages the collector efficiency is only slightly greater than unity. This is in contrast to the emitter efficiency γ and transport factor β, both of which are always less than unity.

The product of the above 3 factors, namely, γ, β and M is termed the *common base current amplification factor*. Its value depends on whether large signal or small signal conditions are considered. The large signal value of the common base current amplification factor will be denoted by α_B, while its value under small signal conditions will be denoted by α_b. These will be discussed further in Section 1.4. Here we shall consider d.c. conditions only. If I_E is the d.c. value of the emitter current in Fig. 1.8, the collector current I_C will be equal to $-\alpha_B I_E$ (the negative sign is used because in Fig. 1.8 all the terminal currents are shown flowing into the device).

Throughout the above discussion it has been assumed that the emitter current is finite. Consider now the case when the emitter current is reduced to zero. If the collector–base junction is reverse-biased, thermally generated holes will flow from the base into the collector, and thermally generated electrons will flow from the collector into the base. The flow of these thermally generated charge carriers constitutes the so-called *collector–base leakage current* I_{CBO}. It is also referred to as the *collector cut-off* current.

Like the saturation current of a junction diode, the collector–base leakage current I_{CBO} increases in magnitude with increasing

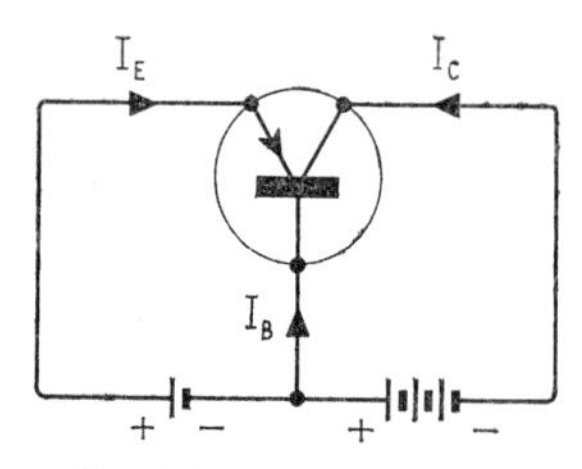

Fig. 1.10. pnp transistor

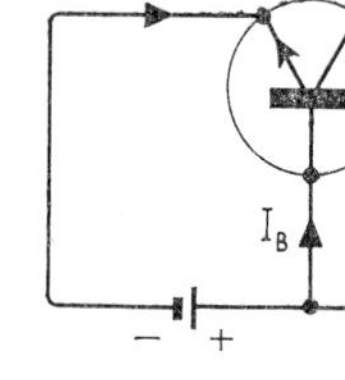

Fig. 1.11. npn transistor

temperature (by a factor of nearly 10 for every 30°C rise). The temperature dependence of I_{CBO} is of particular concern in the design of circuits using germanium transistors as it can cause a substantial shift in the *quiescent operating-point*, unless special precautions are taken. This is discussed further in Section 2.1.

Therefore, the collector current I_C, the emitter current I_E and the collector–base leakage current I_{CBO} are related as follows:

$$I_C = I_{CBO} - \alpha_B I_E \qquad (1.3)$$

Notice that $I_C = I_{CBO}$ when $I_E = 0$ in accordance with the definition of I_{CBO}. Typical values for α_B and I_{CBO} for a germanium junction transistor are:

$$\alpha_B = 0{\cdot}98$$

$$I_{CBO} = -1\mu\text{A at } 20°\text{C} \quad \text{and} \quad -10\mu\text{A at } 50°\text{C}$$

So far only the pnp transistor has been considered. The operation of the npn type can be explained in an analogous manner except that the roles of holes and electrons, p-type and n-type semiconductors, and positive and negative biasing voltages are interchanged.

The representations used when pnp and npn junction transistor types are employed as circuit elements are shown in Figs. 1.10 and

1.11, respectively. In both Figures, the arrow on the emitter lead specifies the direction of actual direct emitter current flow when the emitter–base junction is forward-biased.

1.4. TRANSISTOR CHARACTERISTIC CURVES

The transistor is basically a 3-terminal device as shown in Fig. 1.12, which also includes the terminal currents and voltages. It follows from Fig. 1.12 that the terminal currents I_E, I_B and I_C are related as

$$I_E + I_B + I_C = 0 \tag{1.4}$$

Further, from the definitions of the terminal voltages V_{CB}, V_{CE} and V_{EB} shown in Fig. 1.12 we find that:

$$V_{CB} - V_{CE} - V_{EB} = 0 \tag{1.5}$$

Thus, from Equations 1.4 and 1.5 we see that it is only necessary to

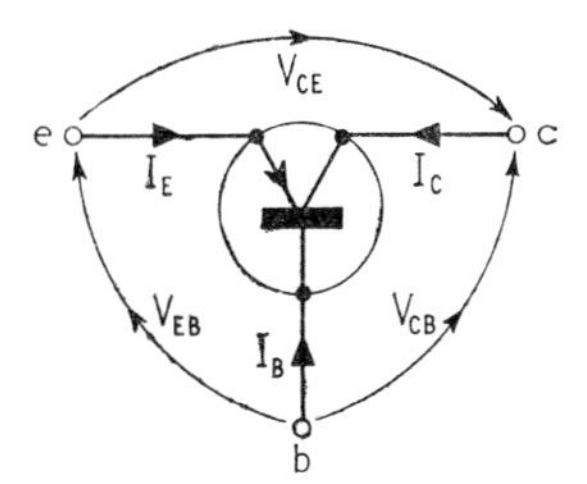

Fig. 1.12. pnp transistor as a 3-terminal device

specify any 2 of the 3 terminal currents and any 2 of the 3 terminal voltages.

The external behaviour of a transistor under static, that is d.c., conditions is usually expressed in the form of one family of characteristic curves to describe the input circuit and another family of characteristic curves to describe the output circuit of the device. These 2 sets of curves then completely define the static behaviour of the transistor. Now a transistor can be operated in any 1 of 3 useful configurations, namely, the *common base*, *common emitter* and *common collector* configurations illustrated in Figs. 1.13–1.15. In Fig. 1.13 the base terminal is arranged to be common to the input and output circuits. Similarly, in Figs. 1.14 and 1.15 the emitter and collector terminals are common, respectively. Therefore, we find that it is possible to develop numerous graphical representations for describing the external behaviour of a transistor under d.c. conditions. However, only a few of these possibilities are found to be particularly useful in the study of transistor circuits. Accordingly, attention will be given only to the common base and common emitter characteristic curves that are widely used in practice.

1.4.1. Common Base Characteristics

Consider the common base configuration of Fig. 1.13 using a pnp junction transistor. In Fig. 1.16 the emitter–base voltage V_{EB} is

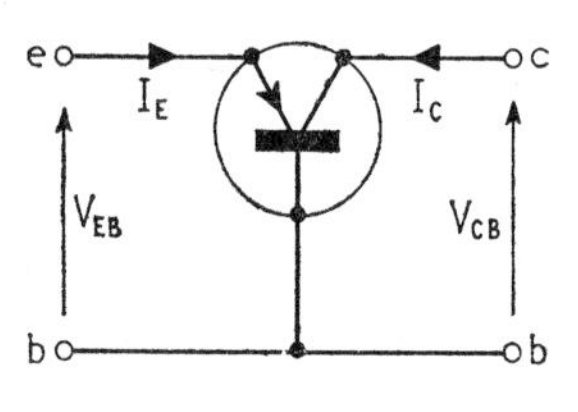

Fig. 1.13. Common base configuration

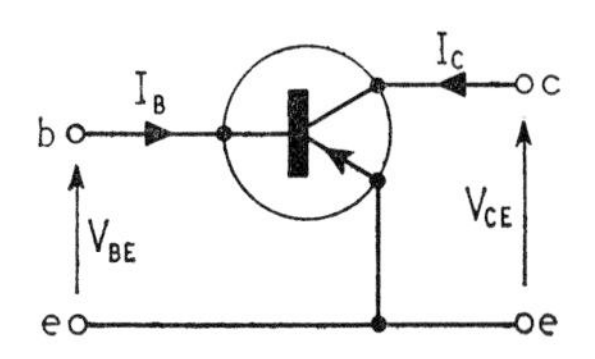

Fig. 1.14. Common emitter configuration

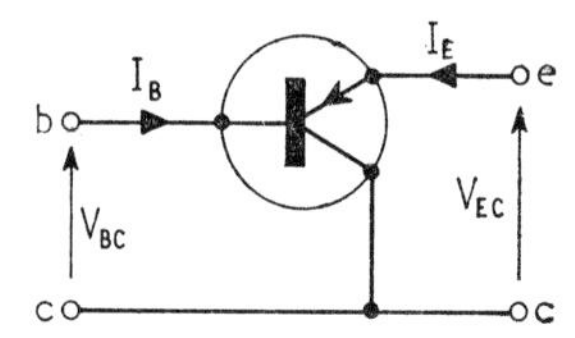

Fig. 1.15. Common collector configuration

shown plotted as a function of the emitter current I_E with the collector–base voltage V_{CB} as a running parameter. The output collector current I_C is shown plotted in Fig. 1.17 as a function of V_{CB} with I_E as a running parameter. Thus Figs. 1.16 and 1.17 illustrate typical input and output static characteristic curves, respectively, of a pnp transistor operated in the common base mode.

From the input characteristics of Fig. 1.16 we see that, for a given value of V_{CB}, the resulting characteristic curve is exponential and very similar to the voltage–current characteristic of a forward-biased pn junction diode (see Fig. 1.7). The slight dependence of the input characteristic on the collector–base voltage V_{CB} is due to the so-called *base width modulation effect*. This effect arises because the collector–base junction is, like an ordinary pn junction diode, associated with a depletion layer which extends mainly into the base region, and whose width increases as the applied collector–base voltage is increased in magnitude.

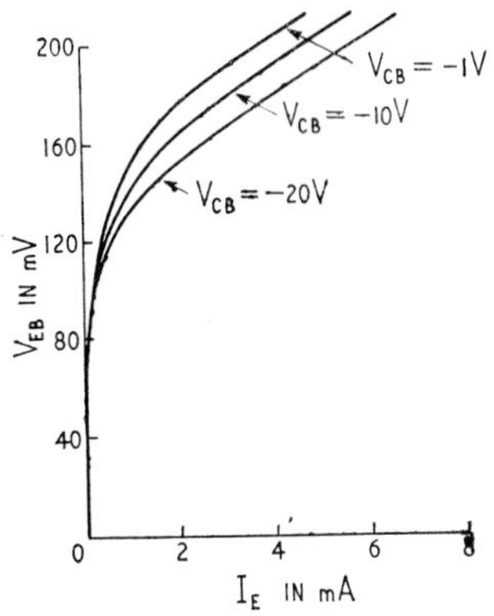

Fig. 1.16. Common base input characteristics

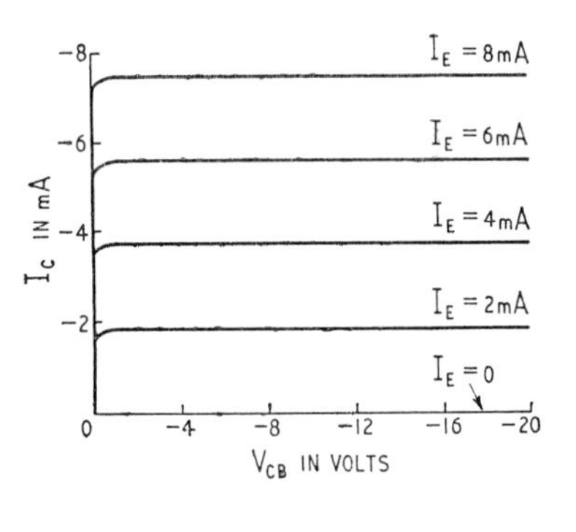

Fig. 1.17. Common base output characteristics

Next consider the family of output characteristics of Fig. 1.17. Here it is observed that:

(1) The characteristic curve corresponding to $I_E = 0$ lies almost coincident with the voltage axis. This is because, by definition, when $I_E = 0$ the collector current I_C is equal to the collector–base leakage current I_{CBO} which, at ordinary temperatures, is of the order of only a few microamps.

(2) For a given value of emitter current, the collector current reaches its limiting value for small collector–base voltages. Thereafter the value of the collector current is very nearly equal to the emitter current because α_B is very close to unity

(3) For a given value of I_E the slope of the linear portion of the output characteristic curve is very small, indicating a very large collector output resistance for the common base configuration. The slight dependence of the collector current on the collector–base voltage is mainly due to the base width modulation effect mentioned earlier. As a result of this effect the base region becomes narrower with the increasing magnitude of collector–base voltage and, therefore, the common base current amplification factor α_B approaches unity more closely. It thus follows from Equation 1.3 that, for a given value of I_E, the collector current I_C will increase slightly as V_{CB} is increased in magnitude.

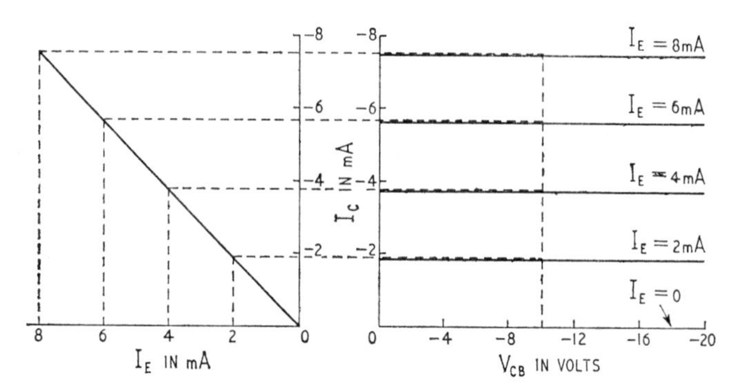

Fig. 1.18. Common base current transfer characteristic

Therefore, when the emitter–base junction is forward-biased and the collector–base junction is reverse-biased, the magnitude of the collector current is mainly controlled by the emitter current, and it is only very slightly sensitive to variations in the collector–base voltage once it exceeds a few tenths of a volt. In other words, the junction transistor is essentially a current-controlled device.

The output characteristics of Fig. 1.17 can be used to determine the common base *current transfer characteristic* which relates the output collector current I_C to the input emitter current I_E for a specified value of collector–base voltage. Thus, using the graphical construction indicated in Fig. 1.18 we can deduce the desired transfer characteristic which applies to a specified collector–base voltage of -10V. Suppose that the transistor is biased such that the emitter current is I_E and the corresponding value of collector current is I_C. The large signal or d.c. value of the common base current amplification factor α_B is deduced from Equation 1.3 to be

$$\alpha_B = -\frac{I_C - I_{CBO}}{I_E} \tag{1.6}$$

For small signal conditions the common base current amplification factor α_b is defined as

$$\alpha_b = -\left(\frac{dI_C}{dI_E}\right)_{V_{CB}=\text{constant}} \tag{1.7}$$

Thus the small signal value α_b is equal to the ratio of a small change in collector current to the small change in emitter current which gives rise to it, with the collector–base voltage maintained constant. α_b must be evaluated at the quiescent operating-point fixed by the specified values of I_C and V_{CB}.

1.4.2. Common Emitter Characteristics

Consider the common emitter configuration of Fig. 1.14. In this case the family of input characteristics is obtained by plotting the base–emitter voltage V_{BE} as a function of the base current I_B with the collector–emitter voltage V_{CE} as a running parameter. In the

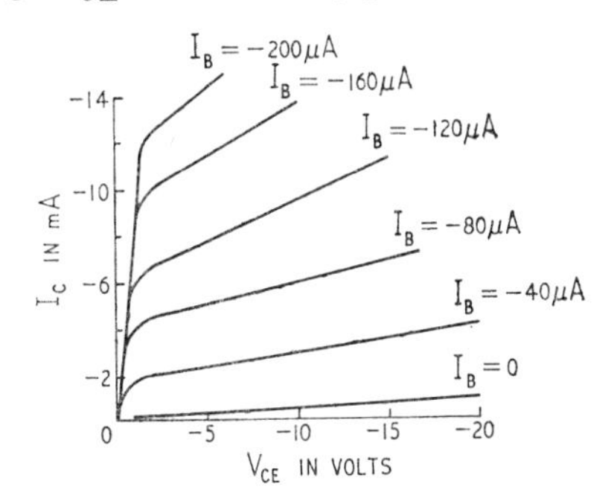

Fig. 1.19. Common emitter output characteristics

output characteristics of Fig. 1.19 the output collector current I_C is plotted against V_{CE} with I_B as a running parameter.

From the output characteristics of Fig. 1.19 we observe that:

(1) The collector current corresponding to $I_B = 0$ is quite appreciable. This can be explained by eliminating I_E between Equations 1.3 and 1.4. Thus we obtain that

$$I_C = \frac{\alpha_B}{1-\alpha_B} . I_B + \frac{I_{CBO}}{1-\alpha_B}$$

$$= \alpha_E \, I_B + I_{CEO} \tag{1.8}$$

where α_E = d.c. value of *common emitter current amplification factor*

$$= \frac{\alpha_B}{1-\alpha_B} \tag{1.9}$$

and I_{CEO} = *collector–emitter leakage current*

$$= \frac{I_{CBO}}{1-\alpha_B} \tag{1.10}$$

From Equation 1.8 we find that when the base current I_B is reduced to zero, that is the base terminal is open circuited, the collector current is equal to the collector–emitter leakage current I_{CEO}. However, as α_B is normally close to unity, Equation 1.10 shows that I_{CEO} can be quite large. In other words, when the base terminal is open-circuited, the transistor amplifies its own leakage current. Typical values for I_{CEO} of a germanium junction transistor are as follows:

$$I_{CEO} = -50\mu A \text{ at } 20°C$$
$$= -500\mu A \text{ at } 50°C$$

(2) In the linear portion of the characteristics, a relatively small base current can result in a large collector current. This is also due to the closeness of α_B to unity, giving a large value for α_E in Equation 1.9. A typical value for α_E is 49.

(3) The linear portion of the output characteristics has a greater slope than for a common base stage indicating a lower output resistance. This is again due to the closeness of α_B to unity.

Consider now the common emitter current transfer characteristic which relates the output collector current I_C to the input base current I_B when the collector–emitter voltage V_{CE} is maintained constant. This characteristic is obtained from the common emitter output characteristic using the graphical construction illustrated in Fig. 1.20, where it is assumed that the specified collector–emitter voltage is -10V. Suppose that the transistor is biased to have a base current of I_B and a corresponding collector current of I_C. The large signal or d.c. value α_E of the common emitter current amplification factor is deduced from Equation 1.8 to be

$$\alpha_E = \frac{I_C - I_{CEO}}{I_B} \tag{1.11}$$

For small signal conditions the common emitter current amplification factor α_e is defined as

$$\alpha_e = \left(\frac{dI_C}{dI_B}\right)_{V_{CE}=\text{constant}} \tag{1.12}$$

Thus α_e is equal to the ratio of a small change in collector current to the small change in base current which gives rise to it, with the collector–emitter voltage kept constant at the specified quiescent value.

In general, α_e and α_E can have different values owing to the non-linear nature of the current transfer curve. However, since the common emitter current transfer characteristic is more non-linear than the common base one (see Figs. 1.18 and 1.20) the values of α_b and α_B for the common base configuration are closer to each

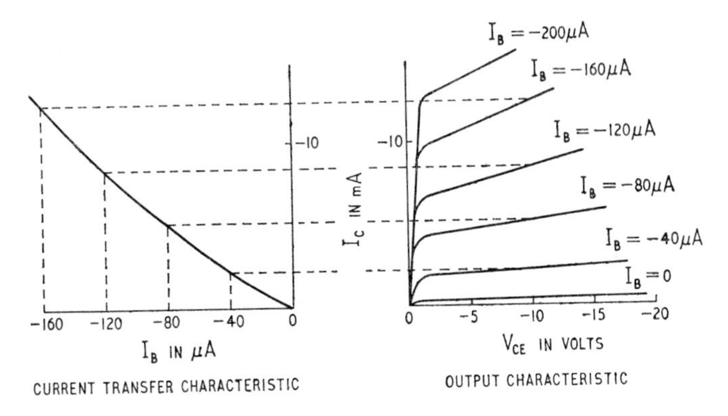

Fig. 1.20. Common emitter current transfer characteristic

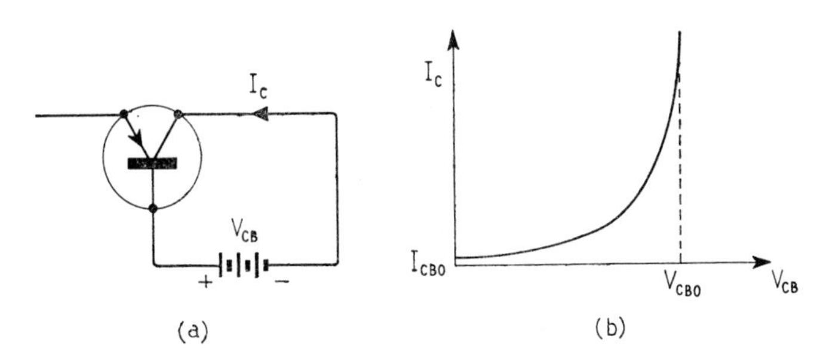

Fig. 1.21. Breakdown of common base with its emitter terminal open circuited

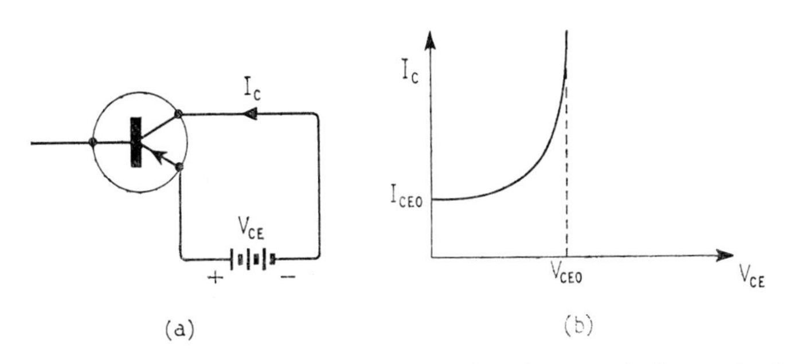

Fig. 1.22. Breakdown of common emitter with its base terminal open circuited

other than the values of α_e and α_E for the common emitter configuration.

The large signal values α_E and α_B of the common emitter and common base current amplification factors are related to each other by Equation 1.9. If in this equation small signal values are substituted for large signal values, we find that α_e and α_b are related as follows:

$$\alpha_e = \frac{\alpha_b}{1-\alpha_b} \tag{1.13}$$

1.5. PERMISSIBLE WORKING REGION

Transistor manufacturers normally specify a maximum value for the collector voltage, collector current and collector power dissipation. The transistor can be expected to operate satisfactorily only when it is ensured that the above maximum values are not exceeded.

The *maximum collector voltage* is determined by the collector breakdown, which can occur in either one of two ways. Consider

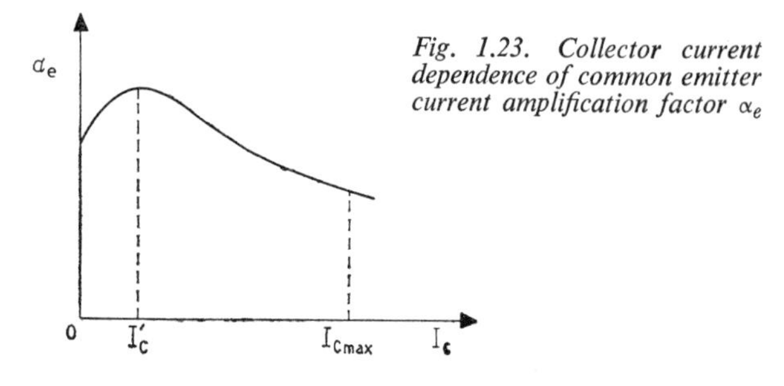

Fig. 1.23. Collector current dependence of common emitter current amplification factor α_e

first a pnp junction transistor having its emitter terminal open-circuited as in Fig. 1.21(a). The resulting collector current will be equal to the collector–base leakage current I_{CBO}. As the externally applied voltage V_{CB} is increased, the collector current increases in magnitude as indicated in Fig. 1.21(b). When the *critical breakdown voltage* V_{CBO} is reached, the collector current begins to rise rapidly and becomes self-maintaining. This breakdown mechanism is due to the avalanche multiplication process which was explained earlier when we were considering the breakdown of a reverse-biased pn junction diode (see page 7). Consider next a pnp junction transistor having its base terminal open-circuited as in Fig. 1.22(a). In this case the resulting collector current is equal to the collector–emitter leakage current I_{CEO}. Here again as the applied voltage is increased, the collector current increases too, and breakdown occurs when the critical voltage V_{CEO} is reached. The cause of this second breakdown mechanism is that as the applied collector voltage increases, the depletion layer associated with the collector–base junction extends more and more deeply into the base region. When

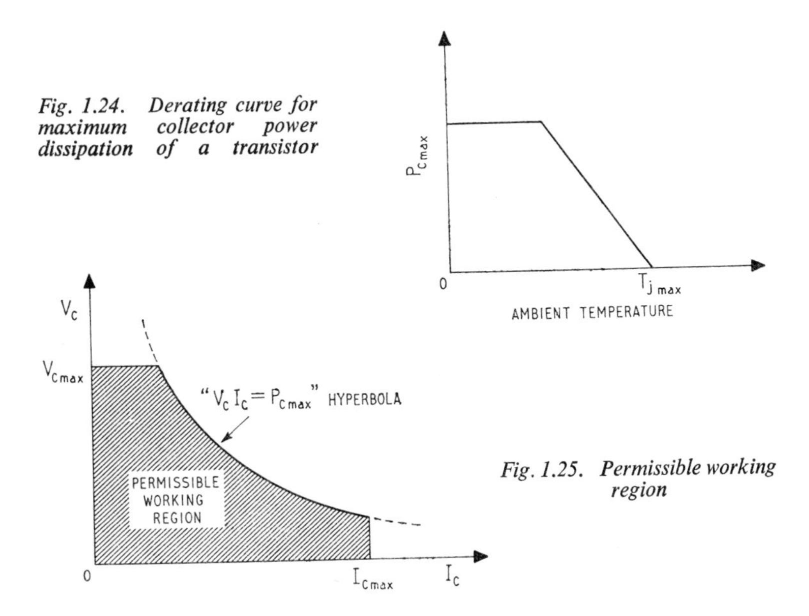

Fig. 1.24. Derating curve for maximum collector power dissipation of a transistor

Fig. 1.25. Permissible working region

the applied voltage reaches V_{CEO}, the depletion layer occupies the base region entirely. Then a path of very low resistance is established across the base region. This phenomenon is known as *punch-through breakdown* and the collector voltage at which it occurs is called *punch-through breakdown voltage*. Therefore, the maximum collector voltage $V_{C\,max}$ is determined by either V_{CBO} or V_{CEO}, whichever is the smaller one.

When the collector current I_C is varied it is found in practice that the common emitter current amplification factor α_e varies in the manner indicated in Fig. 1.23. Here we observe that initially α_e rises reaching a maximum value at I_C' and then it starts to decrease with increasing collector current. Thus a *maximum collector current* $I_{C\,max}$ is specified up to which the reduction in the magnitude of α_e is considered to be tolerable in practice.

The *maximum collector power dissipation* $P_{C\,max}$, which a particular transistor type can handle at any ambient temperature, is determined as a result of life-tests carried out by the manufacturer. For convenience, the power dissipation rating is expressed in terms of the *thermal resistance* θ and the maximum collector junction temperature $T_{j\,max}$ that is permissible. The maximum collector power dissipation $P_{C\,max}$, at any ambient temperature T_{amb}, is

$$P_{C\,max} = \frac{1}{\theta}(T_{j\,max} - T_{amb}) \tag{1.14}$$

Therefore, at higher temperatures the transistor must be derated as shown in Fig. 1.24.

The specified values of $V_{C\,max}$, $I_{C\,max}$ and $P_{C\,max}$ define the permissible working region in the $V_C - I_C$ plane as illustrated in Fig. 1.25. In this diagram the hyperbola corresponds to $V_C I_C = P_{C\,max}$. This hyperbola shifts closer to the origin with increasing ambient temperature.

CHAPTER 2

GRAPHICAL ANALYSIS

2.1. STABILISATION OF OPERATING-POINT

The *quiescent operating-point* of a transistor stage can be established by specifying suitable values for the collector current I_C and collector–emitter voltage V_{CE}. However, the operating-point can shift from its desired position owing to two major causes:

(1) the temperature dependence of the collector–base leakage current I_{CBO},

(2) the wide variations in the large signal value of the common emitter current amplification factor α_E.

Therefore, the objective of the proper d.c. biasing of a transistor stage ought to be not only to establish a specified operating-point

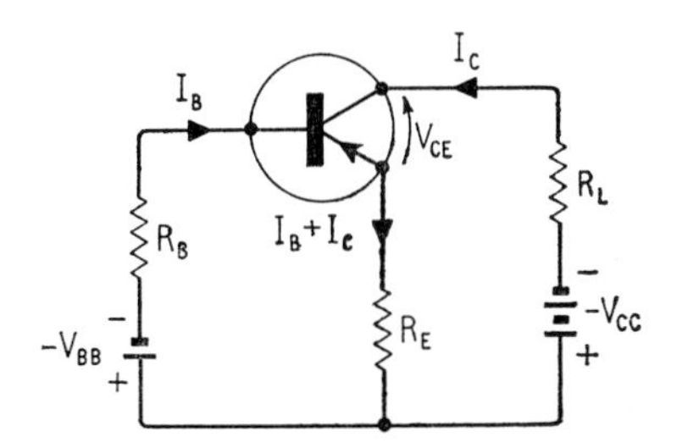

Fig. 2.1. General biasing circuit

but also to maintain it inside a prescribed region in spite of temperature changes and replacement of the transistor.

Consider the biasing circuit shown in Fig. 2.1. This circuit is sufficiently general to enable us to assess the merits of a number of different biasing circuit arrangements often used in practice. Normally, in a germanium transistor the d.c. value of the base–

emitter voltage V_{BE} is of the order of $-0{\cdot}1$ to $-0{\cdot}2$V and it can be neglected if the voltage drop across the emitter resistor is large enough. Applying Kirchoff's Voltage Law to the input loop shown in Fig. 2.1 and neglecting V_{BE} we find

$$-V_{BB} = R_B I_B + R_E(I_C + I_B) \tag{2.1}$$

From Equation 1.8, the collector current I_C and base current I_B are related as

$$I_C = \frac{\alpha_B}{1-\alpha_B} . I_B + \frac{I_{CBO}}{1-\alpha_B} \tag{2.2}$$

where α_B is the large signal value of the common base current amplification factor.

Eliminating I_B between Equations 2.1 and 2.2 we obtain

$$-V_{BB} = \frac{I_C}{\alpha_B}[R_E + R_B(1-\alpha_B)] - \frac{I_{CBO}}{\alpha_B}(R_E + R_B) \tag{2.3}$$

This equation gives the dependence of the collector current I_C of the circuit of Fig. 2.1 on the collector–base leakage current I_{CBO}. A factor which is a measure of the shift in operating-point produced by a change in I_{CBO} is the *stability factor S* defined as

$$S = \frac{dI_C}{dI_{CBO}} \tag{2.4}$$

Here we note that the closer S is to unity the more d.c. stable is the particular biasing circuit. The stability factor of the biasing circuit arrangement of Fig. 2.1 is evaluated by differentiating both sides of Equation 2.3 with respect to I_{CBO}. Since V_{BB} is constant, and assuming that α_B is independent of temperature we find

$$S = \frac{R_E + R_B}{R_E + (1-\alpha_B)\, R_B} \tag{2.5}$$

Consider the following two extreme cases:

(1) When $R_B = 0$, that is, there is zero resistance in the base circuit as in Fig. 2.2, Equation 2.5 gives $S = 1$. This is the most desirable condition.

(2) When $R_E = 0$, that is, there is zero resistance in the emitter circuit as in Fig. 2.3, Equation 2.5 gives $S = 1/(1-\alpha_B)$. Since α_B is normally close to unity, the stability factor of the simple biasing circuit of Fig. 2.3 can be very large and so $R_E = 0$ corresponds to the worst condition.

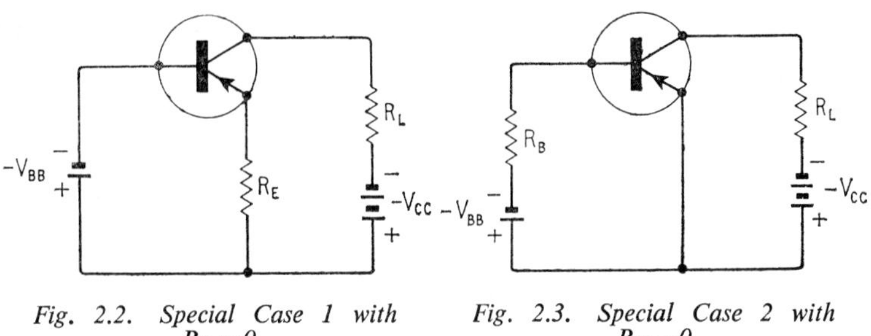

Fig. 2.2. Special Case 1 with $R_B = 0$

Fig. 2.3. Special Case 2 with $R_E = 0$

Therefore, the displacement in the position of the operating-point of Fig. 2.1, as a result of temperature changes, is reduced by increasing the emitter resistance R_E and by reducing the base resistance R_B.

The collector resistance R_L is chosen such that the voltage drop across it, produced by the flow of collector current I_C, leaves the collector at a specified voltage with respect to the emitter. If V_{CE} is the collector–emitter voltage, then applying Kirchoff's Voltage Law to the output loop consisting of the d.c. supply $-V_{CC}$,

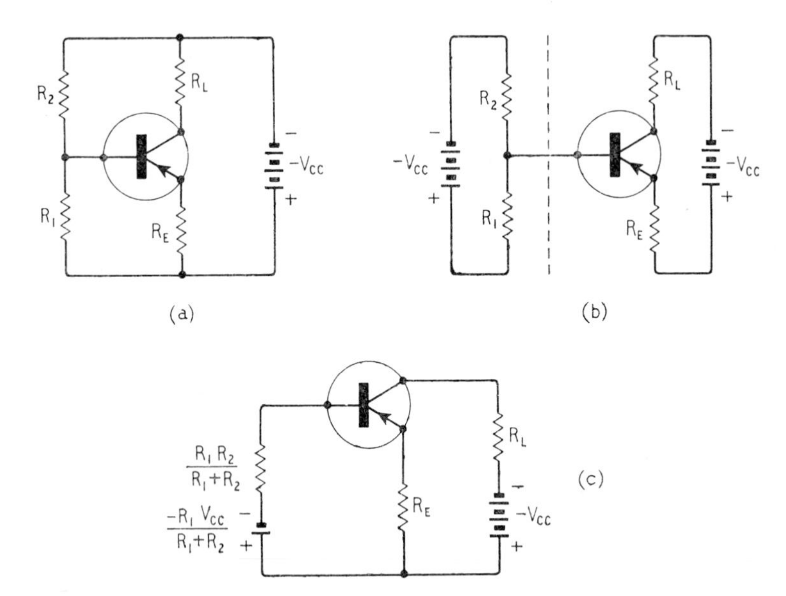

Fig. 2.4. Single-supply biasing-circuit and its reduction to the same form as the general biasing circuit of Fig. 2.1

Thus in the example considered the operating-point of the transistor stage of Fig. 2.1 shifts from the position of

$$(I_C = -1\text{mA}, \quad V_{CE} = -6\text{V})$$

at ordinary temperatures to the position of

$$(I_C = -1{\cdot}08\text{mA}, \quad V_{CE} = -5{\cdot}7\text{V})$$

at the elevated temperature of 50°C. This is illustrated in Fig. 2.5.

2.2. LOAD-LINES

Consider the elementary biasing circuit of Fig. 2.6 where the transistor stage is *directly coupled* to a load of resistance R_L. This circuit behaves in a similar way to that of Fig. 2.3 and they become

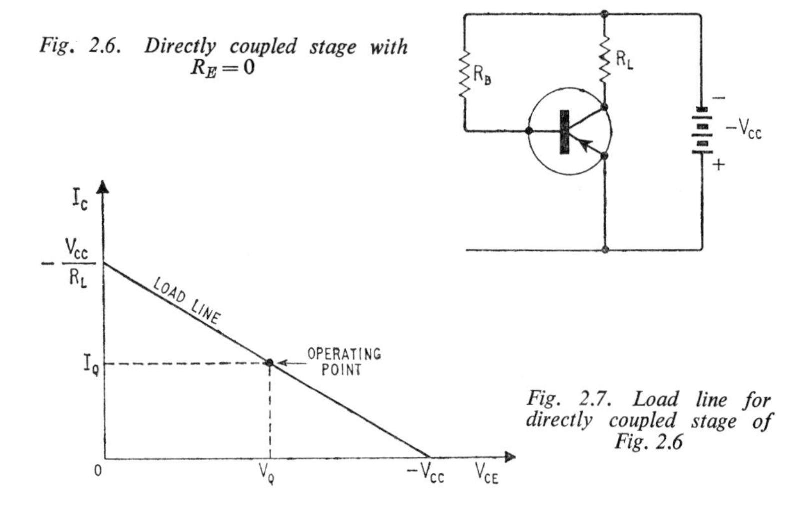

Fig. 2.6. Directly coupled stage with $R_E = 0$

Fig. 2.7. Load line for directly coupled stage of Fig. 2.6

identical when $V_{CC} = V_{BB}$. The circuit of Fig. 2.6 has zero resistance in the emitter circuit. Therefore, putting $R_E = 0$ in Equation 2.6 we find

$$-V_{CC} = V_{CE} + I_C R_L \tag{2.8}$$

Notice that when $I_C = 0$, $V_{CE} = -V_{CC}$, and when $V_{CE} = 0$, $I_C = -V_{CC}/R_L$. Thus Equation 2.8 corresponds to a straight line having a slope equal to $-1/R_L$ as shown in Fig. 2.7. This straight line is usually referred to as the dc or *static load-line* as it applies when operating under d.c. conditions. Suppose that the transistor is biased such that the collector current is I_Q and the collector–emitter voltage is V_Q. Thus I_Q and V_Q fix the quiescent operating-point which will be located on the static load-line as in Fig. 2.7. As the collector current is varied, the quiescent operating-point will move up and down, but it will always remain located on the static load-line.

Consider next the circuit of Fig. 2.6 operating under a.c. conditions. Assuming that the internal resistance of the d.c. supply is zero, the circuit of Fig. 2.6 simplifies to the form shown in Fig. 2.8. The load resistance of the stage, under a.c. conditions, is still equal to R_L. Therefore the stage of Fig. 2.6 has an a.c. or *dynamic load-line* of

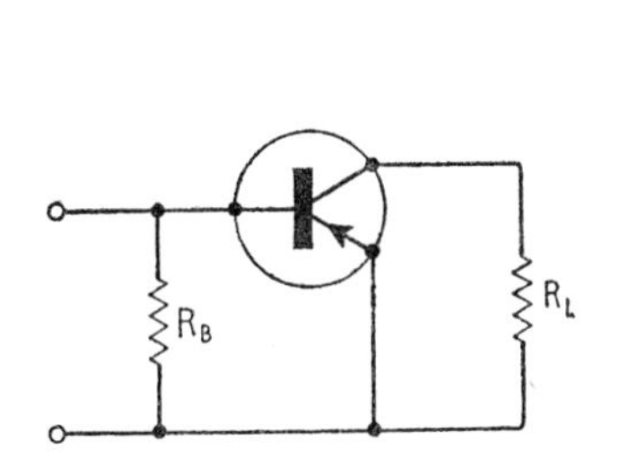

Fig. 2.8. A.c. equivalent circuit of Fig. 2.6

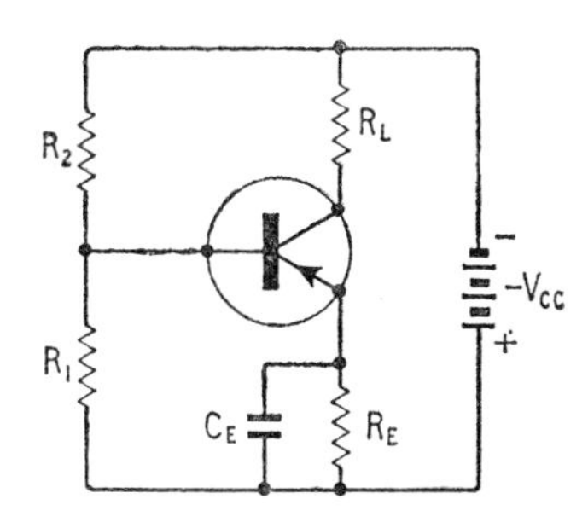

Fig. 2.9. Single-supply biasing-circuit with emitter by-pass capacitor

slope also equal to $-1/R_L$. Thus the static and dynamic load-lines of the circuit of Fig. 2.6 are coincident.

The circuit of Fig. 2.6 has zero resistance in the emitter circuit. In order to investigate the consequences of including a finite resistance in series with the emitter terminal, consider the circuit of Fig. 2.4(a). This is reproduced in Fig. 2.9, where a by-pass capacitor C_E has been connected across the emitter resistor R_E. The capacitor C_E has been added so that, at the operating-frequencies, it short-circuits R_E. Therefore, under a.c. conditions the effective load resistance of the stage is equal to R_L, and so the static and dynamic load-lines of the stage of Fig. 2.9 will have different slopes. Under d.c. conditions the collector current I_C and collector–emitter voltage V_{CE} of the stage of Fig. 2.9 are related by Equation 2.6. From this equation we find that, under d.c. conditions, the effective load resistance of the stage of Fig. 2.9 is very nearly equal to R_L+R_E. This is subject to the condition that $|I_C|\gg|I_B|$ which is normally satisfied. Therefore, the static and

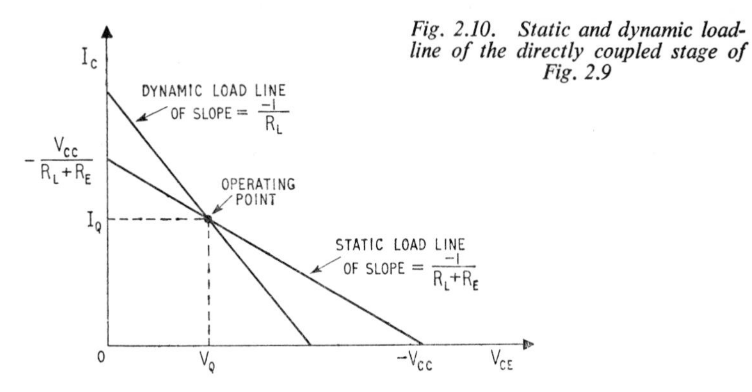

Fig. 2.10. Static and dynamic load-line of the directly coupled stage of Fig. 2.9

Fig. 2.11 (right). Capacitively coupled stage

Fig. 2.12. Static and dynamic load-lines of capacitively coupled stage of Fig. 2.11

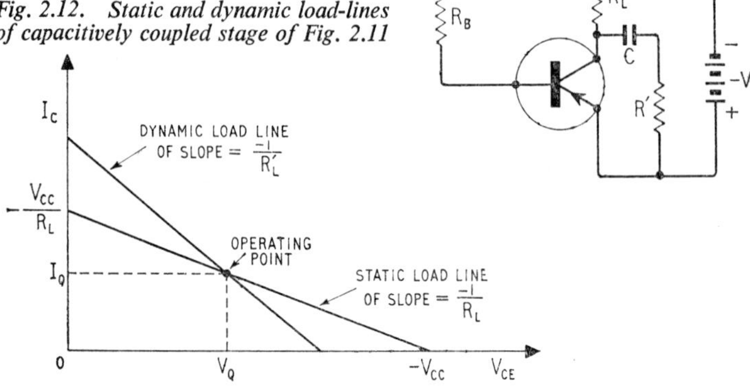

Fig. 2.13 (right). Transformer coupled stage

Fig. 2.14. Static and dynamic load-line of transformer coupled stage of Fig. 2.13

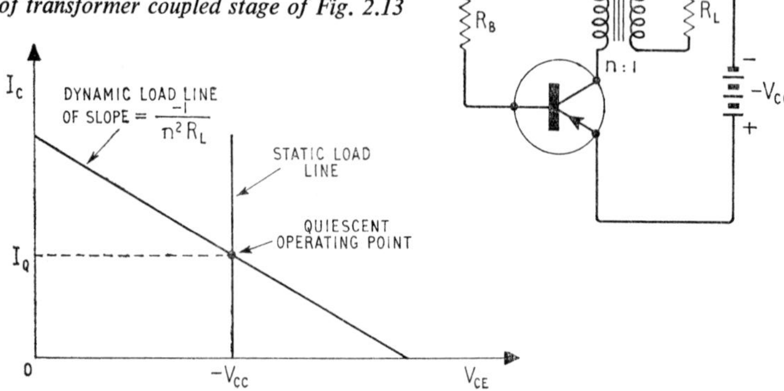

dynamic load-lines of the circuit of Fig. 2.9 have slopes equal to $-1/R_L$ and $-1/(R_L+R_E)$ respectively, and they intersect at the quiescent operating-point as shown in Fig. 2.10.

In the circuits of Figs. 2.6 and 2.9 the load is directly coupled to the transistor stage. Alternatively, the load can be *capacitively coupled* as in Fig. 2.11. The coupling capacitor C is chosen to be sufficiently large such that its impedance is negligible at the operating-frequencies. In such a case we find that under d.c. conditions the load resistance of the stage is equal to R_L. Hence, the static load-line has a slope of $-1/R_L$ and it passes through the quiescent operating-point as in Fig. 2.12. On the other hand, under a.c. conditions the effective load resistance R_L' is equal to the parallel combination of R_L and R', that is

$$R_L' = \frac{R'R_L}{R'+R_L} \tag{2.9}$$

Therefore, the dynamic load-line of the stage of Fig. 2.11 has a slope of $-1/R_L'$ and it intersects the static load-line at the quiescent operating-point as in Fig. 2.12. Notice that, since $R_L > R_L'$, the dynamic load-line of the stage of Fig. 2.11 has a greater slope than the static load-line.

Another common method of coupling a transistor stage to its load is by the use of a transformer as in Fig. 2.13. In this case, the load resistance of the stage, under d.c. conditions, is simply equal to the resistance of the primary winding of the transformer. This resistance is normally quite small and so, for all practical purposes, we can assume that the static load-line of the *transformer coupled* stage of Fig. 2.13 is vertical, with the intercept on the voltage axis equal to the d.c. supply voltage $-V_{CC}$ as in Fig. 2.14. If next we assume that the shunt inductance of the primary winding of the coupling transformer is large, then at the operating-frequencies the effective load resistance of the stage of Fig. 2.13 is equal to the reflected resistance of $n^2 R_L$, where n is the transformer turns-ratio. Therefore, the dynamic load-line will have a slope equal to $-1/n^2 R_L$ and it will intersect the static load-line at the quiescent operating-point as in Fig. 2.14. In this case, the static load-line has a slope greater than that of the dynamic load-line. This is in direct contrast to the case of the capacitively coupled stage of Fig. 2.11.

Having drawn the dynamic load-line for the particular transistor stage we can go on to use the static characteristic curves of the transistor to determine such quantities of interest as power output and non-linear distortion.

2.3. POWER OUTPUT AND NON-LINEAR DISTORTION

In order to illustrate how the response of a given transistor stage can be analysed by graphical means, consider the single stage circuit of Fig. 2.15, where the transistor is operated in its common emitter configuration. It is transformer-coupled to its load at the output terminals and capacitively coupled to the driving voltage source v_s' of series resistance R_s' at the input terminals. Normally the coupling capacitor is sufficiently large for its effect to be neglected at the operating signal frequencies. We thus find that under a.c. conditions the response of the stage can be determined from the

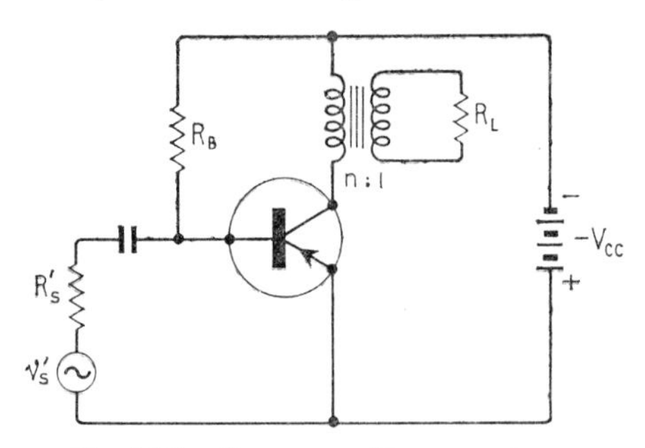

Fig. 2.15. Common emitter stage transformer coupled to its load and capacitively coupled to its source

circuit of Fig. 2.16(a). To simplify this circuit we apply Thevenin's Theorem to the left of the dotted line and so obtain the equivalent circuit of Fig. 2.16(b), where

$$v_s = \frac{R_B v_s'}{R_B + R_s'} \tag{2.10}$$

$$R_s = \frac{R_B R_s'}{R_B + R_s'} \tag{2.11}$$

Since the transistor is operated in the common emitter configuration, it is convenient to use the common emitter static characteristic curves for the graphical analysis. We shall first examine the conditions existing in the output circuit. For this purpose consider Fig. 2.17, showing a family of output characteristic curves for a typical medium-power transistor. On these characteristics we superimpose the static and dynamic load-lines which, for the stage of Fig. 2.15, have slopes equal to $-\infty$ and $-1/n^2 R_L$, respectively. Next we proceed to plot the *dynamic current transfer curve* which

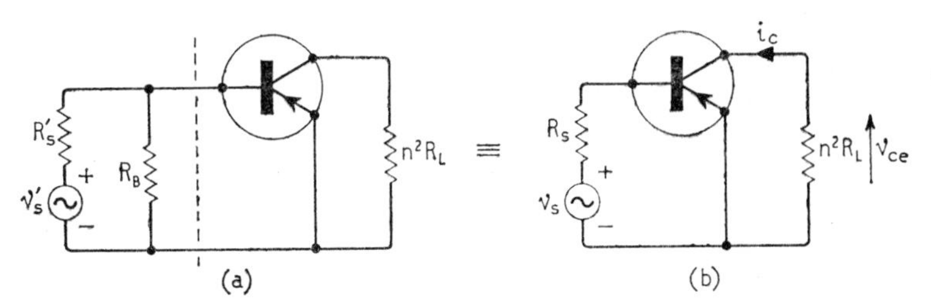

Fig. 2.16. A.c. equivalent circuits of Fig. 2.15. The circuit of (b) is obtained by applying Thevenin's theorem to the circuit of (a)

relates the output collector current to the input base current under a.c. conditions. To construct this dynamic characteristic choose a suitable value for the base current I_B and evaluate the corresponding value of collector current I_C as determined by the intersection of the dynamic load-line and the static output characteristic curve corresponding to I_B. This process is then repeated for other suitable values of base current until we obtain a sufficient number of points to plot the dynamic current transfer curve, Fig. 2.17. It is important to appreciate that the dynamic current transfer curve is a function not only of the transistor type but also of the load resistance, such that if either is changed then we must replot it.

Next we place the dynamic current transfer curve beside the static input characteristic in the manner shown in Fig. 2.18. Here it is assumed that the input characteristic corresponds to the specified quiescent value of collector–emitter voltage V_{CE}. The graphical construction of Fig. 2.18 enables us to investigate, in a convenient way, the effects of varying the voltage source and its series resistance

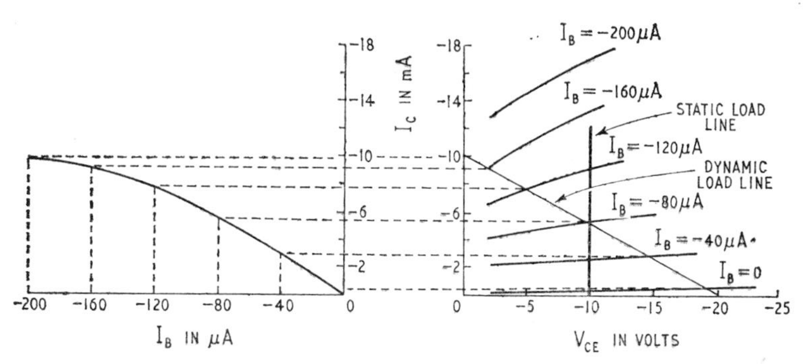

Fig. 2.17. Graphical construction of dynamic current transfer curve

on the output collector current. The effect of the source resistance is taken into account by drawing a source line of slope equal to R_s and passing through the operating-point fixed by the quiescent values of base current and base–emitter voltage.

Assuming that the source voltage is sinusoidal, then

$$v_s = V_s \sin \omega t \tag{2.12}$$

where V_s is the peak value, ω is the angular frequency, and t is time. Owing to the non-linear nature of the static input characteristic and of the dynamic current transfer curve we find that the output collector current is not linearly related to the source voltage, and the output signal is not a faithful reproduction of the input signal. Thus the output collector current contains signal frequencies which were not present in the driving voltage source. This distortion of the output signal is called *non-linear distortion* or *harmonic distortion.*

In order to evaluate the non-linear distortion assume that the instantaneous collector current i_C is related to the source voltage by the following relationship

$$i_C = I_Q + av_s + bv_s^2 \tag{2.13}$$

where I_Q = quiescent value of collector current with $v_s = 0$ and a and b are coefficients to be evaluated.

To be more precise we should include, on the right-hand side of Equation 2.13, a cubic and higher power terms, but if we are seeking results which are accurate to a first order of approximation, then we are justified in using Equation 2.13.

Eliminating v_s between Equations 2.12 and 2.13 we find

$$i_C = I_Q + aV_s \sin \omega t + bV_s^2 \sin^2 \omega t \tag{2.14}$$

However,

$$\sin^2 \omega t = \tfrac{1}{2}(1 - \cos 2\omega t) \tag{2.15}$$

in which case on rewriting Equation 2.14 and collecting terms we have

$$i_C = \left(I_Q + \frac{bV_s^2}{2}\right) + aV_s \sin \omega t - \frac{bV_s^2}{2} . \cos 2\omega t \tag{2.16}$$

We, therefore, see that the output collector current consists of the following three components:

(1) A d.c. component equal to $I_Q + bV_s^2/2$.

(2) An alternating fundamental component of a frequency equal to that of the driving voltage source and of a peak value equal to aV_s.

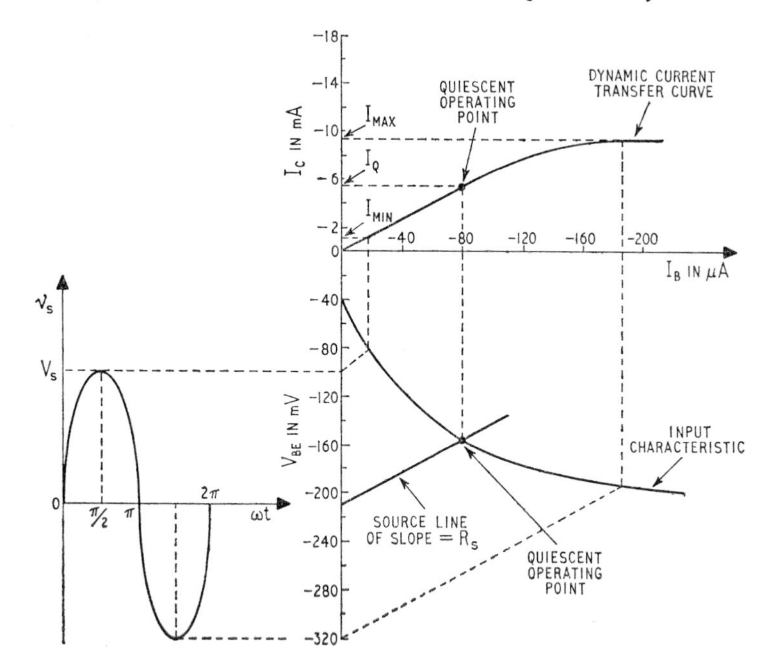

Fig. 2.18. Graphical construction for evaluating non-linear distortion

(3) An alternating second harmonic component of a frequency equal to twice that of the driving voltage source and of a peak value equal to $bV_s^2/2$.

These three components are shown in Fig. 2.19. Notice that non-linear distortion causes an increase in the mean value of the output collector current, the increase being equal to the peak value of the second harmonic component.

The coefficients a and b can be determined from the graphical construction of Fig. 2.18. Thus

$$i_C = I_{\min} \quad \text{when} \quad \omega t = \pi/2$$

and

$$i_C = I_{\max} \quad \text{when} \quad \omega t = 3\pi/2$$

Substituting these two sets of values in Equation 2.14 we obtain, respectively, that

$$I_{\min} = I_Q + aV_s + bV_s^2 \tag{2.17}$$

$$I_{\max} = I_Q - aV_s + bV_s^2 \tag{2.18}$$

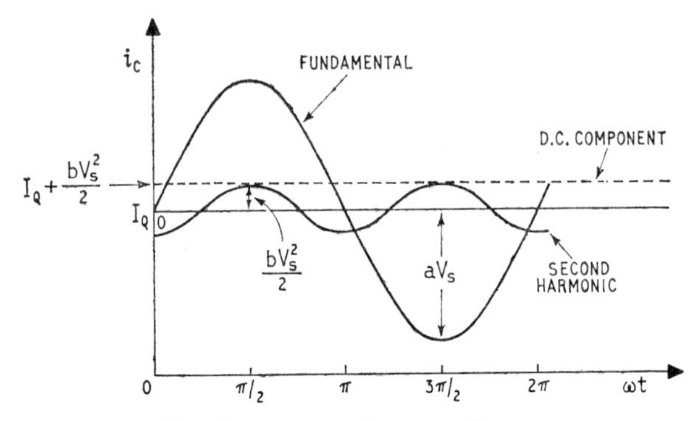

Fig. 2.19. Components of output collector current

Adding these two equations and solving for the peak value of the second harmonic component,

$$\frac{bV_s^2}{2} = \tfrac{1}{4}(I_{\text{max}} + I_{\text{min}} - 2I_Q) \tag{2.19}$$

Next, subtracting Equations 2.17 and 2.18 and then solving for the peak value of the fundamental component

$$aV_s = \tfrac{1}{2}(I_{\text{min}} - I_{\text{max}}) \tag{2.20}$$

However, from Equation 2.16 the alternating fundamental component i_{c1} of the output collector current is equal to $aV_s \sin \omega t$, and if we use Equation 2.20 for the coefficient a we find

$$i_{c1} = \tfrac{1}{2}(I_{\text{min}} - I_{\text{max}}) \sin \omega t \tag{2.21}$$

Therefore, the useful a.c. power P_1 delivered to the load and produced by the fundamental current component i_{c1} is

$$P_1 = \begin{pmatrix}\text{RMS value of fundamental} \\ \text{component of collector current}\end{pmatrix}^2 \times \begin{pmatrix}\text{reflected load} \\ \text{resistance}\end{pmatrix}$$

$$= \left[\frac{1}{\sqrt{2}} \cdot \frac{1}{2}(I_{\text{min}} - I_{\text{max}})\right]^2 . n^2 R_L$$

$$= \frac{n^2 R_L}{8}(I_{\text{min}} - I_{\text{max}})^2 \tag{2.22}$$

Further, the fundamental component v_{ce1} of the output voltage developed across the reflected load resistance $n^2 R_L$ is equal to $n^2 R_L i_{c1}$ that is,

$$\begin{aligned} v_{ce1} &= -n^2 R_L i_{c1} \\ &= -\frac{n^2 R_L}{2}(I_{min} - I_{max}) \sin \omega t \end{aligned} \tag{2.23}$$

The ratio of v_{ce1} to the driving voltage v_s is defined as the *external voltage gain K* of the stage. Hence, Equations 2.12 and 2.23 give

$$\begin{aligned} K &= \frac{v_{ce1}}{v_s} \\ &= -\frac{n^2 R_L}{2V_s}(I_{min} - I_{max}) \end{aligned} \tag{2.24}$$

A factor which is a measure of the non-linear distortion associated with the output waveform is the *second harmonic distortion factor D_2* defined as

$$\begin{aligned} D_2 &= \frac{\text{peak value of second harmonic component}}{\text{peak value of fundamental component}} \\ &= \left| \frac{(I_{max} + I_{min} - 2I_Q)}{2(I_{min} - I_{max})} \right| \end{aligned} \tag{2.25}$$

Thus knowing I_{max}, I_{min} and I_Q it is possible to determine the fundamental power output and the second harmonic distortion factor.

Example 2.2

Given the static input and output characteristic curves of Figs. 2.18 and 2.17, determine the fundamental power output, external voltage gain and second harmonic distortion factor of the stage of Fig. 2.15 when

$$\begin{aligned} n^2 R_L &= 2\text{k}\Omega \\ R_s &= 650\Omega \\ V_{CC} &= 10\text{V} \\ V_s &= 110\text{mV} \end{aligned}$$

From the graphical construction of Figs. 2.17 and 2.18 we deduce the following

$$\begin{aligned} I_Q &= -5{\cdot}4\text{mA} \\ I_{max} &= -9{\cdot}2\text{mA} \\ I_{min} &= -1\text{mA} \end{aligned}$$

Therefore, Equation 2.22 gives the fundamental a.c. power output

$$P_1 = \frac{2\times 10^3}{8}(9{\cdot}2\times 10^{-3} - 10^{-3})^2$$

$$= 16{\cdot}8\times 10^{-3}\ \text{W}$$

$$= 16{\cdot}8\text{mW}$$

Equation 2.24 gives the external voltage gain of the stage to be

$$K = -\frac{2\times 10^3}{2\times 110\times 10^{-3}}(9{\cdot}2\times 10^{-3} - 10^{-3})$$

$$= -\ 74{\cdot}6$$

Finally, the second harmonic distortion factor is deduced from Equation 2.25 to be

$$D_2 = \left|\frac{9{\cdot}2\times 10^{-3} + 10^{-3} - 2\times 5{\cdot}4\times 10^{-3}}{2(9{\cdot}2\times 10^{-3} - 10^{-3})}\right|$$

$$= 0{\cdot}036$$

$$= 3{\cdot}6\%$$

CHAPTER 3

SMALL SIGNAL EQUIVALENT CIRCUITS AND PARAMETERS

In the previous chapter it was shown how the static characteristic curves can be used for evaluating the a.c. power output and gain of a transistor stage. The graphical method of analysis gives reasonably accurate results only if the amplitudes of the applied signals are relatively large. If, however, the amplitudes of the applied signals are small it is found more convenient to replace the transistor by a suitable equivalent circuit which, ideally, should have exactly the same electrical performance as the transistor itself. Thus we shall go on first to develop a number of widely used small signal equivalent circuits and then apply them later for determining the external circuit performance of various transistor configurations.

3.1. SMALL SIGNAL EQUIVALENT CIRCUITS

Consider a junction transistor having its emitter–base junction forward-biased and its collector–base junction reverse-biased. Further, let the transistor be operated in its common base mode as in Fig. 3.1 and have the static characteristics of Fig. 3.2. As a first approximation we can represent the transistor by the elementary equivalent circuit of Fig. 3.3, which is valid if the amplitudes of the applied signals are small. The current source $\alpha_b i_e$ is controlled by the small signal value i_e of the input emitter current and it is, therefore, referred to as a *current controlled current source.* The *control parameter* α_b is the small signal value of the common base current amplification factor. It is defined by Equation 1.7 which is reproduced here for convenience

$$\alpha_b = -\left(\frac{\mathrm{d}I_C}{\mathrm{d}I_E}\right)_{V_{CB}=\text{constant}} \tag{3.1}$$

Thus α_b is proportional to the vertical displacement of the output characteristic curve of Fig. 3.2(b), resulting from a small change in the input emitter current.

The resistor r_i of Fig. 3.3 is equal to the slope of the input characteristic curve of Fig. 3.2(a), that is,

$$r_i = \left(\frac{dV_{EB}}{dI_E}\right)_{V_{CB}=\text{constant}} \tag{3.2}$$

The elementary equivalent circuit of Fig. 3.3 does not account for the relatively small influence of the output collector–base voltage V_{CB}

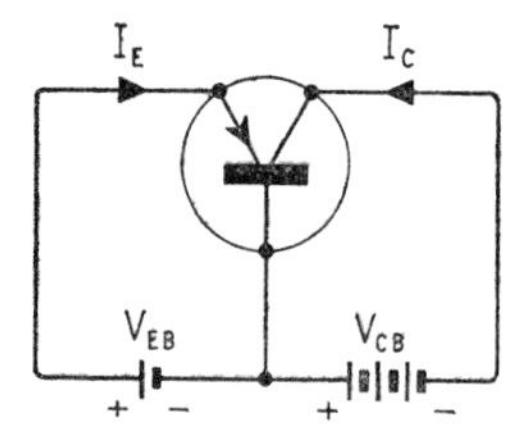

Fig. 3.1. Common base configuration

on the input and output characteristic curves as evidenced in Fig. 3.2. The effect of V_{CB} on the output collector current can be taken into account by placing a resistor r_o across the collector to base terminals as shown in Fig. 3.4(a). The value of r_o is equal to the slope of the output characteristic curve of Fig. 3.2(b), that is

$$r_o = \left(\frac{dV_{CB}}{dI_C}\right)_{I_E=\text{constant}} \tag{3.3}$$

The dependence of the input characteristic curves on the output collector–emitter voltage V_{CB} can be represented by placing a voltage source $\mu_b v_{cb}$ in series with the input emitter terminal as shown in Fig. 3.4(a). This voltage source is controlled by the small signal value v_{cb} of the output collector–base voltage and it is, therefore, referred to as a *voltage controlled voltage source.* The dimensionless control parameter μ_b is defined as

$$\mu_b = \left(\frac{dV_{EB}}{dV_{CB}}\right)_{I_E=\text{constant}} \tag{3.4}$$

Thus μ_b is proportional to the vertical displacement of the input characteristic curve of Fig. 3.2(a) resulting from a small change in the output collector–base voltage.

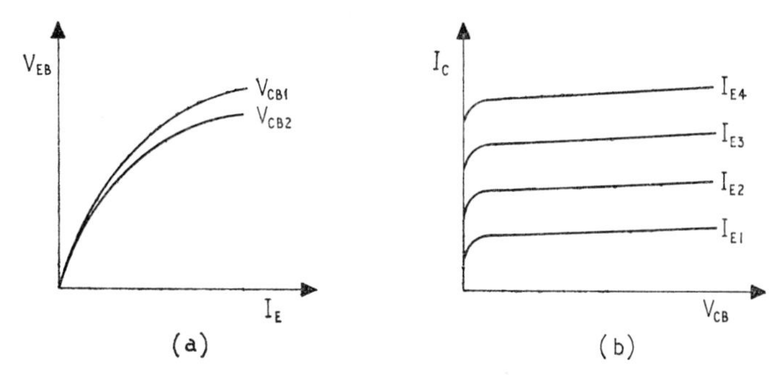

Fig. 3.2. *Common base static characteristics*

The parameters r_i, μ_b, α_b and r_o must be evaluated at the quiescent operating-point fixed by the specified d.c. values of the collector–base voltage and collector current. They have the following typical values:

$$r_i = 28\Omega$$

$$\mu_b = 2 \times 10^{-4}$$

$$\alpha_b = 0{\cdot}98$$

$$r_o = 2\text{M}\Omega$$

The small signal equivalent circuit of Fig. 3.4(a) is usually referred to as the common base *hybrid* or h-*parameters equivalent circuit* because a mixed set of voltages and currents, namely, v_{cb} and i_e have

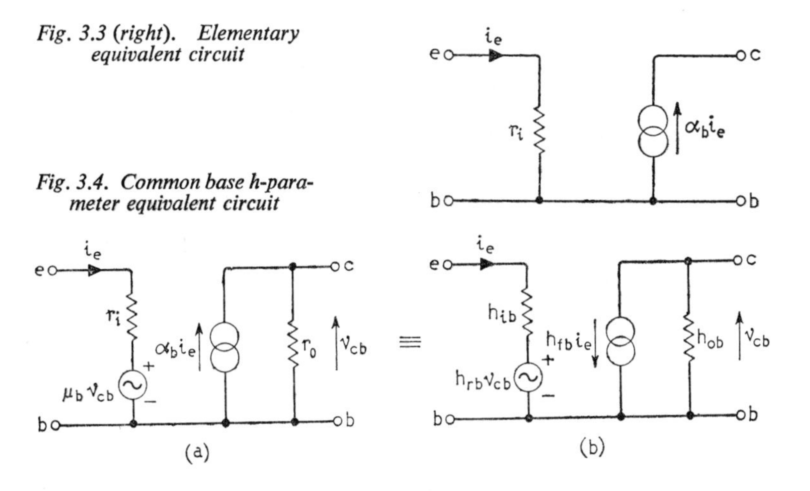

Fig. 3.3 (right). *Elementary equivalent circuit*

Fig. 3.4. *Common base h-parameter equivalent circuit*

been chosen as the independent variables. A more common way of designating the parameters of this equivalent circuit is:

$$\left.\begin{aligned} h_{ib} &= r_i \\ h_{rb} &= \mu_b \\ h_{fb} &= -\alpha_b \\ h_{ob} &= \frac{1}{r_o} \end{aligned}\right\} \quad (3.5)$$

h_{ib}, h_{rb}, h_{fb} and h_{ob} are referred to as the *common base* h-*parameters.* They are defined in the equivalent circuit of Fig. 3.4(b).

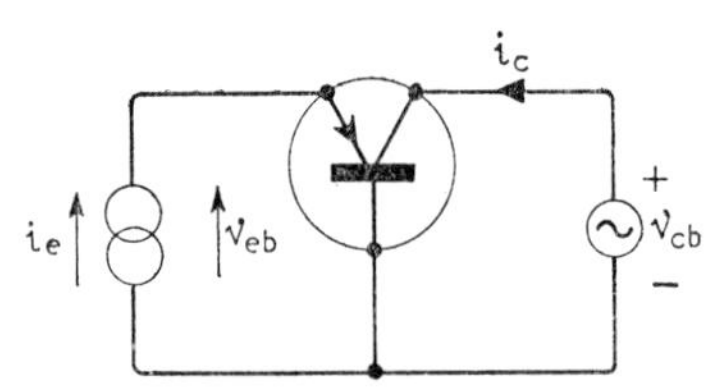

Fig. 3.5. Common base stage under a.c. conditions

In terms of the common base h-parameters we, therefore, can relate the input quantities i_e and v_{eb}, output quantities i_e and v_{cb} of the common base configuration of Fig. 3.5 as follows:

$$\left.\begin{aligned} v_{eb} &= h_{ib}\, i_e + h_{rb}\, v_{cb} \\ i_c &= h_{fb}\, i_e + h_{ob}\, v_{cb} \end{aligned}\right\} \quad (3.6)$$

Equations 3.1 to 3.6 relate the common base h-parameters to the common base static characteristic curves, from which equations they may be evaluated. An alternative and more accurate method is to apply the following definitions for the common base h-parameters. When $v_{cb} = 0$ we obtain from Equation 3.6 the following definitions for h_{ib} and h_{fb}:

$$\left.\begin{aligned} h_{ib} &= \left(\frac{v_{eb}}{i_e}\right)_{v_{cb}=0} \\ h_{fb} &= \left(\frac{i_c}{i_e}\right)_{v_{cb}=0} \end{aligned}\right\} \quad (3.7)$$

On the other hand, when $i_e = 0$ from Equation 3.6 we obtain the following definitions for h_{rb} and h_{ob}:

$$\left.\begin{aligned} h_{rb} &= \left(\frac{v_{eb}}{v_{cb}}\right)_{i_e=0} \\ h_{ob} &= \left(\frac{i_c}{v_{cb}}\right)_{i_e=0} \end{aligned}\right\} \tag{3.8}$$

Notice that the condition $v_{cb} = 0$ corresponds to short circuiting the output terminals of the common base configuration while $i_e = 0$ corresponds to open circuiting its input terminals for a.c. signals. Thus we can make the following deductions:

(1) The parameter h_{ib} is equal to the *short-circuit input resistance* of a common base stage, that is, it is equal to the resistance measured looking into the input terminals when the output terminals are short circuited as in Fig. 3.6.

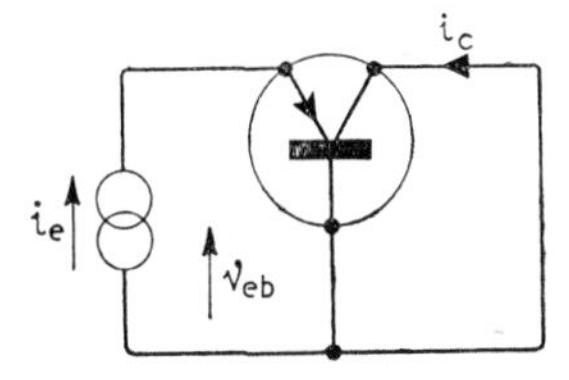

Fig. 3.6. Common base stage with output terminals short circuited

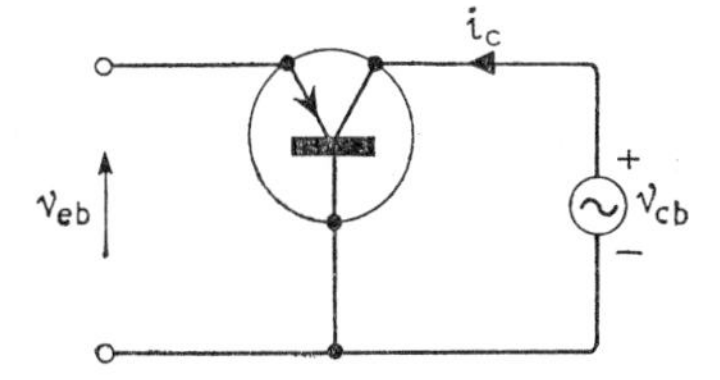

Fig. 3.7. Common base stage with input terminals open circuited

(2) The parameter h_{fb} is the *short-circuit forward current gain* of the common base stage, that is, it is equal to the ratio of the output collector current to the input emitter current when the output terminals are short circuited, see Fig. 3.6.

(3) The parameter h_{rb} is the *open-circuit reverse-voltage gain* of the common base configuration; that is, it is equal to the ratio of input emitter–base voltage to output collector–base voltage when the input terminals are open circuited as in Fig. 3.7.

(4) The parameter h_{ob} is the *open-circuit output conductance* of the common base configuration; that is, it is equal to the conductance measured looking into the output terminals when the input terminals are open circuited as in Fig. 3.7.

Therefore, we see that h_{fb} and h_{rb} are both dimensionless ratios. On the other hand, h_{ib} has the dimensions of a resistance while h_{ob}

has the dimensions of a conductance. Further, the parameter h_{fb} is responsible for signal transmission in the forward direction while h_{rb} gives rise to signal transmission in the reverse direction.

The great advantage of the common base h-parameters is the ease with which they can be all measured accurately and directly in practice. In Section 3.4 it will be shown that the common base configuration has a very low input resistance and a high output resistance. Consequently, it should be an easy matter to simulate an accurate open circuit across the input terminals of the common base configuration, and to simulate an accurate short circuit across its output terminals for a.c. signals. This should then enable an accurate evaluation of all the common base h-parameters by applying the definitions of Equations 3.7 and 3.8.

The disadvantage of the h-parameters equivalent circuit of Fig. 3.4 is an analytical one and it is due to the fact that it involves a current controlled current source as well as a voltage controlled voltage source. In practice, however, it is often found preferable to use an equivalent circuit representation that involves one controlled source

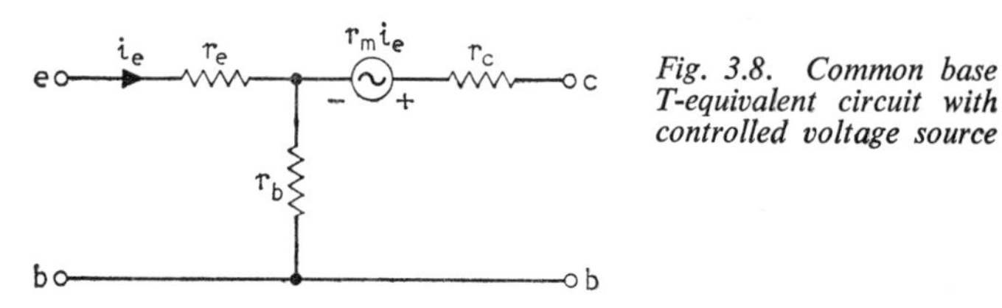

Fig. 3.8. Common base T-equivalent circuit with controlled voltage source

only. In order to develop such an equivalent circuit let us first solve the second line of Equation 3.6 for the output voltage v_{cb} and so obtain

$$v_{cb} = -\frac{h_{fb}}{h_{ob}} i_e + \frac{1}{h_{ob}} i_c \tag{3.9}$$

Next, on eliminating v_{cb} between Equation 3.9 and the first line of Equation 3.6 and collecting terms

$$v_{eb} = \left(h_{ib} - \frac{h_{rb} h_{fb}}{h_{ob}}\right) i_e + \frac{h_{rb}}{h_{ob}} i_c \tag{3.10}$$

Further, Equations 3.10 and 3.9 can be rearranged as follows, respectively:

$$\left.\begin{aligned} v_{eb} &= r_e i_e + r_b(i_e + i_c) \\ v_{cb} &= r_b(i_e + i_c) + r_m i_e + r_c i_c \end{aligned}\right\} \tag{3.11}$$

where the parameters r_e, r_b, r_m and r_c are related to the common base h-parameters by

$$\left.\begin{aligned} r_e &= h_{ib} - \frac{(1+h_{fb})\,h_{rb}}{h_{ob}} \\ r_b &= \frac{h_{rb}}{h_{ob}} \\ r_m &= -\frac{h_{fb}+h_{rb}}{h_{ob}} \\ r_c &= \frac{1-h_{rb}}{h_{ob}} \end{aligned}\right\} \qquad (3.12)$$

Normally, $h_{rb} \ll 1$ and the absolute value of h_{fb} is quite close to unity. Hence, the parameters r_m and r_c may be approximated by the simpler expressions:

$$\left.\begin{aligned} r_m &\simeq -\frac{h_{fb}}{h_{ob}} \\ r_c &\simeq \frac{1}{h_{ob}} \end{aligned}\right\} \qquad (3.13)$$

Equation 3.11 leads directly to the equivalent circuit of Fig. 3.8. Here we observe that the 3 arms of the circuit form a T-section, and that the base terminal is common to the input and output circuits. For these reasons the circuit of Fig. 3.8 is usually referred to as the *common base* T-*equivalent circuit.*

The circuit of Fig. 3.8 involves a *current controlled voltage source* $r_m i_e$. It is sometimes found useful to employ a controlled current source. This can be achieved by applying Norton's Theorem* to the branch consisting of the resistance r_c in series with the voltage source $r_m i_e$, resulting in the T-equivalent circuit of Fig. 3.9, which involves the current controlled current source $a_b i_e$. The control parameter a_b is given by:

$$a_b = \frac{r_m}{r_c} \qquad (3.14)$$

From Equations 3.13 and 3.14 we see that a_b is very nearly equal to $-h_{fb}$ and, since h_{fb} itself is equal to $-\alpha_b$ (see Equation 3.5) the parameter a_b of Fig. 3.9 is very nearly equal to α_b. That is

$$a_b \simeq \alpha_b \qquad (3.15)$$

* See the Appendix.

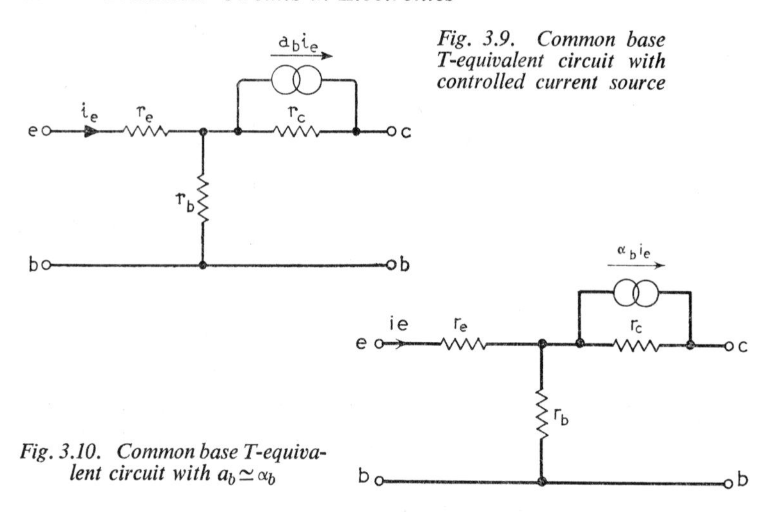

Fig. 3.9. Common base T-equivalent circuit with controlled current source

Fig. 3.10. Common base T-equivalent circuit with $a_b \simeq \alpha_b$

For this reason we shall ignore the distinction between the two quantities a_b and α_b and so only use the latter. Therefore, we can replace the controlled current source $a_b i_e$ of the T-equivalent circuit of Fig. 3.9 by the controlled current source $\alpha_b i_e$ of Fig. 3.10.

Example 3.1

Evaluate the common base T-equivalent circuit parameters for a transistor having the following common base h-parameters:

$$h_{ib} = 28\Omega$$

$$h_{rb} = 2 \times 10^{-4}$$

$$h_{fb} = -0{\cdot}98$$

$$h_{ob} = 0{\cdot}5 \times 10^{-6}\,\text{S}$$

Using Equation 3.12 we find that the common base T-equivalent circuit parameters r_e and r_b have the following values:

$$r_e = 28 - \frac{(1-0{\cdot}98)\,2 \times 10^{-4}}{0{\cdot}5 \times 10^{-6}}$$

$$= 20\Omega$$

$$r_b = \frac{2 \times 10^{-4}}{0{\cdot}5 \times 10^{-6}}$$

$$= 400\Omega$$

Using Equation 3.13 we obtain the following values for r_m and r_c:

$$r_m = \frac{0{\cdot}98}{0{\cdot}5 \times 10^{-6}}$$

$$= 1{\cdot}96\text{M}\Omega$$

$$r_c = \frac{1}{0{\cdot}5 \times 10^{-6}}$$

$$= 2\text{M}\Omega$$

Finally from Equations 3.14 and 3.15 we deduce that $\alpha_b = 0{\cdot}98$.

The equivalent circuits of Figs. 3.8 and 3.10 are suitable for analysing circuits involving the common base configuration. In each of these 2 circuits the active elements, namely, $r_m i_e$ and $\alpha_b i_e$ are controlled by the input emitter current i_e. In the common emitter configuration of Fig. 3.11, however, we see that the input current is equal to the base current i_b. Therefore, it would be desirable to transform the 2 equivalent circuits of Figs. 3.8 and 3.10 such that the active element in each case is controlled by the input base current. For this purpose let us first rearrange the T-equivalent circuit of Fig. 3.8 so that its terminals correspond to those of the

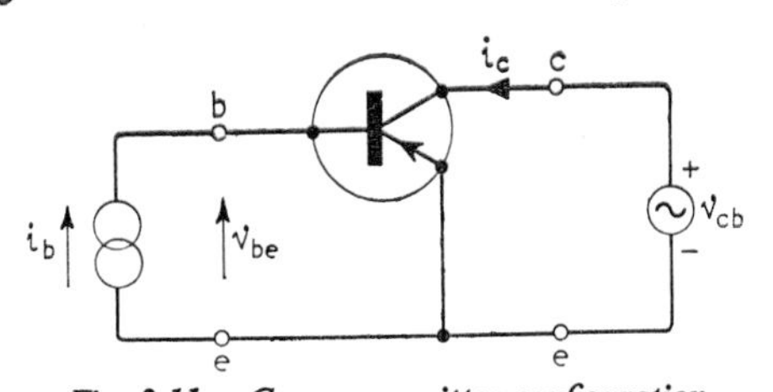

Fig. 3.11. Common emitter configuration

common emitter configuration of Fig. 3.11. This has been done in Fig. 3.12(a). The base current i_b, emitter current i_e and collector current i_c are related as follows:

$$i_e = -(i_b + i_c) \tag{3.16}$$

Therefore, we find that the controlled voltage source $r_m i_e$ is equivalent to 2 controlled voltage sources of $-r_m i_b$ and $-r_m i_c$ connected in series as shown in Fig. 3.12(b). Since the current flowing through each of these two controlled sources is equal to i_c, it follows that a resistance of $-r_m$ can be substituted for the controlled source of $-r_m i_c$ as shown in Fig. 3.12(c). Finally, if we reverse the polarity of the remaining controlled source $-r_m i_b$ we can ignore the minus sign associated with it as shown in Fig. 3.13. In this diagram we have also denoted the resistance $r_c - r_m$ by r_d, that is,

$$r_d = r_c - r_m$$

$$= r_c\left(1 - \frac{r_m}{r_c}\right)$$

$$\simeq r_c(1 - \alpha_b) \tag{3.17}$$

where we have made use of Equations 3.14 and 3.15.

Next, if in Fig. 3.13 we apply Norton's Theorem to the branch consisting of the resistance r_d in series with the controlled voltage

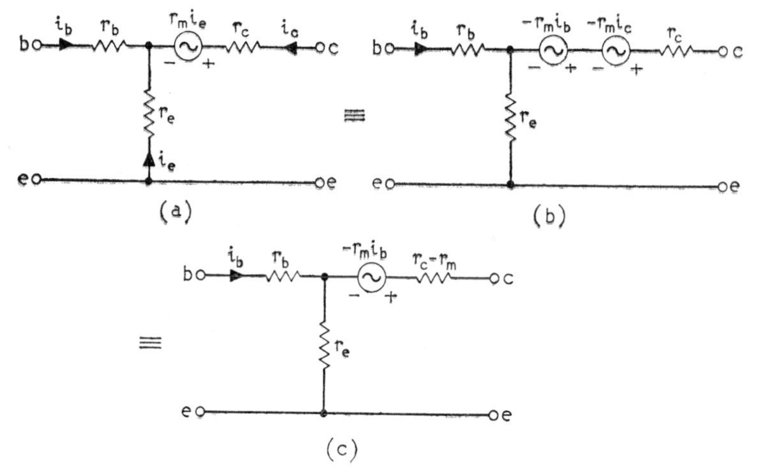

Fig. 3.12. Common emitter T-equivalent circuit

source $r_m i_b$ we obtain the modified circuit representation of Fig. 3.14, where the common emitter current amplification factor α_e is given by

$$\alpha_e = \frac{\alpha_b}{1 - \alpha_b}$$

$$\simeq \frac{r_m}{r_d} \tag{3.18}$$

The equivalent circuits of Figs. 3.13 and 3.14 are known as the *common emitter* T-*equivalent circuits*, and they are useful for analysing the external circuit response of the common emitter configuration.

Comparing, on the one hand, the common base configuration of Fig. 3.5 and its 2 T-equivalent circuits of Figs. 3.8 and 3.10 and, on

Table 3.1. ANALOGOUS QUANTITIES IN COMMON BASE AND COMMON EMITTER T-EQUIVALENT CIRCUITS

Common base	*Common emitter*
i_e	i_b
r_e	r_b
r_b	r_e
r_c	r_d
r_m	$-r_m$
α_b	α_e

the other hand, the common emitter configuration of Fig. 3.11 and its 2 T-equivalent circuits of Figs. 3.13 and 3.14, we can deduce the analogies listed in Table 3.1. This Table is found to be useful when a certain quantity associated with the common base configuration is known and it is desired to determine the corresponding quantity associated with the common emitter configuration and vice versa. Thus let us use the Table for evaluating the *common emitter*

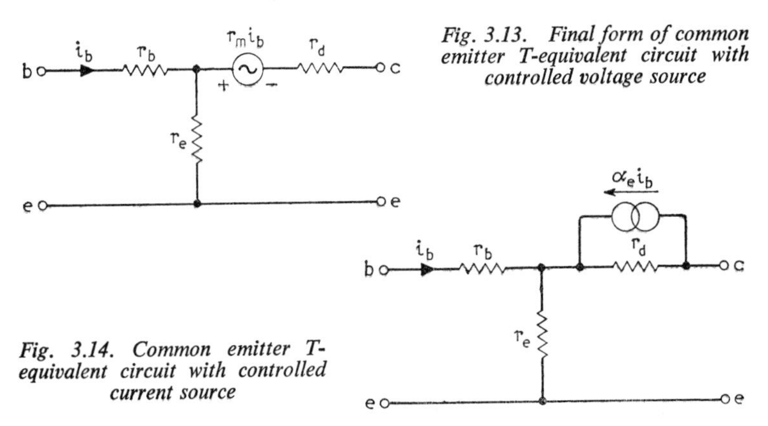

Fig. 3.13. Final form of common emitter T-equivalent circuit with controlled voltage source

Fig. 3.14. Common emitter T-equivalent circuit with controlled current source

h-*parameters.* With reference to the common emitter configuration of Fig. 3.11 we can define its h-parameters as follows:

$$\left.\begin{aligned} v_{be} &= h_{ie}\,i_b + h_{re}\,v_{ce} \\ i_c &= h_{fe}\,i_b + h_{oe}\,v_{ce} \end{aligned}\right\} \qquad (3.19)$$

where h_{ie}, h_{re}, h_{fe} and h_{oe} are the common emitter h-parameters. Equation 3.19 leads to the common emitter h-parameter equivalent circuit of Fig. 3.15. The input base current i_b, input base–emitter voltage v_{be}, output collector current i_c and output collector–emitter voltage v_{ce} are as defined in Fig. 3.11. The second line of Equation 3.13 relates h_{ob} to r_c. The parameter h_{ob} of the common base configuration corresponds to the parameter h_{oe} of the common emitter configuration. Further, from Table 3.1 we see that r_c of the common base T-equivalent circuit corresponds to r_d of the common emitter T-equivalent circuit. Therefore, if in the second line of Equation 3.13 we use r_d in place of r_c, and use h_{oe} in place of h_{ob} we obtain

$$\frac{1}{r_d} \simeq h_{oe} \tag{3.20}$$

Next examine the first line of Equation 3.13. Here we see that the parameters h_{fb} and h_{ob} of the common base configuration correspond

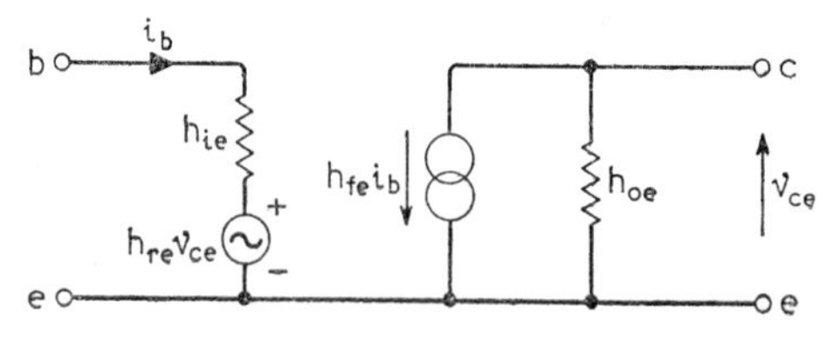

Fig. 3.15. Common emitter h-parameters equivalent circuit

to the parameters h_{fe} and h_{oe} of the common emitter configuration. From Table 3.1 we also see that r_m of the common base T-equivalent circuit corresponds to $-r_m$ of the common emitter T-equivalent circuit. Therefore, if in the first line of Equation 3.13 we use $-r_m$ in place of r_m, h_{fe} in place of h_{fb} and h_{oe} in place of h_{ob} we find

$$r_m \simeq \frac{h_{fe}}{h_{oe}} \tag{3.21}$$

Moving back to the second line of Equation 3.12 we see that the parameters h_{rb} and h_{ob} of the common base configuration correspond to the parameters h_{re} and h_{oe} of the common emitter configuration, respectively. Also from Table 3.1, r_b of the common base T-equivalent circuit corresponds to r_e of the common emitter T-equivalent circuit. Hence, using r_e in place of r_b, h_{re} in place of h_{rb}, and h_{oe} in place of h_{ob} in the second line of Equation 3.12 we obtain

$$r_e = \frac{h_{re}}{h_{oe}} \tag{3.22}$$

Finally, in the first line of Equation 3.12, if we use the common emitter parameters h_{ie}, h_{fe}, h_{re} and h_{oe} in the place of the common base parameters h_{ib}, h_{fb}, h_{rb} and h_{ob}, respectively, and use r_b in place of r_e as deduced from Table 3.1, we find that for the common emitter configuration:

$$r_b = h_{ie} - \frac{(1+h_{fe})\,h_{re}}{h_{oe}} \tag{3.23}$$

Therefore, knowing the common emitter h-parameters, we can determine the common emitter T-equivalent circuit parameters, by using Equations 3.20 to 3.23. Further, from Equations 3.12 and

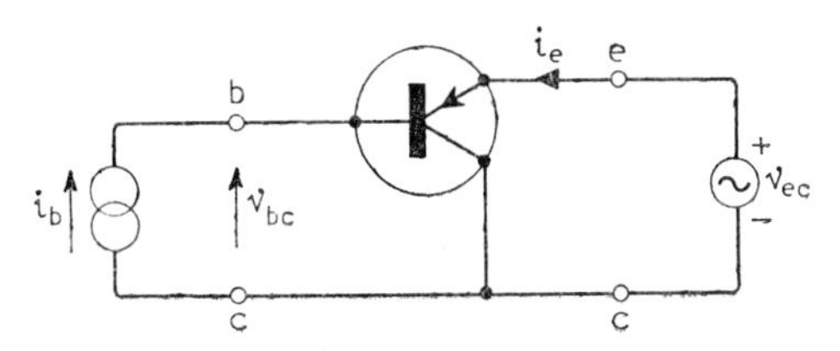

Fig. 3.16. Common collector configuration

3.20 to 3.23 we find that the common emitter h-parameters are related to the common base h-parameters as follows:

$$\left.\begin{aligned} h_{ie} &\simeq \frac{h_{ib}}{1+h_{fb}} \\ h_{fe} &\simeq \frac{-h_{fb}}{1+h_{fb}} \\ h_{re} &\simeq \frac{h_{ib}\,h_{ob}}{1+h_{fb}} - h_{rb} \\ h_{oe} &\simeq \frac{h_{ob}}{1+h_{fb}} \end{aligned}\right\} \tag{3.24}$$

So far we have discussed the equivalent circuit representations that are suitable for the common base and common emitter configurations. From Chapter 1 it is recalled that there exists a third useful way in which the transistor can be operated, and that is the common collector configuration shown in Fig. 3.16. Here we see that the input current is equal to the base current i_b. Therefore, for the common collector configuration we can use either of the 2 common emitter T-equivalent circuits of Figs. 3.13 and 3.14 provided that in each case we rearrange the terminals so as to correspond to those of the common collector configuration of Fig. 3.16. This has been done in the circuits of Figs. 3.17 and 3.18.

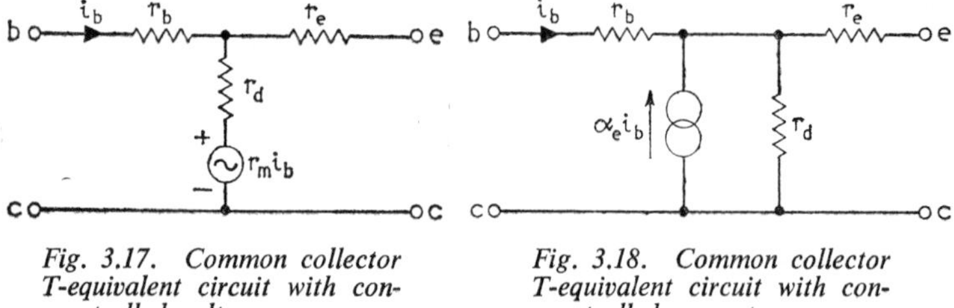

Fig. 3.17. Common collector T-equivalent circuit with controlled voltage source

Fig. 3.18. Common collector T-equivalent circuit with controlled current source

To complete the small signal equivalent circuit representations for the transistor we shall now consider the *common collector h-parameters.* The external behaviour of the common collector configuration of Fig. 3.16 can be defined in terms of its h-parameters h_{ic}, h_{rc}, h_{fc} and h_{oc} as

$$\left.\begin{aligned} v_{bc} &= h_{ic}\, i_b + h_{rc}\, v_{ec} \\ i_e &= h_{fc}\, i_b + h_{oc}\, v_{ec} \end{aligned}\right\} \qquad (3.25)$$

where the input base current i_b, input base–collector voltage v_{bc}, output emitter current i_e and output emitter–collector voltage v_{ec} are defined in Fig. 3.16. Equation 3.25 leads to the *common collector h-parameters equivalent circuit* of Fig. 3.19. The *common collector h-parameters* are related to the common emitter h-parameters by:

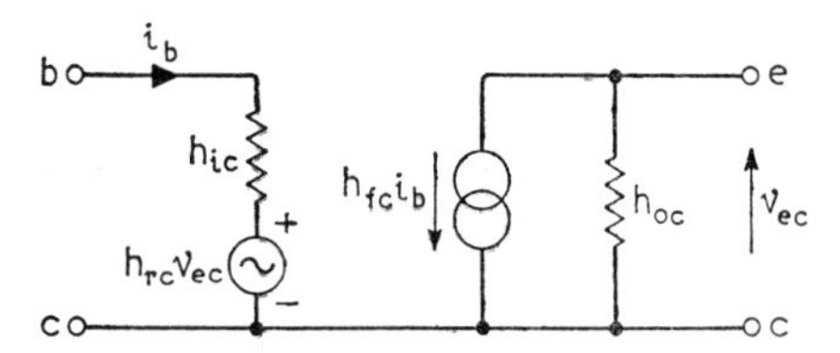

Fig. 3.19. Common collector h-parameters equivalent circuit

$$\left.\begin{aligned} h_{ic} &= h_{ie} \\ h_{rc} &= 1 \\ h_{fc} &= -(1+h_{fe}) \\ h_{oc} &= h_{oe} \end{aligned}\right\} \qquad (3.26)$$

The first and third relationships can be established as follows. From Equation 3.25 we see that in order to evaluate h_{ic} and h_{fc} the output voltage v_{ec} should be reduced to zero, in which case

$$\left.\begin{aligned} h_{ic} &= \left(\frac{v_{bc}}{i_b}\right)_{v_{ec}=0} \\ h_{fc} &= \left(\frac{i_e}{i_b}\right)_{v_{ec}=0} \end{aligned}\right\} \tag{3.27}$$

Further, in Equation 3.19 if v_{ce} is reduced to zero we obtain the following definitions for the common emitter h-parameters

$$\left.\begin{aligned} h_{ie} &= \left(\frac{v_{be}}{i_b}\right)_{v_{ce}=0} \\ h_{fe} &= \left(\frac{i_c}{i_b}\right)_{v_{ce}=0} \end{aligned}\right\} \tag{3.28}$$

Next, from Figs. 3.11 and 3.16

$$v_{bc} = v_{be} + v_{ec} \tag{3.29}$$

Therefore, when v_{ec} is set equal to zero, then from Equation 3.29 it follows that $v_{bc} = v_{be}$; and, since the condition $v_{ec} = 0$ is exactly the same as $v_{ce} = 0$, we find from Equations 3.27 and 3.28 that h_{ic} and h_{ie} are equal.

Consider next Equation 3.16. Dividing both sides of this equation by i_b we obtain

$$\frac{i_e}{i_b} = -\left(1 + \frac{i_c}{i_b}\right)$$

in view of which it follows from Equations 3.27 and 3.28 that h_{fc} and h_{fe} must be related as in Equation 3.26.

The second and fourth relationships of Equation 3.26 can be established by putting $i_b = 0$ in Equation 3.25, when

$$\left.\begin{aligned} h_{rc} &= \left(\frac{v_{bc}}{v_{ec}}\right)_{i_b=0} \\ h_{oc} &= \left(\frac{i_e}{v_{ec}}\right)_{i_b=0} \end{aligned}\right\} \tag{3.30}$$

However, when $i_b = 0$ we find from the T-equivalent circuit of Fig. 3.18 that h_{rc} is equal to $r_d/(r_d + r_e)$ and that h_{oc} is equal to $1/(r_d + r_e)$. From Equation 3.30, since r_d is normally much larger than r_e, we find that h_{rc} and h_{oc} are as given in Equation 3.26.

Example 3.2

Evaluate the common emitter and common collector h-parameters for the transistor whose common base h-parameters are as given in Example 3.1 (page 44).

From Equation 3.24 we find that the common emitter h-parameters have the following values:

$$h_{ie} = \frac{28}{1-0{\cdot}98}$$

$$= 1{,}400\Omega$$

$$h_{fe} = \frac{0{\cdot}98}{1-0{\cdot}98}$$

$$= 49$$

$$h_{re} = \frac{28 \times 0{\cdot}5 \times 10^{-6}}{1-0{\cdot}98} - 2 \times 10^{-4}$$

$$= 5 \times 10^{-4}$$

$$h_{oe} = \frac{0{\cdot}5 \times 10^{-6}}{1-0{\cdot}98} = 25 \times 10^{-6}\,\text{S}$$

Having evaluated the common emitter h-parameters we can go on to apply Equation 3.26 to obtain the following results for the common collector h-parameters:

$$h_{ic} = 1400\Omega$$

$$h_{rc} = 1$$

$$h_{fc} = -(1+49) = -50$$

$$h_{oc} = 25 \times 10^{-6}\,\text{S}$$

3.2. EXTERNAL CIRCUIT PROPERTIES

The transistor is basically a 3-terminal device, as illustrated in Fig. 3.20. Normally, one terminal is used as the input, another one is the output and the remaining terminal is arranged to be common to the input and output circuits. Therefore, the 4-terminal network shown in Fig. 3.21 can be assumed to represent any of the 3 basic transistor configurations.

Let the 4-terminal network be coupled to a voltage source v_s of resistance R_s at its input terminals 11′, and be closed by a load resistance R_L across its output terminals 22′ as shown in Fig. 3.22. The external circuit properties that are of particular interest are

(1) *Input resistance* R_{in}. It is defined as the resistance measured looking into the input terminals 11′ of the 4-terminal network

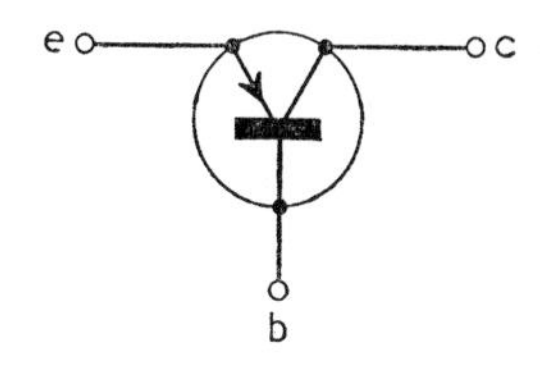

Fig. 3.20. The transistor as a 3-terminal network

when its output terminals are closed with the load resistance R_L as in Fig. 3.23, that is

$$R_{\text{in}} = \frac{v_1}{i_1} \tag{3.31}$$

where v_1 is the input voltage and i_1 is the input current.

(2) *Voltage gain* K_v. It is defined as the ratio of the output voltage v_2 developed across the load to the input voltage v_1 applied to the network, that is:

$$K_v = \frac{v_2}{v_1} \tag{3.32}$$

Fig. 3.21. General 4-terminal network

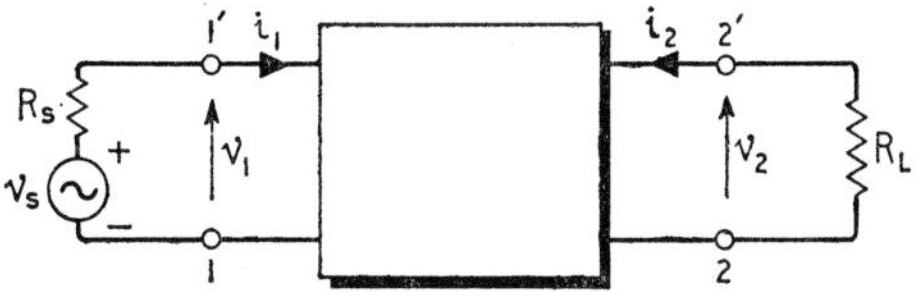

Fig. 3.22. Terminated 4-terminal network

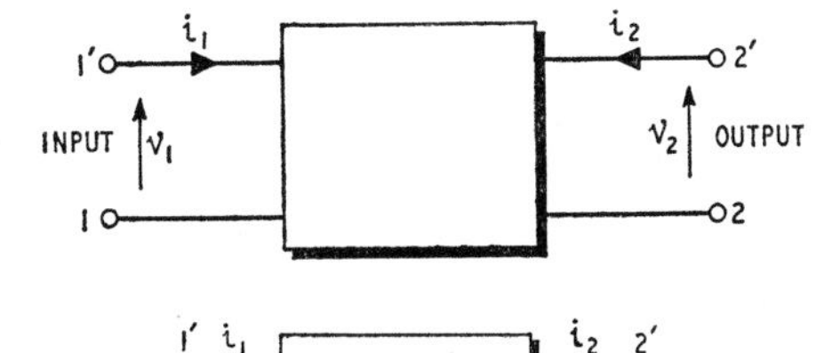

Fig. 3.23. Four-terminal network with R_L across output terminals

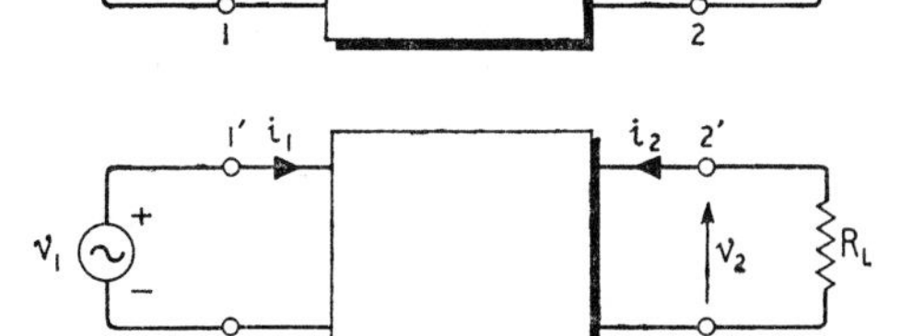

(3) *Current gain* K_i. It is defined as the ratio of the output current i_2, delivered to the load, to the input current i_1 applied to the network, that is

$$K_i = \frac{i_2}{i_1} \tag{3.33}$$

At this stage it is useful to relate the input resistance, voltage gain and current gain. For this purpose we can rewrite Equation 3.33 as follows

$$K_i = \frac{i_2}{v_2} \cdot \frac{v_2}{v_1} \cdot \frac{v_1}{i_1}$$

$$= \frac{\dfrac{v_2}{v_1} \cdot \dfrac{v_1}{i_1}}{\dfrac{v_2}{i_2}} \tag{3.34}$$

However, from Fig. 3.23

$$v_2 = -i_2 R_L \tag{3.35}$$

Therefore, using Equations 3.31 to 3.35 we obtain

$$K_i = -\frac{K_v R_{in}}{R_L} \tag{3.36}$$

Alternatively, this relationship can be expressed as

$$K_v = -\frac{K_i R_L}{R_{in}} \tag{3.37}$$

(4) *Power gain* K_p. It is defined as the ratio of output power, delivered to the load, to input power applied to the network, that is,

$$K_p = \left| \frac{v_2 i_2}{v_1 i_1} \right|$$

$$= |K_v K_i|$$

Using the relationships of Equations 3.36 and 3.37 we find that

$$K_p = \frac{K_v^2 R_{in}}{R_L}$$

$$= \frac{K_i^2 R_L}{R_{in}} \tag{3.38}$$

The power gain K_p is usually expressed in *decibels*, denoted by dB and defined as

$$10\log_{10} K_p = 10\log_{10}\left(\frac{K_v{}^2 R_{\text{in}}}{R_L}\right)$$

$$= 20\log_{10} K_v + 10\log_{10}\left(\frac{R_{\text{in}}}{R_L}\right) \tag{3.39}$$

If the input resistance of the network is equal to the terminating load resistance R_L, then from Equation 3.39 the power gain is equal to $20\log_{10} K_v$ dB.

(5) *Output resistance* R_{out}. This is defined as the resistance measured looking into the output terminals 22′ of the 4-terminal network when its input terminals 11′ are closed with the source resistance R_s as in Fig. 3.24, that is,

$$R_{\text{out}} = \frac{v_2}{i_2} \tag{3.40}$$

(6) *External voltage gain K.* This is defined as the ratio of the output voltage v_2, developed across the load, to the driving source voltage v_s (see Fig. 3.22), that is,

$$K = \frac{v_2}{v_s} \tag{3.41}$$

Now from Fig. 3.22

$$v_s = v_1 + i_1 R_s \tag{3.42}$$

in view of which we can rewrite Equation 3.41 as follows:

$$K = \frac{v_2}{v_1 + i_1 R_s}$$

$$= \frac{\dfrac{v_2}{i_1}}{\dfrac{v_1}{i_1} + R_s}$$

$$= \frac{\dfrac{v_2}{i_2}\cdot\dfrac{i_2}{i_1}}{\dfrac{v_1}{i_1} + R_s} \tag{3.43}$$

Next, using Equations 3.31, 3.33, 3.35 and 3.43 we find that the external voltage gain K is related to the input resistance R_{in} and current gain K_i by

$$K = -\frac{R_L K_i}{R_{in}+R_s} \tag{3.44}$$

It is important to note that, unlike the earlier gain definitions, the external voltage gain K depends on the resistance R_s of the driving source.

We shall now apply the above definitions to the various transistor circuit configurations. The analysis will be first carried out in

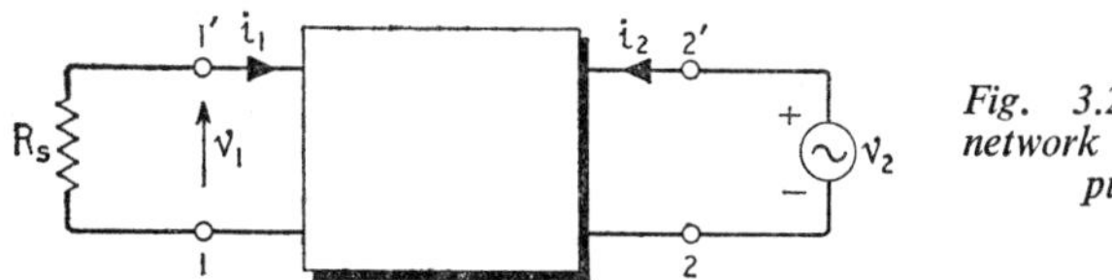

Fig. 3.24. Four-termina network with R_s across input terminals

terms of the h-parameters and then in terms of the T-equivalent circuit representations that were developed in Section 3.1.

3.3. ANALYSIS IN TERMS OF h-PARAMETERS

Consider the 4-terminal network of Fig. 3.22, which can represent any of the 3 basic transistor configurations. If we take h_i, h_r, h_f and h_o to be the general h-parameters of the 4-terminal network of Fig. 3.22, then the input current i_1, input voltage v_1, output current i_2 and output voltage v_2 are related as follows*

$$\left.\begin{aligned} v_1 &= h_i i_1 + h_r v_2 \\ i_2 &= h_f i_1 + h_o v_2 \end{aligned}\right\} \tag{3.45}$$

Eliminating v_2 between Equation 3.35 and the second line of Equation 3.45 and then solving for the current gain K_i, we find

$$K_i = \frac{i_2}{i_1}$$

$$= \frac{h_f}{1+h_o R_L} \tag{3.46}$$

* Equation 3.45 provides only one possible set for relating the input and output quantities. Indeed there are 5 other possibilities. See HAYKIN, S. S., *Junction Transistor Circuit Analysis*, Iliffe Books Ltd. (1962).

Next, from Equation 3.35, the first line of Equation 3.45 and 3.46 we obtain

$$v_1 = h_i i_1 - \frac{h_f h_r R_L}{1+h_o R_L} i_1$$

Therefore, Equation 3.31 gives the input resistance R_{in} in terms of the h-parameters as

$$R_{in} = h_i - \frac{h_f h_r R_L}{1+h_o R_L} \tag{3.47}$$

Notice from this equation that the input resistance depends on the terminating load resistance. It is for this reason that the transistor is often referred to as a *non-unilateral device* in that the conditions at the output terminals influence those existing at the input.

Let us next evaluate the output resistance R_{out} in terms of the h-parameters. From Fig. 3.24

$$v_1 = -i_1 R_s \tag{3.48}$$

From Equations 3.45 and 3.48 we find

$$i_2 = -\frac{h_f h_r}{h_i + R_s} v_2 + h_o v_2$$

which, together with Equation 3.40 leads to the following expression for the output conductance G_{out} (that is, the inverse of the output resistance R_{out}),

$$G_{out} = \frac{1}{R_{out}}$$

$$= h_o - \frac{h_f h_r}{h_i + R_s} \tag{3.49}$$

This expression shows that the output resistance R_{out} depends on the terminating source resistance, emphasising further the non-unilateral nature of a transistor.

Equations 3.46, 3.47 and 3.49 are considered to be adequate because, knowing K_i and R_{in}, Equations 3.37, 3.38 and 3.44 can be used for evaluating the voltage gain K_v, power gain K_p and external voltage gain K, respectively. Further, Equations 3.46, 3.47 and 3.49 can be applied to any of the 3 transistor configurations; all that is required is that we should use the appropriate h-parameters. Thus, for example, if we wish to evaluate the input resistance of a common emitter stage for a specified value of load resistance R_L, we use the values of the common emitter h-parameters h_{ie}, h_{fe}, h_{re}

and h_{oe} in place of the general parameters h_i, h_f, h_r and h_o, respectively, in Equation 3.47. Similarly for the other external circuit properties and the other 2 transistor configurations. This we shall now demonstrate by way of an example.

Example 3.3

A transistor has the following common base h-parameters:

$$h_{ib} = 28\Omega$$

$$h_{rb} = 2 \times 10^{-4}$$

$$h_{fb} = -0{\cdot}98$$

$$h_{ob} = 0{\cdot}5 \times 10^{-6}\text{S}$$

Evaluate the current gain, input resistance, output resistance, voltage gain, power gain and external voltage gain of the terminated common base stage of Fig. 3.25 given that

$$R_s = 100\Omega$$

$$R_L = 10\text{k}\Omega$$

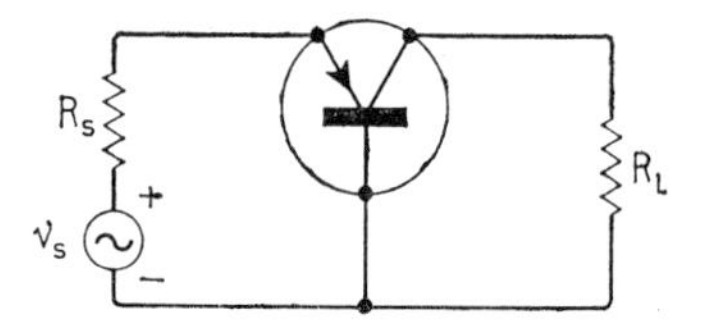

Fig. 3.25. Terminated common base stage

Using the given values of h_{ib}, h_{rb}, h_{fb} and h_{ob} in place of h_i, h_r, h_f and h_o, respectively, in Equations 3.46 and 3.47 we find that for $R_L = 10\text{k}\Omega$

$$K_i = \frac{-0{\cdot}98}{1+0{\cdot}5\times 10^{-6}\times 10^4}$$

$$= -0{\cdot}975$$

$$R_{\text{in}} = 28 + \frac{0{\cdot}98\times 2\times 10^{-4}\times 10^4}{1+0{\cdot}5\times 10^{-6}\times 10^4}$$

$$= 30\Omega$$

Next using Equation 3.49 we find that for $R_s = 100\Omega$ the output resistance is given by

$$\frac{1}{R_{\text{out}}} = 0{\cdot}5 \times 10^{-6} + \frac{0{\cdot}98 \times 2 \times 10^{-4}}{28+100}$$

$$= 2{\cdot}04 \times 10^{-6} \mho$$

Therefore

$$R_{\text{out}} = \frac{1}{2{\cdot}04 \times 10^{-6}}$$

$$= 0{\cdot}49 \text{M}\Omega$$

Finally using Equations 3.37, 3.38 and 3.44 we obtain that:

$$K_v = \frac{0{\cdot}975 \times 10^4}{30}$$

$$= 325$$

$$K_p = 0{\cdot}975 \times 325$$

$$= 317$$

$$K = \frac{10^4 \times 0{\cdot}975}{30+100}$$

$$\simeq 75$$

3.4. ANALYSIS USING THE T-EQUIVALENT CIRCUITS

Consider next how the various T-equivalent circuits of Section 3.1 can be used for evaluating the external circuit properties of the 3 transistor configurations.

3.4.1. The Common Base Stage

Fig. 3.26 shows a common base stage having a load resistance R_L connected across its output terminals and a voltage v_{eb} applied across its input terminals. The input resistance R_{in} is thus equal to v_{eb}/i_e and the current gain K_i is equal to i_c/i_e, where the emitter current i_e and collector current i_c are as defined in Fig. 3.26. In order to evaluate R_{in} and K_i let us replace the transistor with the common base T-equivalent circuit of Fig. 3.8, resulting in the

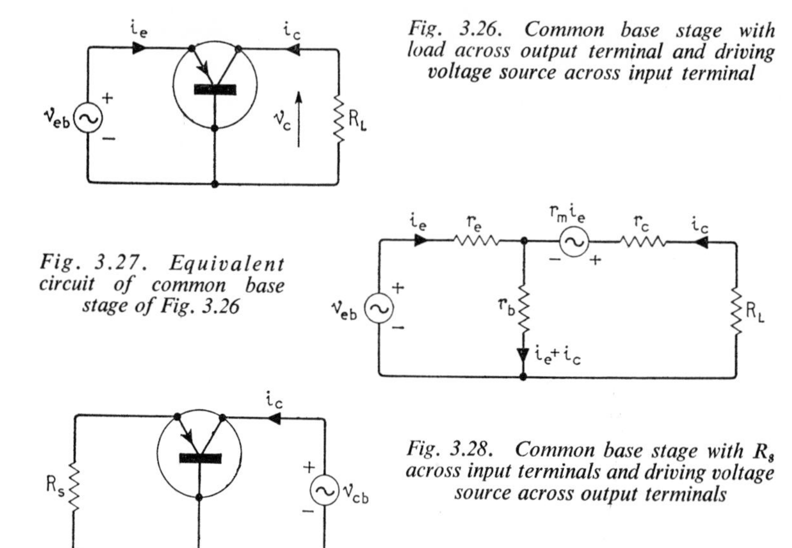

Fig. 3.26. Common base stage with load across output terminal and driving voltage source across input terminal

Fig. 3.27. Equivalent circuit of common base stage of Fig. 3.26

Fig. 3.28. Common base stage with R_s across input terminals and driving voltage source across output terminals

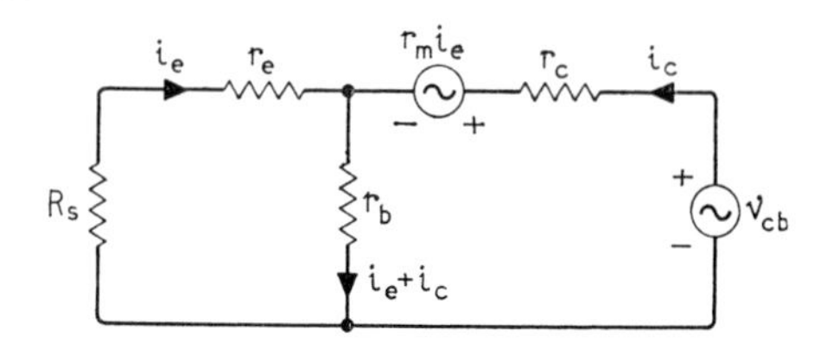

Fig. 3.29. Equivalent circuit of common base stage of Fig. 3.28

circuit representation of Fig. 3.27. Applying Kirchoff's Voltage Law to the input loop of Fig. 3.27 we find

$$\begin{aligned} v_{eb} &= r_e i_e + r_b(i_e + i_c) \\ &= (r_e + r_b)\, i_e + r_b\, i_c \end{aligned} \tag{3.50}$$

and applying it to the output loop we obtain

$$-r_m i_e = r_b(i_e + i_c) + (r_c + R_L)\, i_c \tag{3.51}$$

Collecting terms in Equation 3.51 and solving for the ratio of i_c/i_e we find that the common base current gain is

$$\begin{aligned} K_i &= \frac{i_c}{i_e} \\ &= -\frac{r_b + r_m}{r_b + r_c + R_L} \end{aligned} \tag{3.52}$$

From Equations 3.50 and 3.52 we find that the input resistance of he common base stage is given by

$$R_{\text{in}} = \frac{v_{eb}}{i_e}$$

$$= r_e + r_b \cdot \frac{R_L + r_c - r_m}{R_L + r_c + r_b} \tag{3.53}$$

For evaluating the output resistance, consider Fig. 3.28, where the common base stage is terminated with the source resistance R_s at its input terminals and a voltage v_{cb} is applied across its output terminals. The output resistance is then equal to v_{cb}/i_c, where i_c is the resulting collector current as defined in Fig. 3.28. Replacing the transistor with the common base T-equivalent circuit of Fig. 3.8 we obtain the circuit representation of Fig. 3.29. Applying Kirchoff's Voltage Law to the input loop gives:

$$0 = (R_s + r_e)\, i_e + r_b(i_e + i_c)$$

Collecting terms and solving for i_e we find

$$i_e = -\frac{r_b}{R_s + r_e + r_b} i_c \tag{3.54}$$

Applying Kirchoff's Voltage Law to the output loop of Fig. 3.29 gives

$$v_{cb} - r_m i_e = r_c i_c + r_b(i_e + i_c)$$

Collecting terms we obtain

$$v_{cb} = (r_c + r_b)\, i_c + (r_b + r_m)\, i_e \tag{3.55}$$

From Equations 3.54 and 3.55 we find that the output resistance of the common base stage is

$$R_{\text{out}} = r_c - r_b \frac{r_m - R_s - r_e}{r_e + R_s + r_b} \tag{3.56}$$

Finally, by using the relationships of Equations 3.37, 3.52 and 3.53 we find that the voltage gain of the common base stage is given by

$$K_v = \frac{(r_m + r_b)\, R_L}{r_e(r_c + r_b + R_L) + r_b(r_c - r_m + R_L)} \tag{3.57}$$

Table 3.2. EXTERNAL CIRCUIT PROPERTIES OF COMMON BASE STAGE

	Accurate	*Approximate*
R_{in}	$r_e+r_b\dfrac{r_c-r_m+R_L}{r_c+r_b+R_L}$	$r_e+r_b(1-\alpha_b)$
R_{ou}	$r_c-r_b\dfrac{r_m-R_s-r_e}{r_e+R_s+r_b}$	r_c
K_i	$-\dfrac{r_b+r_m}{r_b+r_c+R_L}$	$-\alpha_b$
K_v	$\dfrac{(r_m+r_b)\,R_L}{r_e(r_c+r_b+R_L)+r_b(r_c-r_m+R_L)}$	$\dfrac{\alpha_b\,R_L}{r_e+r_b(1-\alpha_b)}$

All the above results relating to the common base stage have been collected together in Table 3.2.

Example 3.4

The transistor considered in Examples 3.1 and 3.3 has the following common base T-equivalent circuit parameters:

$$r_e = 20\Omega$$

$$r_b = 400\Omega$$

$$r_c = 2\text{M}\Omega$$

$$r_m = 1{\cdot}96\text{M}\Omega$$

Calculate the current gain, input resistance, output resistance and voltage gain of this transistor when it is operated as a common base stage, given that

$$R_s = 100\Omega$$

$$R_L = 10\text{k}\Omega$$

Equation 3.52 gives the common base current gain to be

$$K_i = \frac{-(400+1{\cdot}96\times 10^6)}{400+2\times 10^6+10^4}$$

$$= \quad 0{\cdot}975$$

The input resistance, for $R_L = 10\text{k}\Omega$, is deduced from Equation 3.53 to be

$$R_{\text{in}} = 20+400\frac{10^4+2\times10^6-1{\cdot}96\times10^6}{10^4+2\times10^6+400}$$

$$= 30\Omega$$

From Equation 3.56 we find that the output resistance is

$$R_{\text{out}} = 2\times10^6-400\frac{1{\cdot}96\times10^6-100-20}{20+100+400}$$

$$= 0{\cdot}49\times10^6\Omega$$

Finally, Equation 3.37 gives the voltage gain to be:

$$K_v = \frac{0{\cdot}975\times10^4}{30}$$

$$= 325$$

Notice that, for the same transistor, the h-parameters and T-equivalent circuit methods of analysis lead to the same values for the external circuit properties of the transistor and so they should.

3.4.2. The Common Emitter Stage

Fig. 3.30 shows a common emitter stage connected to a load resistance R_L at its output terminals and having a voltage v_{be} applied across its input terminals. The input resistance of the common emitter stage is equal to v_{be}/i_b and its current gain is equal to i_c/i_b, where the input base current i_b and output collector current i_c are as defined in Fig. 3.30.

The circuit response of the transistor stage of Fig. 3.30 is best analysed using the common emitter T-equivalent circuit of Fig. 3.13. Therefore, replacing the transistor with this equivalent circuit we obtain Fig. 3.31. Applying Kirchoff's Voltage Law to the input loop we find

$$v_{be} = r_b i_b + r_e(i_b+i_c)$$

$$= (r_b+r_e)\, i_b + r_e i_c \tag{3.58}$$

while applying it to the output loop we obtain

$$r_m i_b = (r_d+R_L)\, i_c + r_e(i_b+i_c)$$

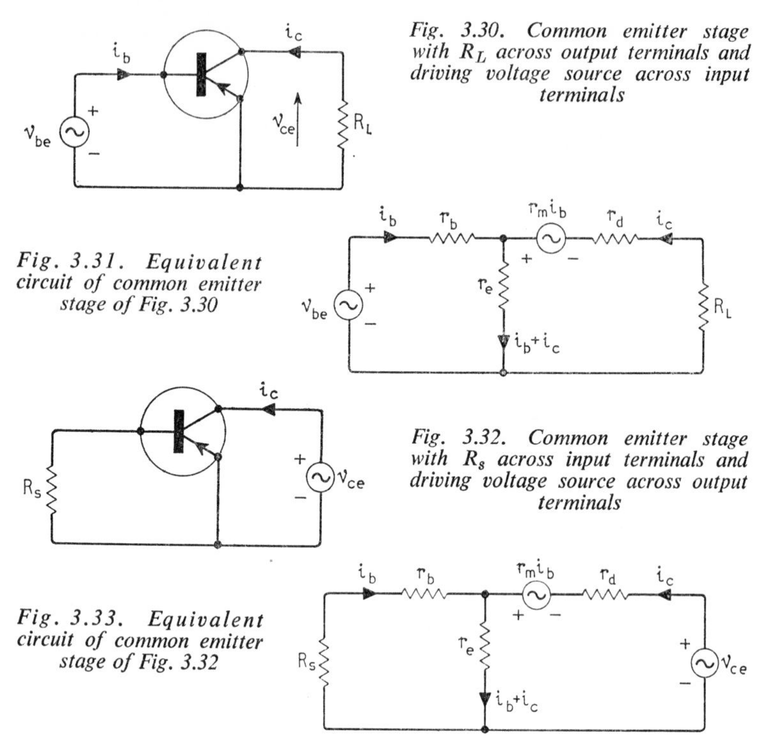

Fig. 3.30. Common emitter stage with R_L across output terminals and driving voltage source across input terminals

Fig. 3.31. Equivalent circuit of common emitter stage of Fig. 3.30

Fig. 3.32. Common emitter stage with R_s across input terminals and driving voltage source across output terminals

Fig. 3.33. Equivalent circuit of common emitter stage of Fig. 3.32

Collecting terms and solving for the current ratio i_c/i_b we find that the current gain of the common emitter stage is

$$K_i = \frac{i_c}{i_b}$$

$$= \frac{r_m - r_e}{r_d + R_L + r_e} \tag{3.59}$$

From Equations 3.58 and 3.59 we find that the input resistance of the common emitter stage is given by

$$R_{\text{in}} = \frac{v_{be}}{i_b}$$

$$= r_b + r_e \cdot \frac{r_c + R_L}{r_d + r_e + R_L} \tag{3.60}$$

where we have made use of the fact that r_d is equal to $r_c - r_m$ (see Equation 3.17).

Let us next evaluate the output resistance of the common emitter configuration. For this purpose consider Fig. 3.32 where the input terminals of the common emitter stage have been closed with the source resistance R_s and voltage v_{ce} is applied across its output terminals. The output resistance is thus equal to v_{ce}/i_c. To evaluate it replace the transistor with the common emitter T-equivalent circuit of Fig. 3.13, resulting in Fig. 3.33. Applying Kirchoff's Voltage Law to the input loop gives

$$0 = (R_s + r_b)\, i_b + r_e(i_b + i_c)$$

Collecting terms and solving for i_b we obtain

$$i_b = -\frac{r_e}{r_e + r_b + R_s} i_c \tag{3.61}$$

Next, applying Kirchoff's Voltage Law to the output loop of Fig. 3.33 gives

$$v_{ce} + r_m i_b = r_d i_c + r_e(i_b + i_c)$$

Collecting terms

$$v_{ce} = (r_d + r_e)\, i_c + (r_e - r_m)\, i_b \tag{3.62}$$

From Equations 3.61 and 3.62 we find that the output resistance of the common emitter configuration is

$$R_{\text{out}} = \frac{v_{ce}}{i_c}$$

$$= r_d + r_e \frac{r_m + r_b + R_s}{r_e + r_b + R_s} \tag{3.63}$$

where again we have made use of Equation 3.17.

In order to find the voltage gain of the common emitter stage we use Equations 3.37, 3.59 and 3.60 and so obtain that

$$K_v = \frac{-(r_m - r_e)\, R_L}{r_e(r_c + R_L) + r_b(r_d + r_e + R_L)} \tag{3.64}$$

The above results for the common emitter stage have been collected together in Table 3.3.

Example 3.5

A transistor having the common base T-equivalent circuit parameters of Example 3.4 (page 62) is operated as a common emitter stage having $R_s = 100\Omega$ and $R_L = 10\text{k}\Omega$. Find the current gain, input resistance, output resistance and voltage gain of this stage.

Table 3.3. EXTERNAL CIRCUIT PROPERTIES OF COMMON EMITTER STAGE

	Accurate	*Approximate*
R_{in}	$r_b+r_e\dfrac{r_c+R_L}{r_d+r_e+R_L}$	$r_b+r_e(1+\alpha_e)$
R_{out}	$r_d+r_e\dfrac{r_m+r_b+R_s}{r_e+r_b+R_s}$	r_d
K_i	$\dfrac{r_m-r_e}{r_d+r_e+R_L}$	α_e
K_v	$\dfrac{-(r_m-r_e)\,R_L}{r_e(r_c+R_L)+r_b(r_d+r_e+R_L)}$	$\dfrac{-\alpha_e\,R_L}{r_b+r_e(1+\alpha_e)}$

First, apply Equation 3.17 to evaluate the parameter r_d

$$r_d = 2\times 10^6 - 1{\cdot}96\times 10^6 = 4\times 10^4\Omega$$

Next, from Equation 3.59 we find that the common emitter current gain has the following value when $R_L = 10^4\Omega$

$$K_i = \frac{1{\cdot}96\times 10^6-20}{4\times 10^4+10^4+20}$$

$$= 39{\cdot}2$$

From Equation 3.60 we find that the input resistance is

$$R_{in} = 400+20\frac{2\times 10^6+10^4}{4\times 10^4+20+10^4}$$

$$= 1{,}200\Omega$$

To evaluate the output resistance we apply Equation 3.63:

$$R_{out} = 4\times 10^4+20\frac{1{\cdot}96\times 10^6+400+100}{20+400+100}$$

$$= 115\times 10^3\Omega$$

Finally from Equation 3.37 we find that the voltage gain is

$$K_v = -\frac{39{\cdot}2\times 10^4}{1200}$$

$$= -326$$

The minus sign associated with the voltage gain indicates that the output voltage signal of a common emitter stage is 180° out of phase relative to the input signal. Such a phase reversal is not present in the common base stage as is observed from Example 3.4.

3.4.3. The Common Collector Stage

Fig. 3.34 shows a common collector stage having a load resistance R_L connected across its output terminals and a voltage v_{bc} applied across its input terminals. The input resistance of the stage is equal to

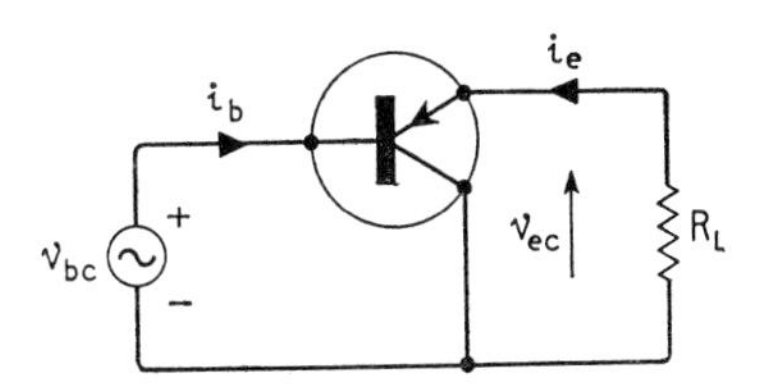

Fig. 3.34. Common collector stage with R_L across output terminals and driving voltage source across input terminals

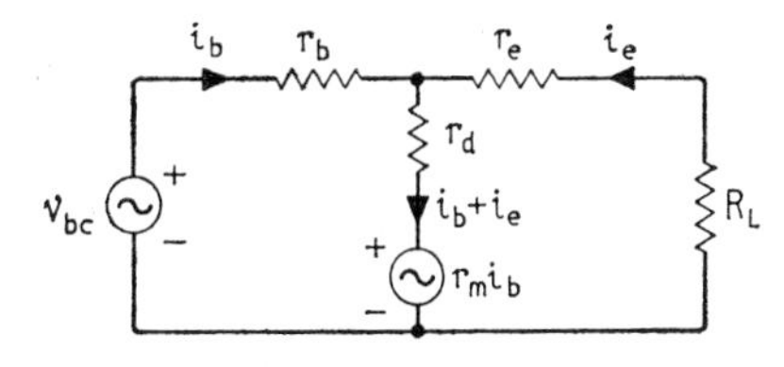

Fig. 3.35. Equivalent circuit of common collector stage of Fig. 3.34

v_{bc}/i_b and its current gain is equal to i_e/i_b. As pointed out earlier in Section 3.1 the common collector stage is best analysed using the common emitter T-equivalent circuit but with its terminals rearranged so as to correspond to those of the common collector stage. This was done in the equivalent circuit of Fig. 3.17. Therefore, using this equivalent circuit to replace the transistor in Fig. 3.34, we obtain Fig. 3.35. Applying Kirchoff's Voltage Law to the input loop then gives

$$v_{bc} - r_m i_b = r_b i_b + r_d(i_b + i_e)$$

Collecting terms and using the relationship of Equation 3.17 we find that

$$v_{bc} = (r_b + r_c) i_b + r_d i_e \tag{3.65}$$

Next, applying Kirchoff's Voltage Law to the output loop of Fig. 3.35 leads to the following

$$-r_m i_b = r_d(i_b + i_e) + (r_e + R_L) i$$

Collecting terms and solving for the current ratio i_e/i_b we find that the current gain of the common collector stage is

$$K_i = \frac{i_e}{i_b}$$

$$= \frac{-r_c}{r_d+r_e+R_L} \tag{3.66}$$

where again we have made use of Equation 3.17. From Equations 3.65 and 3.66 we find that the input resistance of the common collector stage is

$$R_{\text{in}} = \frac{v_{bc}}{i_b}$$

$$= r_b + r_c \frac{r_e+R_L}{r_d+r_e+R_L} \tag{3.67}$$

Consider next the evaluation of the common collector output resistance. In Fig. 3.36 the common collector stage has the source resistance R_s connected across its input terminals and a voltage v_{ec} applied across its output terminals. Replacing the transistor with

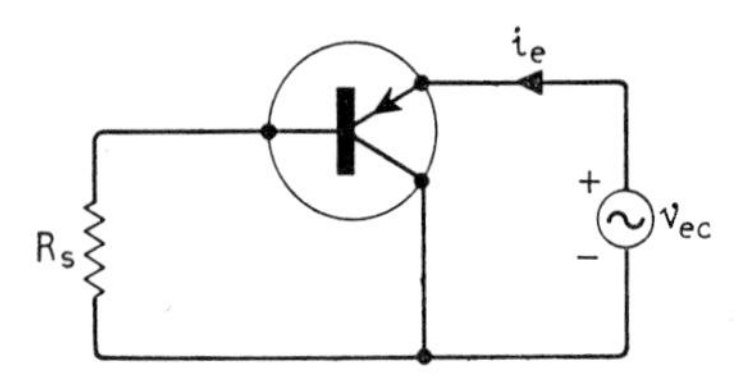

Fig. 3.36. Common collector stage with R_s across input terminals and driving voltage source across output terminals

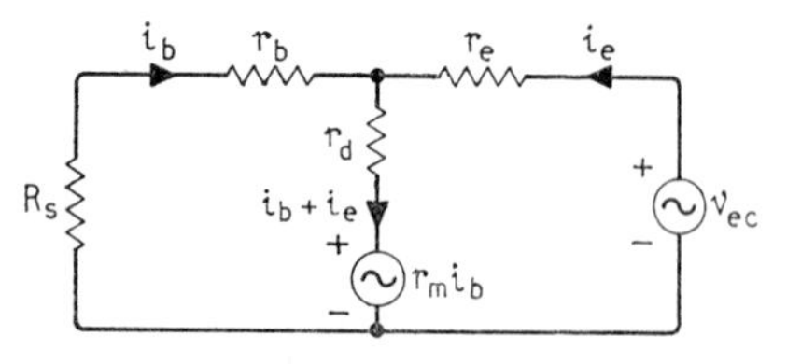

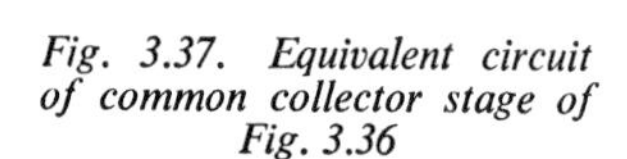

Fig. 3.37. Equivalent circuit of common collector stage of Fig. 3.36

the T-equivalent circuit of Fig. 3.17, we obtain the circuit representation of Fig. 3.37. Applying Kirchoff's Voltage Law to the input loop of this circuit gives:

$$-r_m i_b = (R_s+r_b)\, i_b + r_d(i_b+i_e)$$

Collecting terms and solving for the base current i_b and using the relationship of Equation 3.17 we find that

$$i_b = -\frac{r_d\, i_e}{r_c+R_s+r_b} \tag{3.68}$$

Table 3.4. EXTERNAL CIRCUIT PROPERTIES OF COMMON COLLECTOR STAGE

	Accurate	*Approximate*
R_{in}	$r_b+r_c\dfrac{r_e+R_L}{r_d+r_e+R_L}$	$r_b+(1+\alpha_e)(r_e+R_L)$
R_{out}	$r_e+r_d\dfrac{R_s+r_b}{r_c+R_s+r_b}$	$r_e+(1-\alpha_b)(r_b+R_s)$
K_i	$\dfrac{-r_c}{r_d+r_e+R_L}$	$-(1+\alpha_e)$
K_v	$\dfrac{r_c R_L}{r_c(r_e+R_L)+r_b(r_d+r_e+R_L)}$	$\dfrac{(1+\alpha_e)R_L}{r_b+(1+\alpha_e)(r_e+R_L)}$

Next, applying Kirchoff's Voltage Law to the output loop of Fig. 3.37 gives:

$$v_{ec}-r_m i_b = r_e i_e + r_d(i_b+i_e)$$

Collecting terms and using Equation 3.17:

$$v_{ec} = (r_e+r_d)\,i_e + r_c i_b \tag{3.69}$$

From Equations 3.68 and 3.69 we find that the output resistance of the common collector stage is given by:

$$R_{out} = \frac{v_{ec}}{i_e}$$

$$= r_e + r_d\frac{R_s+r_b}{r_c+R_s+r_b} \tag{3.70}$$

Using the relationships of Equations 3.37, 3.66 and 3.67 we find that the voltage gain of the common collector stage is

$$K_v = \frac{r_c R_L}{r_c(r_e+R_L)+r_b(r_d+r_e+R_L)} \tag{3.71}$$

The above results relating to the common collector configuration have been collected in Table 3.4.

Example 3.6

A transistor, whose T-equivalent circuit parameters are as given in Example 3.4 (page 62) is used as a common collector stage having

$R_s = 100\Omega$ and $R_L = 10\text{k}\Omega$. Evaluate its current gain, input resistance, output resistance and voltage gain.

From Equation 3.66 we find that, when $R_L = 10^4\Omega$, the common collector current gain is

$$K_i = \frac{-2 \times 10^6}{4 \times 10^4 + 20 + 10^4}$$

$$= -40$$

Next, applying Equation 3.67 we obtain the following value for the input resistance

$$R_{\text{in}} = 400 + 2 \times 10^6 \frac{20 + 10^4}{4 \times 10^4 + 20 + 10^4}$$

$$= 4{\cdot}004 \times 10^5\Omega$$

To evaluate the output resistance for $R_s = 100\Omega$ we apply Equation 3.70:

$$R_{\text{out}} = 20 + 4 \times 10^4 \frac{100 + 400}{2 \times 10^6 + 100 + 400}$$

$$= 30\Omega$$

From Equation 3.37 we find that the voltage gain is

$$K_v = \frac{40 \times 10^4}{4{\cdot}004 \times 10^5}$$

$$\simeq 1$$

Notice the closeness of the common collector voltage gain to unity. Further, since K_v is positive then, as for the common base stage, the input and output voltage signals of the common collector stage are in phase.

3.5. GRAPHICAL REPRESENTATIONS OF EXTERNAL CIRCUIT PROPERTIES OF THE THREE BASIC CONFIGURATIONS

In the previous section we developed expressions for the input resistance, output resistance, current gain and voltage gain for each of the three basic transistor configurations in terms of the T-equivalent circuit parameters. It is informative to plot the variations of these external circuit properties as the terminating load and source resistances are varied between the limiting values of zero and infinity. Thus we shall find these graphical representations are useful for comparing the relative circuit behaviours of the three basic configurations.

3.5.1. Input Resistance

For the common base configuration we see from Equation 3.53 that when $R_L = 0$, that is, its output terminals are short circuited, the corresponding value $R_{\text{in}\,SC}$ of the input resistance is

$$R_{\text{in}\,SC} = r_e + r_b \cdot \frac{r_c - r_m}{r_c + r_b} \tag{3.72}$$

On the other hand, when $R_L = \infty$, that is, the output terminals are open circuited, from Equation 3.53 we find that the corresponding value $R_{\text{in}\,OC}$ of the input resistance is given by:

$$R_{\text{in}\,OC} = r_e + r_b \tag{3.73}$$

From Equations 3.72 and 3.73 we deduce that the open-circuit input resistance $R_{\text{in}\,OC}$ is greater than the short-circuit input resistance $R_{\text{in}\,SC}$. Therefore, the input resistance of the common base stage increases with increasing load resistance.

Consider next the common emitter configuration whose input resistance is defined by Equation 3.60. Putting $R_L = 0$ in this equation we find that the short-circuit input resistance $R_{\text{in}\,SC}$ of the common emitter stage is

$$R_{\text{in}\,SC} = r_b + r_e \frac{r_c}{r_d + r_e} \tag{3.74}$$

Putting $R_L = \infty$ in Equation 3.60 we find that, for the common emitter configuration

$$R_{\text{in}\,OC} = r_b + r_e \tag{3.75}$$

From Equations 3.74 and 3.75 we can deduce that for the common emitter configuration $R_{\text{in}\,SC}$ is greater than $R_{\text{in}\,OC}$. Therefore, the input resistance decreases with increasing load resistance.

For the common collector stage we find from Equation 3.67 that when $R_L = 0$ the short-circuit input resistance is

$$R_{\text{in}\,SC} = r_b + r_e \frac{r_c}{r_d + r_e} \tag{3.76}$$

When $R_L = \infty$ the open-circuit input resistance of the common collector stage is deduced from Equation 3.67 to be

$$R_{\text{in}\,OC} = r_b + r_c \tag{3.77}$$

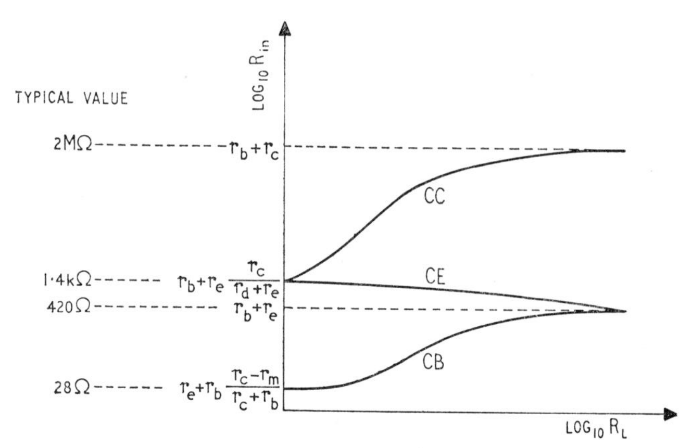

Fig. 3.38. Variation of input resistance with load resistance for the 3 transistor configurations

Thus from Equations 3.76 and 3.77 we deduce that for the common collector stage $R_{\text{in}\,OC}$ is much greater than $R_{\text{in}\,SC}$. Hence, the input resistance increases with increasing load resistance.

Having determined the limiting values for the input resistance of each of the three basic configurations let us next consider how they compare relative to each other. From Equations 3.73 and 3.75 we see that the common base and common emitter configurations have equal open-circuit input resistances. On the other hand, Equations 3.74 and 3.76 show that the common emitter and common collector configurations have equal short-circuit input resistances.

Therefore, it follows that the relative variations of the input resistances of the common base, common emitter and common collector stages are as illustrated in Fig. 3.38. Notice from this diagram that, for any finite load resistance, the input resistance of the common collector stage is always greater than that of the common emitter stage, and the input resistance of the common emitter stage is always greater than that of the common base stage. These deductions are further confirmed by the values calculated in Examples 3.4, 3.5 and 3.6. For convenience, in Fig. 3.38 the logarithm of R_{in} has been plotted against the logarithm of R_L, and typical values have been given.

3.5.2. Output Resistance

The common base stage has its output resistance defined by Equation 3.56. When the source resistance is reduced to zero, that is, the input terminals are short circuited, from Equation 3.56 we find that the short-circuit output resistance is

$$R_{\text{out}\,SC} = r_c - r_b \frac{r_m - r_e}{r_b + r_e} \tag{3.78}$$

When $R_s = \infty$, that is, the input terminals are open circuited then from Equation 3.56 we see that the open-circuit output resistance is

$$R_{\text{out}\,OC} = r_c + r_b \tag{3.79}$$

Therefore, from Equations 3.78 and 3.79 we deduce, that, for the common base configuration $R_{\text{out}\,OC}$ is greater than $R_{\text{out}\,SC}$. In other words, the output resistance of the common base stage increases with increasing source resistance.

Consider next the variation of the common emitter output resistance. This is defined by Equation 3.63. Putting $R_s = 0$ in this expression we find that the short-circuit output resistance

$$R_{\text{out}\,SC} = r_d + r_e \frac{r_m + r_b}{r_e + r_b} \tag{3.80}$$

When $R_s = \infty$ we find from Equation 3.63 that the open-circuit output resistance is given by:

$$R_{\text{out}\,OC} = r_d + r_e \tag{3.81}$$

From Equations 3.80 and 3.81 we note that, for the common emitter configuration, $R_{\text{out}\,SC}$ is greater than $R_{\text{out}\,OC}$. Therefore, the output resistance of the common emitter stage decreases with increasing source resistance.

For the common collector stage we see from Equation 3.70 that when $R_s = 0$ its output resistance has the following short-circuit value:

$$R_{\text{out}\,SC} = r_e + r_d \frac{r_b}{r_c + r_b} \tag{3.82}$$

When $R_s = \infty$ from Equation 3.70 it follows that the open-circuit output resistance is

$$R_{\text{out}\,OC} = r_e + r_d \tag{3.83}$$

Notice from Equations 3.82 and 3.83 that $R_{\text{out}\,OC}$ is greater than $R_{\text{out}\,SC}$. Therefore, the output resistance of the common collector stage increases with increasing load resistance.

Let us next examine the relative variation of the output resistances of the 3 different configurations. First, we see that since r_d is equal to $r_c - r_m$ (see Equation 3.17) then Equations 3.78 and 3.80 show that the short-circuit output resistances of the common base and common emitter configuration are equal. Next, from Equations 3.81 and 3.83 it follows that the open-circuit output resistances of the common emitter and common collector stages are equal. Consequently, the relative variations of the output resistances of

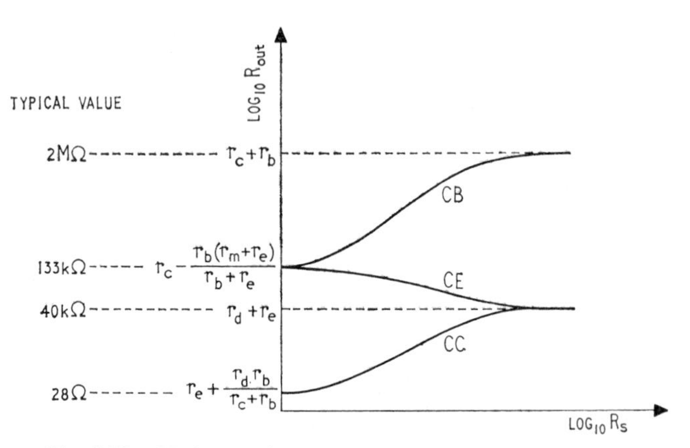

Fig. 3.39. Variation of output resistance with source resistance for the three transistor configurations

the common base, common emitter and common collector configurations are as illustrated in Fig. 3.39. Notice that for any finite source resistance the common base stage has a higher output resistance than the common emitter stage, and the common emitter stage has a higher output resistance than the common collector stage. These deductions are in perfect agreement with the results calculated in the numerical examples, Examples 3.4 to 3.6.

3.5.3. Current Gain

Consider again the common base stage. Its current gain is given by Equation 3.52. It has its maximum value when the output terminals are short circuited, that is, $R_L = 0$. Equation 3.52 then gives the short-circuit current gain to be

$$K_{i\,SC} = -\frac{r_b + r_m}{r_b + r_c} \tag{3.84}$$

When the output terminals are open-circuited, that is, $R_L = \infty$ the current gain is reduced to zero.

Next, for the common emitter stage we see from Equation 3.59 that when $R_L = 0$ its short-circuit current gain is

$$K_{i\,SC} = \frac{r_m - r_e}{r_d + r_e} \tag{3.85}$$

and again the common emitter current gain is reduced to zero when its output terminals are open-circuited.

For the common collector stage we deduce from Equation 3.66 that its short-circuit current gain is

$$K_{i\,SC} = \frac{-r_c}{r_d + r_e} \tag{3.86}$$

and that its open-circuit current gain is zero.

Therefore, it follows that the absolute values of the current gains of all 3 basic configurations decrease with increasing load resistance,

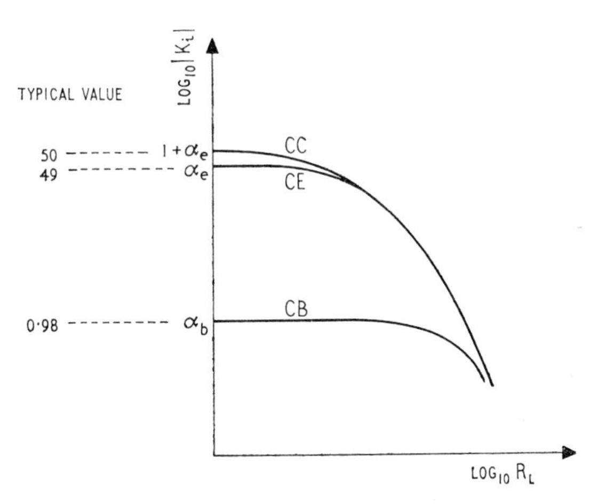

Fig. 3.40. Variation of current gain with load resistance for the three transistor configurations

ultimately becoming zero at infinite load resistance. Further, since each of r_m, r_d and r_c is greater than either of r_b and r_e (see the element values of Example 3.1) then from Equations 3.84 to 3.86 it follows that the short-circuit current gain of the common base stage is equal to $-\alpha_b$, that of the common emitter stage is equal to α_e and that of the common collector stage is equal to $-(1+\alpha_e)$. In these deductions we have made use of Equations 3.14, 3.15 and 3.18.

Normally the common base current amplification factor α_b is very close to unity and the common emitter current amplification factor α_e is much larger than unity. Consequently, the relative variations of the current gains of the three basic configurations will be as indicated in Fig. 3.40. Notice that the absolute values of the common emitter and common collector current gains are very nearly equal, and that the current gain of the common base stage is always less than unity in magnitude.

3.5.4. Voltage Gain

Equations 3.57, 3.64 and 3.71 show that when the output terminals are short circuited, that is, the load resistance R_L is reduced to zero, the voltage gain of each of the three basic configurations is reduced to zero too.

Next, when the output terminals of the common base stage are open circuited then from Equation 3.57 it follows that the corresponding voltage gain is:

$$K_{v\,OC} = \frac{r_m + r_b}{r_e + r_b} \tag{3.87}$$

When the output terminals of the common emitter stage are open circuited we find from Equation 3.64 that the open-circuit voltage gain is

$$K_{v\,OC} = \frac{-(r_m - r_e)}{r_e + r_b} \tag{3.88}$$

Putting $R_L = \infty$ in Equation 3.71 we find that the open-circuit voltage gain of the common collector stage has the following value:

$$K_{v\,OC} = \frac{r_c}{r_c + r_b} \tag{3.89}$$

Now each of r_m and r_c is much greater than either of r_e or r_b. Therefore, we find from Equations 3.87 to 3.89 that the open-circuit voltage gains of the common base and common emitter stages have very nearly equal absolute values, and that the open-circuit voltage gain of the common collector stage is very nearly equal to unity. We thus deduce that the relative variations of the voltage gains of the 3 stages are as shown schematically in Fig. 3.41. Notice that for all load resistance values the common emitter and common base stages have very nearly equal voltage gains, and that the voltage gain of the common collector stage is always less than unity.

3.5.5. Power Gain

In Section 3.2 it was shown that the power gain K_p of any transistor stage is equal to the absolute value of the product of its voltage and current gains. This being so we can make the following observations.

(1) Since the voltage gain of any transistor stage is zero for zero load resistance, and its current gain is zero for infinite load resistance, it follows that the power gain of any transistor stage must be zero at zero and infinite load resistances.

(2) Since only the common emitter stage provides a current gain as well as voltage gain greater than unity in absolute value, it follows that of all the 3 different transistor configurations the common emitter stage provides the highest power gain for any finite load resistance.

Having, in this section, discussed the various circuit properties of the 3 basic transistor configurations we can now assess their circuit

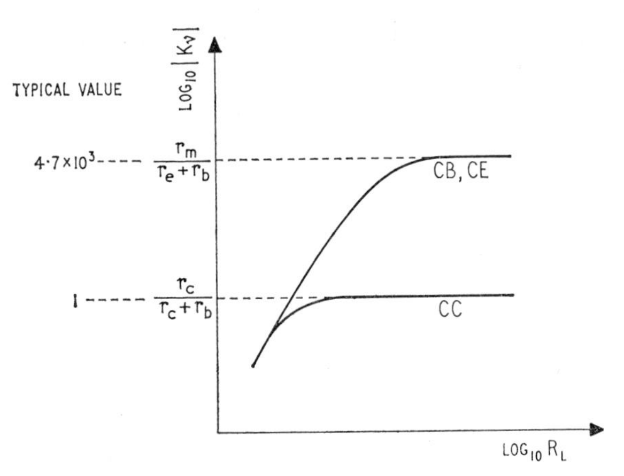

Fig. 3.41. Variation of voltage gain with load resistance for the three transistor configurations

potentialities. First, the common base stage is characterised in having a very low input resistance and a high output resistance. Therefore, this stage is particularly suited for coupling a driving source of low series resistance to a load of high resistance. On the other hand, the common collector stage has a high input resistance and a very low output resistance and it is, therefore, best suited for coupling a driving source of high resistance to a load of low resistance. In such a case, the common collector stage acts basically as a buffer between the source and the load. Finally, the input resistance and output resistance of the common emitter stage lie between their respective values for the common base and common collector stages. This indicates that there will be the least amount of difficulty when cascading a number of common emitter stages so as to obtain increased gain. This will be considered in greater detail in Section 3.7.

3.6. SIMPLIFIED ANALYSIS OF TRANSISTOR CIRCUITS

In Sections 3.3 and 3.4 we presented two different methods for evaluating the various external circuit properties of the three basic transistor configurations. Throughout the work no approximations were made at all. However, it often happens in practice that the terminating load resistance is such that certain justified approximations can be applied, thereby simplifying the analysis a great deal. To illustrate this, consider Equations 3.46 and 3.47 which define the current gain and input resistance, respectively, of the transistor stage

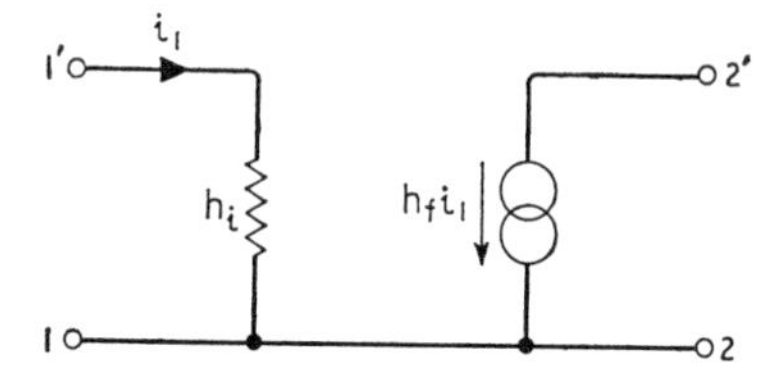

Fig. 3.42. Simplified h-parameters equivalent circuit

in terms of its h-parameters. If the terminating load resistance R_L is much smaller than $1/h_o$ then from these two equations it follows that, to a first order of approximation, the input resistance of the stage is equal to h_i and its current gain is equal to h_f. Therefore the h-parameters equivalent circuit reduces to the simple form shown in Fig. 3.42. Compare this circuit with the elementary equivalent circuit of Fig. 3.3. Thus from Equation 3.44 we obtain directly that the external voltage gain of the stage of Fig. 3.22 has the following approximate value:

$$K \simeq -\frac{h_f R_L}{h_i + R_s} \tag{3.90}$$

Consider next simplifying the circuit analysis in terms of the T-equivalent circuit parameters. We can simplify the various expressions that we developed in Section 3.4 if we first take into account the relative orders of magnitude of the T-equivalent circuit parameters. Thus from the circuit element values given earlier in Example 3.1 we find that

$$r_c \gg r_d \gg r_b \gg r_e \tag{3.91(a)}$$

If we further assume that the terminating load and source resistances satisfy the following inequalities

$$\left.\begin{aligned} R_L &\ll r_d \\ r_c \gg R_s &\gg r_b \end{aligned}\right\} \tag{3.91(b)}$$

then we find that

(1) Equations 3.52, 3.53, 3.56 and 3.57, which apply to the common base stage, approximate to the forms in Table 3.2.
(2) Equations 3.59, 3.60, 3.63 and 3.64, which apply to the common emitter stage, approximate as shown in Table 3.3.
(3) Equations 3.66, 3.67, 3.70 and 3.71, which apply to the common collector stage, approximate as shown in Table 3.4.

3.7. COMMON EMITTER–COMMON EMITTER CASCADE

In order to illustrate how the earlier results obtained for a single transistor stage can be extended to the case of multi-stage amplifiers,

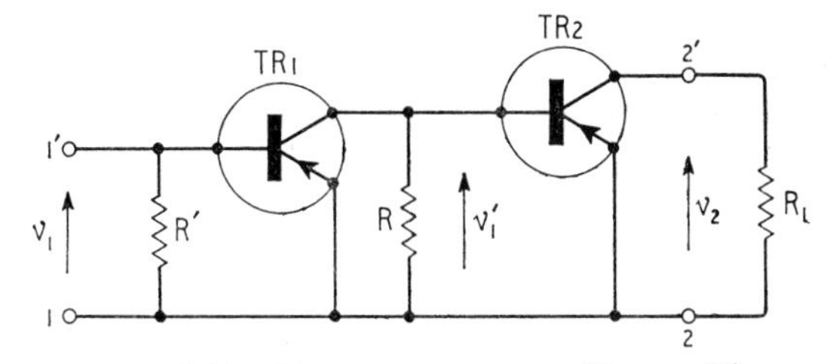

Fig. 3.43. Two-stage common emitter amplifier

consider the circuit of Fig. 3.43 showing two common emitter stages connected in cascade. Evaluation of the overall response is best carried out starting from the load resistance end and working backwards towards the driving source.

Thus from Equation 3.37 it follows that the voltage gain K_{v2} of the second common emitter stage TR2 of Fig. 3.43 is

$$K_{v2} = -\frac{K_{i2} R_L}{R_{\text{in }2}} \tag{3.92}$$

where K_{i2} is the current gain of TR2 and $R_{\text{in }2}$ is its input resistance.

Next, from Fig. 3.43 we observe that the effective load resistance R_{L1} of the first common emitter stage TR1 is equal to the parallel combination of the interstage resistor R and the input resistance $R_{\text{in}\,2}$ of the stage TR2, that is,

$$R_{L1} = \frac{RR_{\text{in}\,2}}{R+R_{\text{in}\,2}} \tag{3.93}$$

Therefore, from Equation 3.37 it follows that the voltage gain K_{v1} of TR1 is

$$K_{v1} = -\frac{K_{i1}\,R_{L1}}{R_{\text{in}\,1}} \tag{3.94}$$

where $R_{\text{in}\,1}$ and K_{i1} are the input resistance and current gain of TR1, respectively.

The overall voltage gain K_v of the two-stage amplifier of Fig. 3.43 is equal to the voltage ratio of v_2/v_1 which we can express as follows:

$$K_v = \frac{v_2}{v_1}$$

$$= \frac{v_2}{v_1'} \cdot \frac{v_1'}{v_1} \tag{3.95}$$

From Fig. 3.43 we see that the voltage ratio v_2/v_1' is equal to the voltage gain K_{v2} of the stage TR2 while the voltage ratio v_1'/v_1 is equal to the voltage gain K_{v1} of the first stage TR1.

Therefore,

$$K_v = K_{v1}\,K_{v2} \tag{3.96}$$

This result can be generalised in that in a multi-stage amplifier, consisting of any number of stages connected in cascade, the overall voltage gain is equal to the product of the voltage gains of all the individual stages.

Using Equations 3.94 to 3.96 we find that, for the two-stage amplifier of Fig. 3.43, the overall voltage gain is

$$K_v = \frac{K_{i1}\,K_{i2}\,R_{L1}\,R_L}{R_{\text{in}\,1}\,R_{\text{in}\,2}} \tag{3.97}$$

Example 3.7

The two-stage common emitter amplifier of Fig. 3.43 uses 2 identical transistors, each one having the following common emitter h-parameters:

$$h_{ie} = 1400\Omega$$
$$h_{re} = 5 \times 10^{-4}$$
$$h_{fe} = 49$$
$$h_{oe} = 25\mu S$$

Evaluate the overall voltage gain, given that $R = R_L = 10k\Omega$.

For the second common emitter stage, having the common emitter h-parameters of $h_{ie2}, h_{re2}, h_{fe2}, h_{oe2}$, Equation 3.47 gives

$$R_{\text{in }2} = h_{ie2} - \frac{h_{re2}\, h_{fe2}\, R_L}{1 + h_{oe2}\, R_L}$$
$$= 1400 - \frac{5 \times 10^{-4} \times 49 \times 10^4}{1 + 25 \times 10^{-6} \times 10^4}$$
$$= 1200\Omega$$

and Equation 3.46 gives its current gain K_{i2} to be

$$K_{i2} = \frac{h_{fe2}}{1 + h_{oe2}\, R_L}$$
$$= \frac{49}{1 + 25 \times 10^{-6} \times 10^4}$$
$$= 39{\cdot}2$$

Equation 3.93 gives the effective load resistance of the first stage to be

$$R_{L1} = \frac{1{\cdot}2 \times 10}{1{\cdot}2 + 10}$$
$$= 1{\cdot}07k\Omega$$

Next, assuming the common emitter h-parameters of the first stage to be h_{ie1}, h_{re1}, h_{fe1} and h_{oe1}, using Equation 3.47 we find that its input resistance is given by

$$R_{\text{in }1} = h_{ie1} - \frac{h_{re1}\, h_{fe1}\, R_{L1}}{1 + h_{oe1}\, R_{L1}}$$
$$= 1400 - \frac{5 \times 10^{-4} \times 49 \times 1{\cdot}07 \times 10^3}{1 + 25 \times 10^{-6} \times 1{\cdot}07 \times 10^3}$$
$$= 1370\Omega$$

Equation 3.46 gives its current gain to be

$$K_{i1} = \frac{h_{fe1}}{1+h_{oe1}R_{L1}}$$

$$= \frac{49}{1+25\times10^{-6}\times1{\cdot}07\times10^{3}}$$

$$= 47{\cdot}7$$

Finally, Equation 3.97 gives the required overall voltage gain to be

$$K_v = \frac{47{\cdot}7\times39{\cdot}2\times1{\cdot}07\times10^{3}\times10^{4}}{1{,}370\times1{,}200}$$

$$= 12{\cdot}2\times10^{3}$$

3.8. PARAMETER VARIATIONS AND THEIR EFFECTS ON EXTERNAL CIRCUIT RESPONSE

Owing to the non-linear nature of the various transistor static characteristic curves as evidenced by Figs. 1.16 to 1.21, it is to be

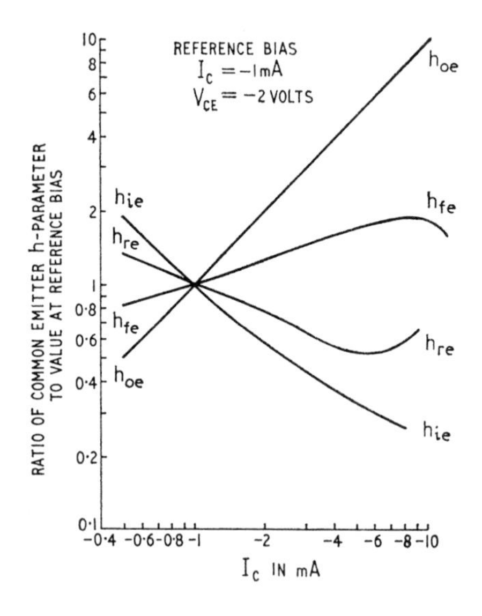

Fig. 3.44. Variation of common emitter h-parameters with collector current

expected that the various transistor parameters will vary as the quiescent operating-point is allowed to change. Thus Fig. 3.44 shows the variations of the common emitter h-parameters of a typical transistor with the collector current I_C, while Fig. 3.45 illustrates their variations with the collector–emitter voltage V_{CE}.

In each case it is found convenient to express the particular h-parameter relative to its value at a nominal operating-point which, in the case of Figs. 3.44 and 3.45 is specified at $I_C = -1\text{mA}$, $V_{CE} = -2\text{V}$. Thus the curves of Figs. 3.44 and 3.45 can be regarded as correction factors for the transistor as they enable the evaluation of the common

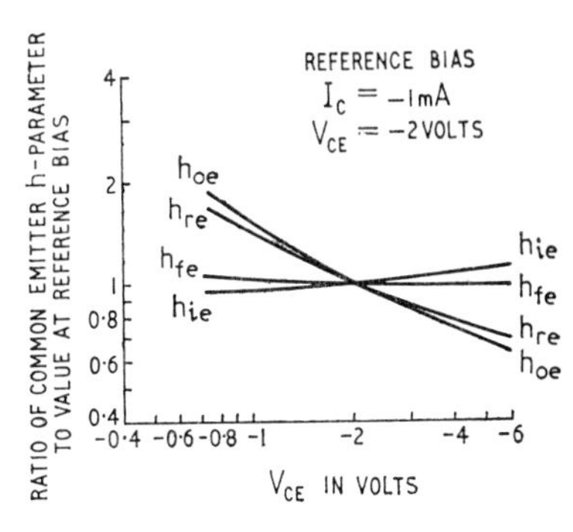

Fig. 3.45. Variation of common emitter h-parameters with collector-emitter voltage

emitter h-parameters at any other operating-point. To illustrate this, consider the following example.

Example 3.8

Determine the common emitter h-parameters of a transistor having the parameter variations of Figs. 3.44 and 3.45 if the operating-point is specified to be $I_C = -4\text{mA}$, $V_{CE} = -6\text{V}$. It is also given that at $I_C = -1\text{mA}$, $V_{CE} = -2\text{V}$:

$$h_{ie} = 1400\Omega$$
$$h_{re} = 5 \times 10^{-4}$$
$$h_{fe} = 49$$
$$h_{oe} = 25 \times 10^{-6}\text{S}$$

From Fig. 3.44 we find that the h_{ie} correction factor at $I_C = -4\text{mA}$ is equal to 0·4, and from Fig. 3.45 we find that the h_{ie} correction factor at $V_{CE} = -6\text{V}$ is equal to 1·05. Therefore, at $I_C = -4\text{mA}$, $V_{CE} = -6\text{V}$, h_{ie} has the following value

$$h_{ie} = 1400 \times 0{\cdot}4 \times 1{\cdot}05$$
$$= 590\Omega$$

Similarly we deduce from Figs. 3.44 and 3.45 that

$$h_{re} = 5 \times 10^{-4} \times 0{\cdot}55 \times 0{\cdot}69$$
$$= 1{\cdot}9 \times 10^{-4}$$

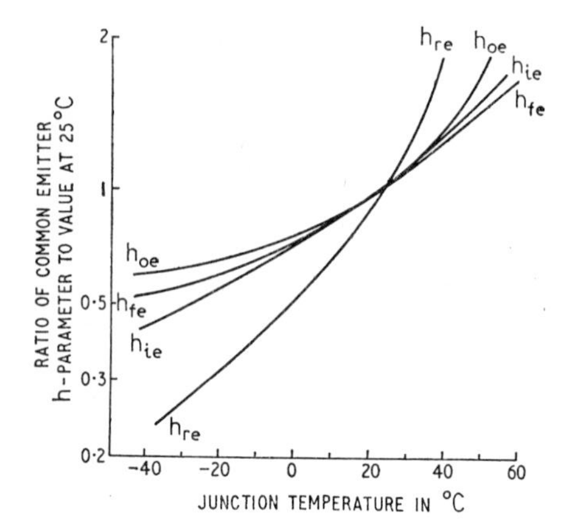

Fig. 3.46. Variation of common emitter h-parameters with unction temperature

$$h_{fe} = 49 \times 1{\cdot}5 \times 1{\cdot}01$$
$$= 74$$
$$h_{oe} = 25 \times 10^{-6} \times 3{\cdot}8 \times 0{\cdot}62$$
$$= 59 \times 10^{-6}\text{S}$$

The transistor parameters also depend on the ambient temperature. This can be seen from Fig. 3.46, which shows the variations of the common emitter h-parameters of a typical germanium junction transistor with temperature.

Lastly it should be mentioned that, because of manufacturing tolerances, the various transistor parameters can vary quite considerably from one transistor sample to another.

An obvious consequence of these parameter variations is that they will influence the external circuit response of the transistor stage. This effect is best demonstrated by way of an example.

Example 3.9

A transistor has the following spread in the values of its common base h-parameters as a result of manufacturing tolerances:

Parameter	*Minimum Value*	*Average Value*	*Maximum Value*
h_{ib} (in ohms)	26	28	30
h_{rb}	10^{-4}	2×10^{-4}	3×10^{-4}
h_{fb}	$-0{\cdot}97$	$-0{\cdot}98$	$-0{\cdot}99$
h_{ob} (in siemens)	$0{\cdot}3 \times 10^{-6}$	$0{\cdot}5 \times 10^{-6}$	$0{\cdot}8 \times 10^{-6}$

Calculate the minimum, average and maximum values of the power gain of this transistor when it is operated as a common base stage with $R_L = 100\text{k}\Omega$.

Using the minimum values of h-parameters given above in Equations 3.46, 3.47 and 3.38, we find that the corresponding value of current gain, input resistance and power gain of the stage are as follows:

$$K_i = \frac{-0{\cdot}97}{1+0{\cdot}3\times 10^{-6}\times 10^5}$$

$$= -0{\cdot}94$$

$$R_{\text{in}} = 26 + \frac{0{\cdot}97\times 10^{-4}\times 10^5}{1+0{\cdot}3\times 10^{-6}\times 10^5}$$

$$= 35{\cdot}4\Omega$$

$$K_p = \frac{(-0{\cdot}94)^2\times 10^5}{35{\cdot}4}$$

$$= 2{,}500$$

Similarly, using the given average values for the h-parameters in the above equations we obtain the following results:

$$K_i = -0{\cdot}933$$

$$R_{\text{in}} = 46{\cdot}7\Omega$$

$$K_p = 1860$$

Finally, using the given maximum values for the h-parameters we find that

$$K_i = -0{\cdot}916$$

$$R_{\text{in}} = 57{\cdot}5\Omega$$

$$K_p = 1460$$

We, therefore, observe that parameter variations from one transistor sample to another can result in relatively large changes in the external circuit properties of a transistor stage.

3.9. HIGH FREQUENCY EFFECTS

The basic limitation to transistor action at high frequencies is the relatively low speed at which the minority charge carriers diffuse across the base region. This leads to transit time effects which, in turn, cause the common base current amplification factor α_b to become frequency dependent. In order to illustrate this effect, assume that the output terminals of a common base stage are short circuited and that a relatively narrow pulse of current (shown in Fig. 3.47) is applied to the emitter terminal. The holes, in a pnp junction transistor, travel across the base region by a process of diffusion. Consequently, the resulting collector current waveform is rounded off and it is no longer a faithful reproduction of the input pulse as can be seen from Fig. 3.47. The rounding off effect

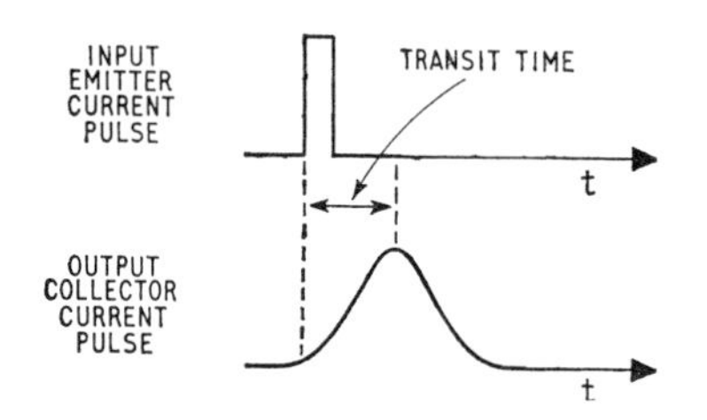

Fig. 3.47. Pulse response of output collector current

is due to the random nature of the diffusion process in that all the charge carriers do not reach the collector terminal simultaneously despite the fact that they are all injected into the base region practically at the same time. It is, therefore, inferred that as a result of the random diffusion process, the higher frequency components of the input current pulse are heavily attenuated on reaching the output collector terminal. In other words, the common base current amplification factor α_b will fall off with increasing frequency. To a first order of approximation the frequency dependence of α_b can be expressed as

$$\alpha_b = \frac{\alpha_{b0}}{1+\dfrac{jf}{f_{\alpha b}}} \tag{3.98}$$

where f is the operating frequency

α_{b0} = the low frequency value of α_b

$f_{\alpha b}$ = the *common base alpha cut-off frequency* which is to be defined presently.

From Equation 3.98 it follows that the modulus of α_b is

$$|\alpha_b| = \frac{\alpha_{b0}}{\left[1+\left(\dfrac{f}{f_{\alpha b}}\right)^2\right]^{\frac{1}{2}}} \tag{3.99}$$

and that its phase angle is

$$\angle\alpha_b = -\tan^{-1}\left(\frac{f}{f_{\alpha b}}\right) \tag{3.100}$$

When the operating-frequency is equal to $f_{\alpha b}$ we find from Equations 3.99 and 3.100 that the modulus and phase angle of α_b are equal to $\alpha_{bo}/\sqrt{2}$ and $-45°$, respectively. Thus $f_{\alpha b}$ is defined as the frequency at which the modulus of α_b drops to $1/\sqrt{2}$ of its low frequency value of α_{bo}.

It must be emphasised that Equation 3.98 is only approximate since an accurate analysis of the high frequency dependence of α_b must take account of the distributed nature of the diffusion process. This is beyond the scope of the present book. However, provided that the operating-frequency does not exceed $f_{\alpha b}/5$, then the expression of Equation 3.98 is sufficiently accurate for most practical purposes.

The alpha cut-off frequency can be improved in either one of 2 ways.

(1) By reducing the width of the base region as then the holes diffuse across the base region in a shorter time.

(2) By grading the impurity distribution in the base region as in *drift transistors.* In such transistors the impurity distribution is made to be exponential, having a maximum value at the emitter end of the base region and dropping to a small value at the collector end. As a result of this non-uniform impurity distribution, an electric field is established in such a way as to accelerate the flow of minority charge carriers across the base region.

Before going on to the discussion of the other factors which affect the high frequency response of junction transistors, we shall find it informative to study the frequency dependence of the common emitter amplification factor α_e in view of Equation 3.98. Now α_b and α_e are related by Equation 3.18. Therefore, eliminating α_b between Equations 3.18 and 3.98 we obtain that:

$$\alpha_e = \frac{\alpha_b}{1-\alpha_b}$$

$$= \frac{\dfrac{\alpha_{bo}}{1+\dfrac{jf}{f_{\alpha b}}}}{1-\dfrac{\alpha_{bo}}{1+\dfrac{jf}{f_{\alpha b}}}}$$

Simplifying this expression we find

$$\alpha_e = \frac{\alpha_{bo}}{1-\alpha_{bo}+\dfrac{jf}{f_{\alpha b}}}$$

Dividing the numerator and denominator by the term $1-\alpha_{bo}$ we obtain

$$\alpha_e = \frac{\alpha_{bo}}{1-\alpha_{bo}} \cdot \frac{1}{1+\dfrac{jf}{f_{\alpha b}(1-\alpha_{bo})}}$$

$$= \frac{\alpha_{eo}}{1+\dfrac{jf}{f_{\alpha e}}} \tag{3.101}$$

where α_{eo} = low frequency value of the common emitter current amplification factor α_e

$$= \frac{\alpha_{bo}}{1-\alpha_{bo}} \tag{3.102}$$

$f_{\alpha e}$ = *common emitter alpha cut-off frequency*

$$= f_{\alpha b}(1-\alpha_{bo}) \tag{3.103}$$

Notice that the product of α_{eo} and $f_{\alpha e}$ is equal to $\alpha_{bo} f_{\alpha b}$.

From Equation 3.101 it follows that the modulus of α_e is

$$|\alpha_e| = \frac{\alpha_{eo}}{\left[1+\left(\dfrac{f}{f_{\alpha e}}\right)^2\right]^{\frac{1}{2}}} \tag{3.104}$$

and its phase angle is

$$\angle\alpha_e = -\tan^{-1}\left(\frac{f}{f_{\alpha e}}\right) \tag{3.105}$$

Here again when $f = f_{\alpha e}$ the modulus and phase angle of α_e are equal to $\alpha_{eo}/\sqrt{2}$ and $-45°$, respectively. Thus $f_{\alpha e}$ is defined as the frequency at which the modulus of the common emitter current amplification factor α_e drops to $1/\sqrt{2}$ of its low frequency value α_{eo}.

Normally, α_{bo} is quite close to unity with the result that the factor $1-\alpha_{bo}$ can be very small compared with unity. It, therefore, follows from Equations 3.102 and 3.103 that α_{eo} is much greater than α_{bo}

while $f_{\alpha e}$ is much smaller than $f_{\alpha b}$. The relative frequency dependences of α_b and α_e are best demonstrated by considering a numerical example.

Example 3.10

A transistor has the following common base parameters

$$\alpha_{bo} = 0{\cdot}98$$

$$f_{\alpha b} = 10\text{MHz}$$

Plot the frequency dependence of the modulus and phase angle of each of α_b and α_e.

Consider first the common base case. From Equation 3.99 we find that the modulus of α_b is

$$|\alpha_b| = \frac{0{\cdot}98}{\left[1+\left(\frac{f}{10}\right)^2\right]^{\frac{1}{2}}}$$

This is shown plotted in Fig. 3.48 where it is assumed that the operating-frequency f is in MHz. Equation 3.100 gives the phase angle of α_b to be

$$\angle\alpha_b = -\tan^{-1}\left(\frac{f}{10}\right)$$

This is shown plotted in Fig. 3.49.

Next, consider the common emitter case. From Equations 3.102 and 3.103 we find that

$$\alpha_{eo} = \frac{0{\cdot}98}{1-0{\cdot}98}$$

$$= 49$$

$$f_{\alpha e} = 10(1-0{\cdot}98)$$

$$= 0{\cdot}2\text{MHz}$$

$$= 200\text{kHz}$$

Therefore, Equation 3.104 gives the modulus of α_e to be

$$|\alpha_e| = \frac{49}{\left[1+\left(\frac{f}{0{\cdot}2}\right)^2\right]^{\frac{1}{2}}}$$

This is shown plotted in Fig. 3.48. From Equation 3.105 we obtain that the phase angle of α_e is

$$\angle\alpha_e = -\tan^{-1}\left(\frac{f}{0{\cdot}2}\right)$$

which is shown plotted in Fig. 3.49.

We thus observe from the curves of Figs. 3.48 and 3.49 that the common base configuration makes the most effective use of the useful frequency range of the transistor.

The second factor which affects the high frequency response of a junction transistor is its *extrinsic base resistance* $r_{bb'}$. This is the

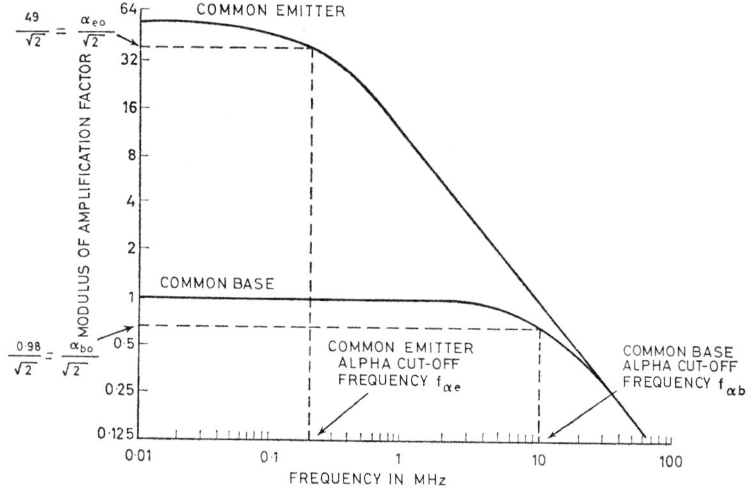

Fig. 3.48. Frequency variation of moduli of common base and common emitter amplification factors

resistance of the bulk of the base material and it constitutes part of the element r_b of the common emitter T-equivalent circuit of Fig. 3.14. The resistance $r_{bb'}$ appears between the external base terminal b and the internal base terminal b' as illustrated in Fig. 3.50, where α_e is as defined in Equation 3.101.

The third factor which affects the high frequency response of a junction transistor is that, due to the transit time effects mentioned earlier, a sudden change in the emitter current can produce only a gradual change in the distribution of the holes set up inside the base region. This effect can be approximately simulated by placing a capacitor $C_{b'e}$ between the emitter terminal and the internal base

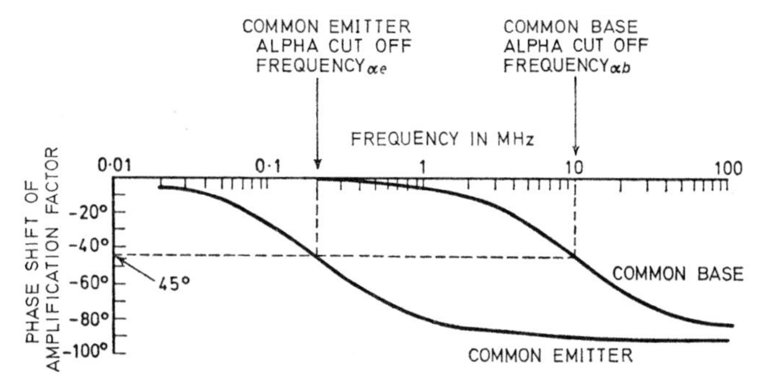

Fig. 3.49. Frequency variation of phase shifts of common base and common emitter amplification factors

terminal as in Fig. 3.50. The capacitance $C_{b'e}$ is termed the *emitter diffusion capacitance* and it is given by

$$C_{b'e} = \frac{1}{2\pi f_{\alpha e} r_{b'e}} \tag{3.106}$$

where $r_{b'e}$ is to be defined later (see Equation 3.108).

The fourth major factor influencing the high frequency performance of a junction transistor is the *depletion layer capacitance* associated with the reverse-biased collector–base junction. The depletion layer

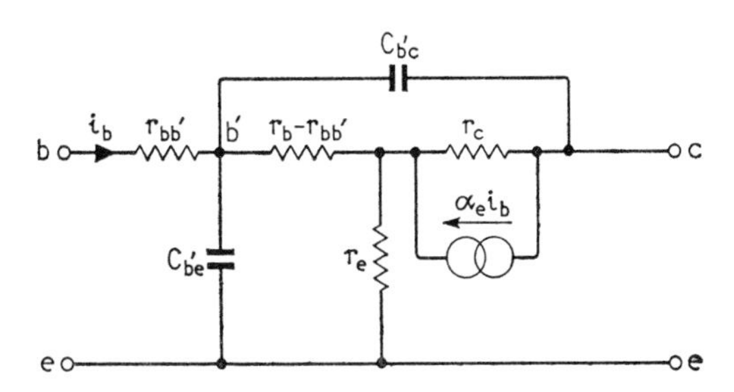

Fig. 3.50. Common emitter T-equivalent circuit at high ferquencies

capacitance is produced by the presence of unneutralised and oppositely charged impurity atoms on the two sides of the depletion layer. The effect of this capacitance can be taken into account by placing a capacitance equal to $C_{b'c}$ between the collector and internal base terminals as in Fig. 3.50. The forward-biased emitter–base junction is also associated with a depletion layer capacitance, but it is normally swamped by the larger value of the emitter diffusion capacitance $C_{b'e}$.

However, in drift transistors the effect of the emitter depletion layer capacitance can become significant, particularly at low values of emitter currents.

3.9.1. Hybrid-π Equivalent Circuit

The equivalent circuit of Fig. 3.50 can be modified into the equivalent form shown in Fig. 3.51. This latter equivalent circuit is usually referred to as the *hybrid-π equivalent circuit* and it is useful for analysing the high frequency response of common emitter circuits. Notice that all the elements of the hybrid-π equivalent circuit are independent of frequency. This is in direct contrast to the equivalent circuit of Fig. 3.50. For a junction transistor having a uniform base impurity distribution, the hybrid-π equivalent circuit parameters have the following typical values:

$$\left.\begin{aligned} r_{bb'} &= 100\Omega \\ r_{b'e} &= 1\text{k}\Omega \\ r_{b'c} &= 2\text{M}\Omega \\ r_{ce} &= 40\text{k}\Omega \\ g_m &= 38\text{mS} \\ C_{b'e} &= 800\text{pF} \\ C_{b'c} &= 10\text{pF} \end{aligned}\right\} \tag{3.107}$$

On the other hand, in a drift transistor the resistors $r_{b'c}$ and r_{ce} are so large that their shunting effects can be justifiably neglected as in Fig. 3.53. Typical values for a drift transistor are as follows:

$$\begin{aligned} r_{bb'} &= 60\Omega \\ r_{b'e} &= 1\text{k}\Omega \\ g_m &= 38\text{mS} \\ C_{b'e} &= 100\text{pF} \\ C_{b'c} &= 2\text{pF} \end{aligned}$$

From the above values we see that the capacitances $C_{b'e}$ and $C_{b'c}$ of a drift transistor are much smaller than their corresponding values in a transistor with a uniform base impurity distribution. This indicates that the drift transistor provides a better frequency response than the transistor with uniform base impurity distribution.

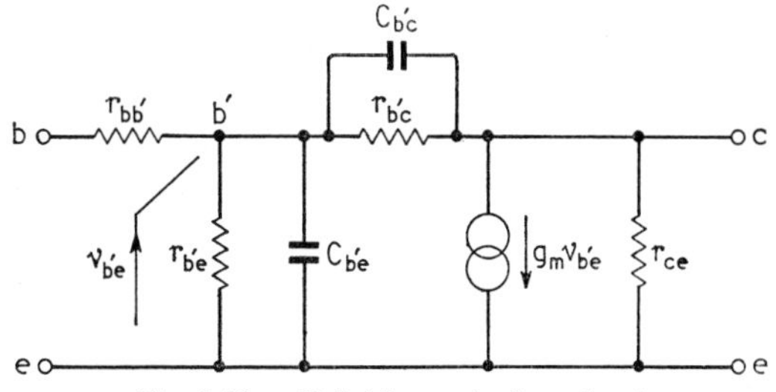

Fig. 3.51. Hybrid-π equivalent circuit

The resistive elements of the hybrid-π equivalent circuit are related to the common emitter h-parameters as follows:

$$\left.\begin{aligned} h_{ie} &= r_{bb'} + r_{b'e} \\ h_{re} &= \frac{r_{b'e}}{r_{b'c}} \\ h_{fe} &= g_m r_{b'e} \\ h_{oe} &= \frac{1}{r_{ce}} + \frac{g_m r_{b'e}}{r_{b'c}} \end{aligned}\right\} \tag{3.108}$$

3.9.2. Frequency Dependence of External Voltage Gain

In order to illustrate how the hybrid-π equivalent circuit of Fig. 3.51 can be used for evaluating the high frequency response of a transistor

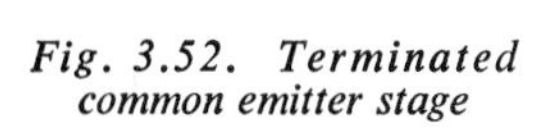

Fig. 3.52. Terminated common emitter stage

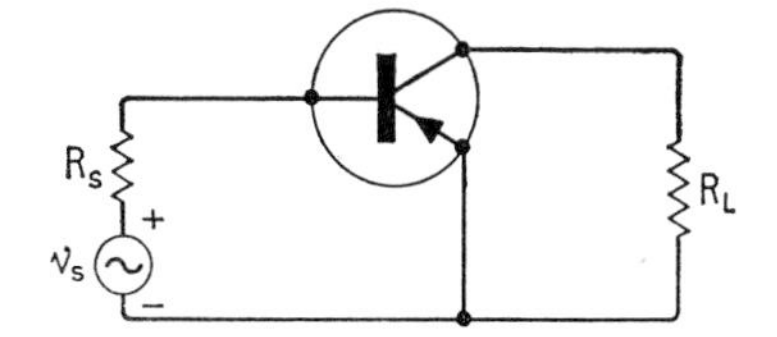

stage, consider the common emitter stage of Fig. 3.52, where it is coupled to a load resistance R_L at the output terminals and it is driven from a voltage source v_s of series resistance R_s. Provided that the load resistance R_L is not too large then, to a first order of approximation we can ignore the resistive elements $r_{b'c}$ and r_{ce} of the hybrid-π equivalent circuit of Fig. 3.51. Therefore, we obtain the circuit representation of Fig. 3.53 for the common emitter stage of

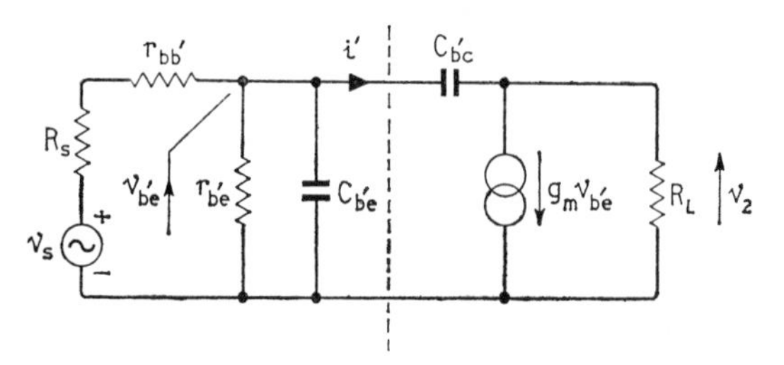

Fig. 3.53. Simplified high frequency equivalent circuit for common emitter stage of Fig. 3.52

Fig. 3.52. From the circuit of Fig. 3.53 we find that the output voltage v_2 is

$$v_2 \simeq -g_m R_L v_{b'e} \tag{3.109}$$

This equation is subject to the condition that, at the operating frequency $1/\omega C_{b'c}$ is large compared to the load resistance R_L.

However, the voltage drop across the capacitance $C_{b'c}$ is equal to $v_{b'e} - v_2$ which, in view of Equation 3.109, becomes

$$v_{b'e} - v_2 \simeq v_{b'e}(1 + g_m R_L) \tag{3.110}$$

Therefore, the current i' flowing through the element $C_{b'c}$ and shown defined in Fig. 3.53 is given as follows:

$$\begin{aligned} i' &= \mathrm{j}\omega C_{b'c}(v_{b'e} - v_2) \\ &\simeq \mathrm{j}\omega C_{b'c}(1 + g_m R_L)\, v_{b'e} \end{aligned} \tag{3.111}$$

We thus find that the admittance Y' measured looking to the right of the dotted terminals in Fig. 3.53 is

$$\begin{aligned} Y' &= \frac{i'}{v_{b'e}} \\ &\simeq \mathrm{j}\omega C_{b'c}(1 + g_m R_L) \end{aligned} \tag{3.112}$$

The admittance Y' appears directly across the capacitive element $C_{b'e}$ of the hybrid-π equivalent circuit. Hence the effect of the capacitance $C_{b'c}$ can be taken into account by placing a capacitance equal to $C_{b'c}(1 + g_m R_L)$ across $C_{b'e}$, thereby modifying the circuit representation of Fig. 3.53 into the simpler form shown in Fig. 3.54. The equivalent capacitance C_{eq} is given by

$$\mathbf{C_{eq} = C_{b'e} + C_{b'c}(1 + g_m R_L)} \tag{3.113}$$

In most junction transistors the capacitance $C_{b'e}$ is much greater than $C_{b'c}$ and so Equation 3.113 approximates to

$$C_{eq} \simeq C_{b'e} + C_{b'c} g_m R_L \tag{3.114}$$

Normally the load resistance R_L of the stage is such that $g_m R_L$ is much greater than unity. Hence, the effect of the capacitance $C_{b'c}$ is

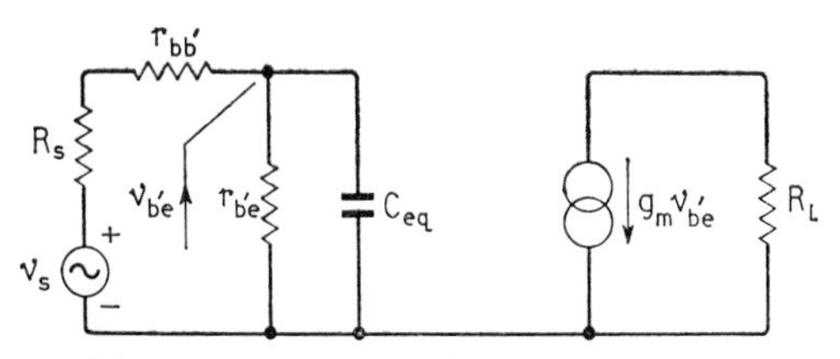

Fig. 3.54. Simplified equivalent circuit for common emitter stage of Fig. 3.52

to reflect an increased capacitance across the internal base-to-emitter terminals and so increase the input capacitance of the common emitter stage. This effect is usually referred to as the *Miller effect.*

The high frequency response of the common emitter stage of Fig. 3.52 can now be obtained from the simplified equivalent circuit of Fig. 3.54. Thus, evaluating the internal voltage drop $v_{b'e}$ we find that

$$v_{b'e} = \frac{v_s}{1+(r_{bb'}+R_s)(g_{b'e}+j\omega C_{eq})} \tag{3.115}$$

where $g_{b'e} = 1/r_{b'e}$.

Next, eliminating $v_{b'e}$ between Equations 3.109 and 3.115 and then evaluating the voltage ratio v_2/v_s we find that, at higher frequencies, the external voltage gain K_{hf} of the common emitter stage is given by

$$K_{hf} = \frac{v_2}{v_s}$$

$$= \frac{-g_m R_L}{1+(r_{bb'}+R_s)(g_{b'e}+j\omega C_{eq})} \tag{3.116}$$

On factorising out the real term of the denominator of Equation 3.116 we can rewrite it as follows:

$$K_{hf} = \frac{K_{mf}}{1+\dfrac{jf}{f_h}} \tag{3.117}$$

where K_{mf} = mid-frequency value of common emitter external voltage gain

$$= -\frac{g_m R_L}{1+g_{b'e}(r_{bb'}+R_s)} \tag{3.118}$$

and

$$f_h = \frac{1}{2\pi C_{eq}}\left(g_{b'e}+\frac{1}{r_{bb'}+R_s}\right) \tag{3·119}$$

Evaluating the modulus of the external voltage gain we obtain

$$|K_{hf}| = \frac{K_{mf}}{\left[1+\left(\frac{f}{f_h}\right)^2\right]^{\frac{1}{2}}} \tag{3.120}$$

From this expression we find that $|K_{hf}| = K_{mf}/\sqrt{2}$ at $f=f_h$. Thus f_h is termed the *upper cut-off frequency* at which the modulus of the external voltage gain drops to $1/\sqrt{2}$ of its mid-frequency value.

3.9.3. Frequency dependence of input impedance

Consider next the variation of the short-circuit input impedance $Z_{\text{in}\,SC}$ of the common emitter stage with frequency. When the

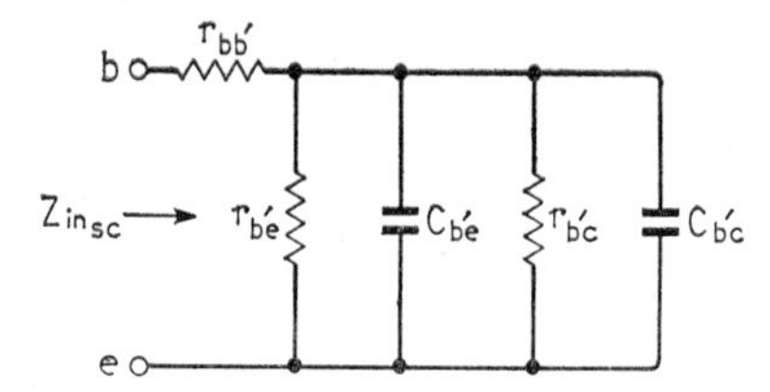

Fig. 3.55. Hybrid-π equivalent circuit with output short circuited

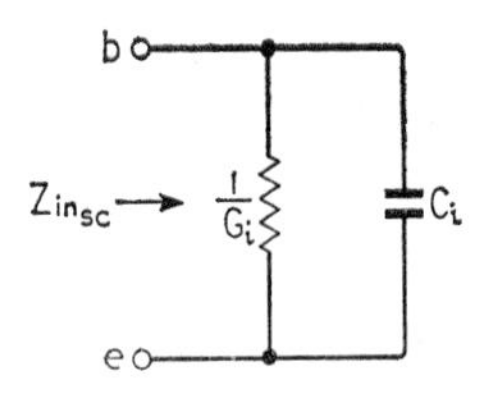

Fig. 3.56. Network for representing short-circuit input impedance

output terminals are short circuited we find that the hybrid-π equivalent circuit of Fig. 3.51 reduces to the form shown in Fig. 3.55. Therefore

$$Z_{\text{in}\,SC} = r_{bb'}+\frac{1}{\left(\frac{1}{r_{b'e}}+\frac{1}{r_{b'c}}\right)+j\omega(C_{b'e}+C_{b'c})} \tag{3.121}$$

However, $r_{b'c} \gg r_{b'e}$ and $C_{b'e} \gg C_{b'c}$ (see the typical values given earlier) Hence Equation 3.121 simplifies to

$$Z_{\text{in}\,SC} \simeq r_{bb'}+\frac{r_{b'e}}{1+j\omega C_{b'e} r_{be}} \tag{3.122}$$

Since $\omega = 2\pi f$ and $C_{b'e}r_{b'e} = 1/2\pi f_{\alpha e}$ (see Equation 3.106) we can rewrite Equation 3.122 as follows:

$$Z_{\text{in }SC} \simeq r_{bb'} + \frac{r_{b'e}}{1+\dfrac{\text{j}f}{f_{\alpha e}}} \tag{3.123}$$

In practice it is found convenient to represent $Z_{\text{in }SC}$ as the parallel combination of a conductance G_i and a capacitance C_i, as in Fig. 3.56. Therefore, we can express $Z_{\text{in }SC}$ as

$$\frac{1}{Z_{\text{in }SC}} = G_i + \text{j}2\pi f C_i \tag{3.124}$$

Thus, Equations 3.123 and 3.124 give that

$$G_i + \text{j}2\pi f C_i \simeq \frac{1+\dfrac{}{f_{\alpha e}}}{(r_{be'}+r_{bb'}) + \text{j}\dfrac{f}{f_{\alpha e}}r_{bb'}} \tag{3.125}$$

Rationalising the right-hand side we obtain

$$G_i + \text{j}2\pi f C_i \simeq \frac{\left(1+\dfrac{\text{j}f}{f_{\alpha e}}\right)\left[(r_{b'e}+r_{bb'}) - \text{j}\dfrac{f}{f_{\alpha e}}r_{bb'}\right]}{(r_{b'e}+r_{bb'})^2 + \left(\dfrac{f}{f_{\alpha e}}r_{bb'}\right)^2}$$

$$= \frac{(r_{b'e}+r_{bb'}) + \left(\dfrac{f}{f_{\alpha e}}\right)^2 r_{bb'} + \dfrac{\text{j}f}{f_{\alpha e}}.r_{b'e}}{(r_{b'e}+r_{bb'})^2 + \left(\dfrac{f}{f_{\alpha e}}r_{bb'}\right)^2} \tag{3.126}$$

Equating real and imaginary parts

$$G_i = \frac{r_{b'e}+r_{bb'} + \left(\dfrac{f}{f_{\alpha e}}\right)^2 r_{bb'}}{(r_{b'e}+r_{bb'})^2 + \left(\dfrac{f}{f_{\alpha e}}r_{bb'}\right)^2}$$

$$C_i = \frac{r_{b'e}/2\pi f_{\alpha e}}{(r_{b'e}+r_{bb'})^2 + \left(\dfrac{f}{f_{\alpha e}}r_{bb'}\right)^2} \tag{3.127}$$

These two relationships clearly demonstrate the frequency dependence of the short-circuit input impedance of the common emitter stage. Thus G_i increases with increasing frequency; it approaches the value of $(r_{bb'}+r_{b'e})^{-1}$ at frequencies low compared with $f_{\alpha e}$ while it approaches $1/r_{bb'}$ at very high frequencies. The input capacitance C_i approaches zero at high frequencies and the value of

$$\frac{r_{b'e}}{2\pi f_{\alpha e}}(r_{b'e}+r_{bb'})^{-2}$$

at low frequencies.

3.9.4. Frequency Dependence of Output Admittance

We shall next determine the frequency variation of the output admittance of the common emitter stage when its input terminals

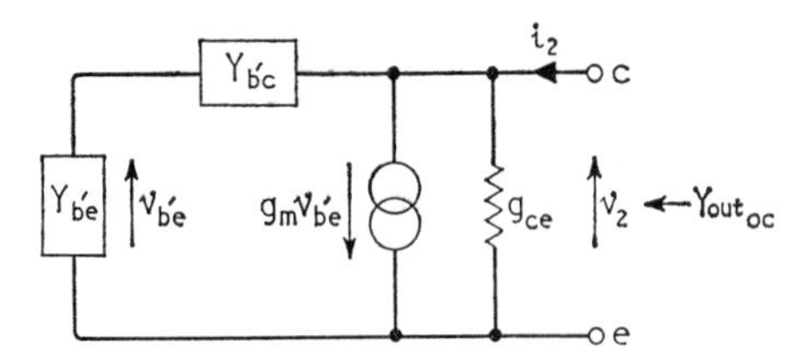

Fig. 3.57. Hybrid-π equivalent circuit with input open circuited

are open circuited as in Fig. 3.57. $Y_{b'e}$ represents the admittance of the elements $r_{b'e}$ and $C_{b'e}$ in parallel while $Y_{b'c}$ represents the admittance of $r_{b'c}$ and $C_{b'c}$ in parallel, that is,

$$\left.\begin{aligned} Y_{b'e} &= \frac{1}{r_{b'e}}+j\omega C_{b'e} \\ Y_{b'c} &= \frac{1}{r_{b'c}}+j\omega C_{b'c} \end{aligned}\right\} \tag{3.128}$$

If a voltage v_2 is applied across the output terminals then the internal base to emitter voltage drop $v_{b'e}$ is

$$v_{b'e} = \frac{Y_{b'c}}{Y_{b'c}+Y_{b'e}}v_2$$

Since $Y_{b'e} \gg Y_{b'c}$ because $r_{b'c} \gg r_{b'e}$ and $C_{b'e} \gg C_{b'c}$ then

$$v_{b'e} \simeq \frac{Y_{b'c}}{Y_{b'e}}v_2 \tag{3.129}$$

From Fig. 3.57 we find that the output current i_2 is given by

$$i_2 \simeq g_{ce} v_2 + g_m v_{b'e} + Y_{b'c} v_2$$

$$\simeq \left(g_{ce} + \frac{g_m Y_{b'c}}{Y_{b'e}} + Y_{b'c} \right) v_2 \tag{3.130}$$

Therefore, the open-circuit output admittance $Y_{\text{out}\ OC}$ of the common emitter stage is

$$Y_{\text{out}\ OC} \simeq g_{ce} + \frac{g_m Y_{b'c}}{Y_{b'e}} + Y_{b'c} \tag{3.131}$$

Substituting the relationships of Equation 3.128 we obtain

$$Y_{\text{out}\ OC} \simeq g_{ce} + \frac{g_m r_{b'e}}{r_{b'c}} \cdot \frac{1 + \mathrm{j}\omega C_{b'c} r_{b'c}}{1 + \mathrm{j}\omega C_{b'e} r_{b'e}} + \frac{1}{r_{b'c}} + \mathrm{j}\omega C_{b'c} \tag{3.132}$$

Let

$$f_c = \frac{1}{2\pi C_{b'c} r_{b'c}} \tag{3.133}$$

Then in terms of f_c and $f_{\alpha e}$ of Equation 3.106 we can rewrite Equation 3.132 as

$$Y_{\text{out}\ OC} \simeq g_{ce} + \mathrm{j}2\pi f C_{b'c} + \frac{g_m r_{b'e}}{r_{b'c}} \cdot \frac{1 + \dfrac{\mathrm{j}f}{f_c}}{1 + \dfrac{\mathrm{j}f}{f_{\alpha e}}} \tag{3.134}$$

where $\omega = 2\pi f$ and we have made use of the inequality $g_{ce} \gg 1/r_{b'c}$ (see the typical values given earlier for the elements of the hybrid-π equivalent circuit in Equation 3.107).

Rationalising the third term on the right-hand side of Equation 3.134 we obtain

$$Y_{\text{out}\ OC} \simeq g_{ce} + \mathrm{j}2\pi f C_{b'c} + \frac{g_m r_{b'e}}{r_{b'c}} \cdot \frac{\left(1 + \dfrac{\mathrm{j}f}{f_c}\right)\left(1 - \dfrac{\mathrm{j}f}{f_{\alpha e}}\right)}{\left(1 + \dfrac{\mathrm{j}f}{f_{\alpha e}}\right)\left(1 - \dfrac{\mathrm{j}f}{f_{\alpha e}}\right)}$$

$$= g_{ce} + \mathrm{j}2\pi f C_{b'c} + \frac{g_m r_{b'e}}{r_{b'c}} \cdot \frac{1 + \dfrac{f^2}{f_{\alpha e} f_c} + \mathrm{j}f\left(\dfrac{1}{f_c} - \dfrac{1}{f_{\alpha e}}\right)}{1 + \left(\dfrac{f}{f_{\alpha e}}\right)^2} \tag{3.135}$$

Suppose that G_o is the conductive component and C_o is the capacitive component of $Y_{\text{out }OC}$ (see Fig. 3.58). Then we can express $Y_{\text{out }OC}$ as

$$Y_{\text{out }OC} = G_o + \text{j}2\pi f C_o \tag{3.136}$$

Equating the real and imaginary parts in Equations 3.135 and 3.136 we obtain

$$\left.\begin{aligned} G_o &\simeq g_{ce} + \frac{g_m r_{b'e}}{r_{b'c}} \cdot \frac{1+\dfrac{f^2}{f_{\alpha e} f_c}}{1+\left(\dfrac{f}{f_{\alpha e}}\right)^2} \\ C_o &\simeq C_{b'c} + \frac{g_m r_{b'e}}{2\pi r_{b'c}} \cdot \frac{\left(\dfrac{1}{f_c}-\dfrac{1}{f_{\alpha e}}\right)}{1+\left(\dfrac{f}{f_{\alpha e}}\right)^2} \end{aligned}\right\} \tag{3.137}$$

We can, therefore, deduce that:

(1) The output conductance G_o approaches the value of

$$g_{ce} + \frac{g_m r_{b'e}}{r_{b'c}}$$

at low frequencies and the value of

$$g_{ce} + \frac{g_m r_{b'e}}{r_{b'c}} \cdot \frac{f_{\alpha e}}{f_c}$$

at high frequencies. However, $f_{\alpha e}$ is normally greater than f_c (see Example 3.11). Thus G_o increases with increasing frequency.

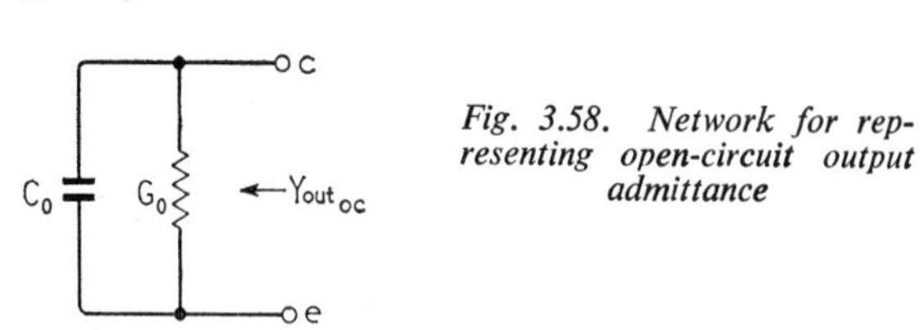

Fig. 3.58. Network for representing open-circuit output admittance

(2) The output capacitance C_o approaches the value of

$$C_{b'c} + \frac{g_m r_{b'e}}{2\pi r_{b'c}}\left(\frac{1}{f_c}-\frac{1}{f_{\alpha e}}\right)$$

at low frequencies and the value of $C_{b'c}$ at high frequencies.

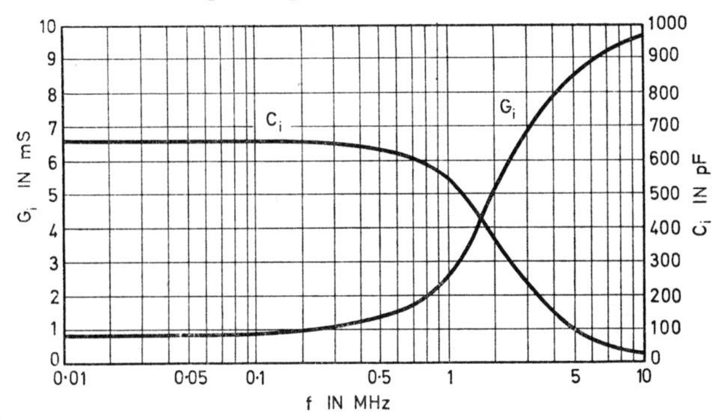

Fig. 3.59. Frequency dependence of G_i and C_i

Example 3.11

Determine and plot the frequency dependence of the short-circuit input impedance and open-circuit output admittance of a common emitter stage having the hybrid-π parameters given in Equation 3.107.

From Equations 3.106 and 3.133 we find that, respectively,

$$f_{\alpha e} = \frac{1}{2\pi C_{b'e} r_{b'e}}$$

$$= \frac{1}{2\pi \times 800 \times 10^{-12} \times 10^{3}} \text{Hz}$$

$$= 200\text{kHz}$$

$$f_c = \frac{1}{2\pi \times 10 \times 10^{-12} \times 2 \times 10^{6}} \text{Hz}$$

$$= 8\text{kHz}$$

Therefore, Equation 3.127 gives the input conductance G_i and input capacitance C_i to be

$$G_i = \frac{1{\cdot}1 + 2{\cdot}5f^2}{1{\cdot}21 + 0{\cdot}25f^2} \text{mS}$$

$$C_i = \frac{800}{1{\cdot}21 + 0{\cdot}25f^2} \text{pF}$$

where the freauency f is in MHz. These are shown plotted in Fig. 3.59.

Next, we find from Equation 3.137 that the output conductance G_o and output capacitance C_o are:

$$G_o = 20+19\frac{1+625f^2}{1+25f^2}\,\mu\text{S}$$

$$C_o = 10+\frac{363}{1+25f^2}\,\text{pF}$$

where f is in MHz. These are shown plotted in Fig. 3.60.

From these graphs we draw the important conclusion that, provided the frequency range of interest is sufficiently narrow, then

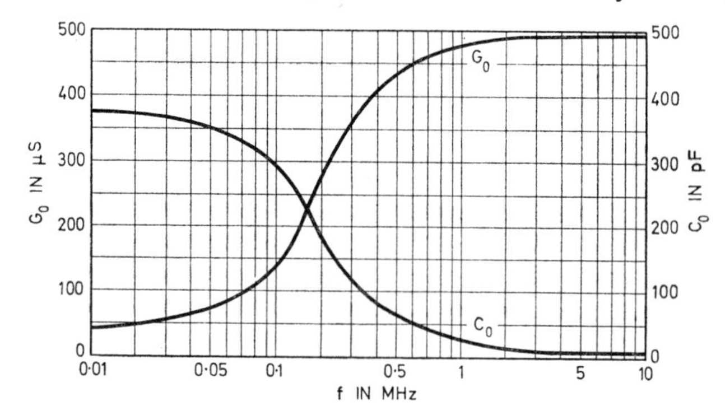

Fig. 3.60. Frequency dependence of G_0 and C_0

for all practical purposes each of C_i, G_i, C_o and G_o can be regarded as constant. This we shall find very useful when we study tuned amplifiers in Chapter 4.

3.10. NOISE

In a broad sense, the term *noise* applies to any undesired disturbance that obscures the signal that is being amplified. There are various kinds of noise but we shall only consider the *fluctuation noise* that is due to the random motion of electrons. Thus in any electrical conductor the electrons are in a continuous state of thermal agitation at a temperature above absolute zero resulting in a minute, but detectable, electric current known as *thermal noise*. A conductor of resistance R is associated with a mean square noise voltage equal to

$$v_n^2 = 4\text{k}TBR \tag{3.138}$$

where $\mathbf{k}$ = Boltzmann's constant

T = Absolute temperature

B = Noise bandwidth

We therefore obtain the noise equivalent circuit of Fig. 3.61 for an electrical conductor.

Suppose that an amplifier is driven from a voltage source of series resistance R_s. The applied signal will be accompanied by noise

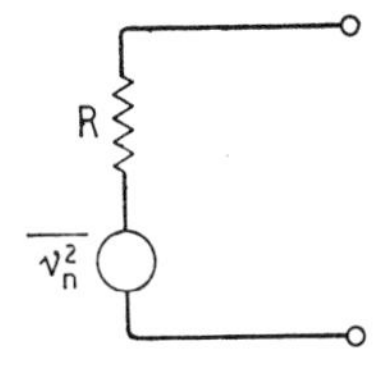

Fig. 3.61. Noise equivalent circuit of electrical conductor

having the mean square noise voltage of Equation 3.138 with R replaced by R_s. In addition the amplifier itself will introduce a further amount of noise.

Noise is generated within a transistor mainly as a result of 5 independent and random causes. In the case of a pnp transistor, for example, we find that

1. The holes travel across the base region by the random process of diffusion. Therefore, there is noise associated with the collection of injected holes.
2. Some of the injected holes recombine with electrons in the base region in a random fashion. Hence there is noise associated with the part of injected emitter current that is subject to recombination.
3. The bulk resistance $r_{bb'}$ of the base material is associated with thermal noise.
4. The thermal generation of charge carriers which results in the collector–base leakage current I_{CBO} is a random process. Noise contributed by this source can be neglected if the quiescent collector current is large compared to $I_{CBO}/(1-\alpha_B)$.
5. At low frequencies surface leakage effects result in the so-called $1/f$ *noise* which varies inversely with frequency f. Normally this source of noise becomes noticeable below a few kHz.

The combined effect of all the above sources of noise is to set a limit on how weak a signal the amplifier can handle. In this respect it is found convenient to use the term *noise figure* to express the noise quality of an amplifier. Noise figure F of a transistor stage is defined as

$$F = \frac{N_a}{N_i} \tag{3.139}$$

where N_a = noise power measured at the output terminals with the driving source and transistor both assumed noisy,

N_i = output noise power with the source assumed noisy and the transistor assumed noiseless.

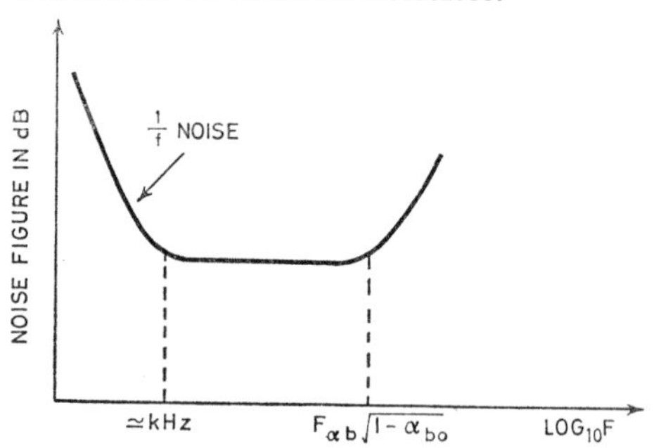

Fig. 3.62. Frequency dependence of transistor noise figure

Notice that with the above definition F is independent of the load resistance but is dependent on the source resistance. Normally, it is expressed in decibels, as $10 \log_{10} F$.

Theoretical considerations show that the noise figure of a transistor is practically independent of the circuit configuration in which it is used, that is, it is the same for the common base, common emitter and common collector configurations.[1] Further, the noise figure, expressed in decibels, varies with frequency in the manner shown in Fig. 3.62.

[1] LO, A. W., ENDRES, R. O., ZAWELS, J., WALDHAUER, F. D., and CHENG, C. C., *Transistor Electronics*, Prentice-Hall Inc. (1955).

CHAPTER 4

AMPLIFIER CIRCUITS

Amplifier circuits can be classified functionally into four major groups:

(1) d.c. amplifiers,

(2) wideband amplifiers,

(3) tuned amplifiers,

(4) power amplifiers.

In this chapter we shall discuss the circuit features of each of these amplifier types in the absence of external feedback. Application of feedback will be considered in Chapter 5.

4.1. D.C. AMPLIFIERS

In *direct current* or d.c. amplifiers the various transistor stages are directly coupled to each other as illustrated in the two-stage common emitter amplifier of Fig. 4.1. They are useful for amplifying input signals which vary slowly with time, that is, signals having very low frequency components. A problem of particular concern in the design of d.c. amplifiers is that of *drift*, which refers to the shift of quiescent operating-point of a transistor stage. It is principally caused by the temperature dependence of the following transistor parameters:

(1) the collector-base leakage current I_{CBO},

(2) the common emitter current amplification factor α_E,

(3) the base–emitter voltage drop V_{BE}.

Drift is troublesome because it is not possible to distinguish variation of the d.c. amplifier output current, due to the above thermal effects, from the input signal. The performance of d.c. amplifiers is conveniently described in terms of the *residual drift voltage* defined as follows. It is the voltage which, if applied to the input terminal of the first stage, will produce the same change in the output current of the complete d.c. amplifier as is produced by the temperature dependence of transistor parameters. We shall find this term useful in Section 4.1.2.

For the d.c. amplifier to be useful it is necessary to minimise the temperature dependence of the output current as far as possible. In general, this objective can be realised by the addition of temperature-sensitive circuit elements (e.g. junction diodes, thermistors) at appropriate points inside the amplifier so as to vary the quiescent operating-point of the various stages in such a way as to compensate

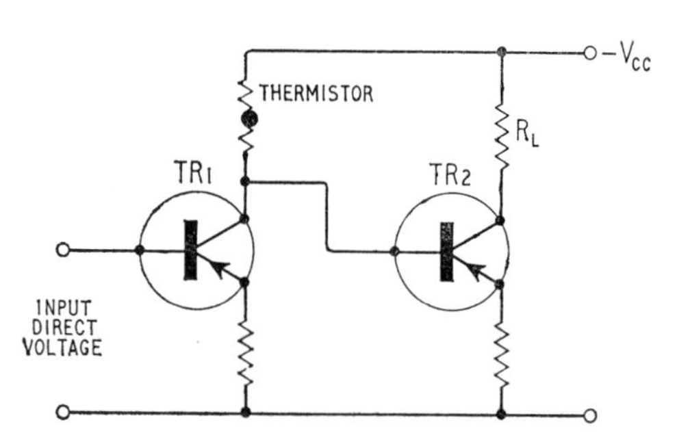

Fig. 4.1. Two-stage d.c. amplifier with compensation

for the temperature dependence of the output current. As an example consider the two-stage d.c. amplifier of Fig. 4.1 where a thermistor is used in the collector circuit of stage TR1. When the temperature is increased the collector current of TR1 increases in magnitude and the resistance of the thermistor decreases. Therefore, the voltage drop across the thermistor and the emitter voltage of TR2 are maintained sensibly constant. It is thus possible to minimise the temperature dependence of the output collector current of TR2.

4.1.1. The Composite Transistor

A circuit configuration that is useful in d.c. amplifiers is the *composite transistor** shown in Fig. 4.2, which involves two transistors. This circuit configuration has a short circuit current gain K_{iSC} that is very close to unity in magnitude and that is not sensitive to transistor

* This composite transistor is often referred to as the *Darlington composite transistor* after its originator.

parameter variations. To determine K_{iSC} replace each transistor with the common base version of the simplified h-parameter equivalent circuit of Fig. 3.42. Thus we obtain the circuit representation of Fig. 4.3 where h_{ib1}, and h_{fb1} are the common base h-parameters of transistor TR1, and h_{ib2} and h_{fb2} refer to transistor TR2.

By definition K_{iSC} is equal to the ratio of the output current i_2 to the input current i_1 when the output terminals are short circuited as in Fig. 4.3, where we see that

$$i_2 = h_{fb1} i_1 + h_{fb2}(1 + h_{fb1}) i_1 \tag{4.1}$$

Therefore, the short-circuit current gain of the composite transistor of Fig. 4.2 is

$$\begin{aligned} K_{iSC} &= \frac{i_2}{i_1} \\ &= h_{fb1} + h_{fb2} + h_{fb1} h_{fb2} \end{aligned} \tag{4.2}$$

To demonstrate the closeness of K_{iSC} to unity, assume identical transistors each having a value of $-0{\cdot}96$ for its h_{fb} parameter.

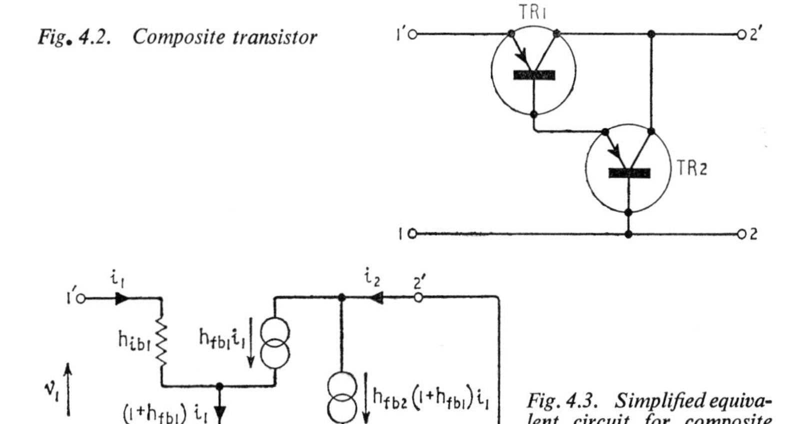

Fig. 4.2. Composite transistor

Fig. 4.3. Simplified equivalent circuit for composite transistor of Fig. 4.2. when output terminals 2 2′ are short circuited

From Equation 4.2 we find that $K_{iSC} = -0{\cdot}998$ which is certainly closer to unity than the h_{fb} of either transistor. Further, K_{iSC} is not so sensitive to transistor parameter variations. To illustrate this suppose that, due to changing circumstances, h_{fb1} and h_{fb2} both drop to $-0{\cdot}9$. The modified value of K_{iSC} is deduced from

Equation 4.2 to be $-0{\cdot}99$. We, therefore, see that a drop of nearly 6% in the value of h_{fb1} and h_{fb2} causes K_{iSC} to decrease by only 0·8%.

Finally, from the circuit of Fig. 4.2 we deduce that

$$v_1 = h_{ib1} i_1 + (1 + h_{fb1}) h_{ib2} i_1 \tag{4.3}$$

and so the short-circuit input resistance $R_{\text{in }SC}$ of the composite transistor of Fig. 4.2 is

$$R_{\text{in }SC} = \frac{v_1}{i_1}$$

$$= h_{ib1} + (1 + h_{fb1}) h_{ib2} \tag{4.4}$$

However, h_{fb1} is normally quite close to -1. Therefore, $R_{\text{in }SC}$ is nearly equal to h_{ib1}.

4.1.2. Emitter Coupled Pair

Another circuit configuration that is widely used in d.c. amplifier design is the *emitter coupled pair* of Fig. 4.4 which involves two transistors having their emitter terminals directly connected to each

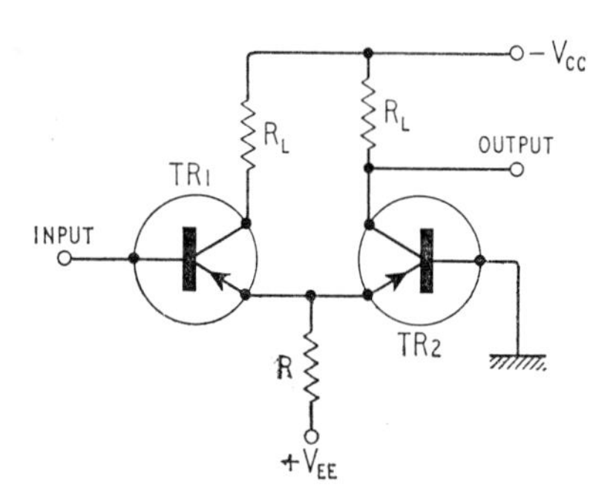

Fig. 4.4. Emitter coupled pair

other; hence the name of the circuit. The great advantage of this circuit is that thermal effects are appreciably reduced at the output because changes in corresponding parameters of the two stages tend to cancel out.

In the following analysis we shall assume a small load resistance R_L and identical transistors. Replacing each transistor with the simplified h-parameters equivalent circuit of Fig. 3.42 we obtain the circuit of Fig. 4.5, where h_{ie} and h_{fe} are the common emitter h-parameters. The input voltages δV_1 and δV_2 signify the residual

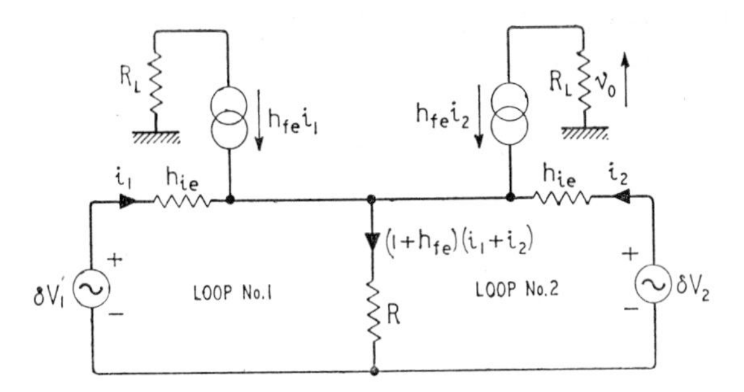

Fig. 4.5. Equivalent circuit of emitter coupled pair

drift voltages of transistors TR1 and TR2, respectively. Applying Kirchoff's Voltage Law to loop No. 1 gives

$$\delta V_1 = h_{ie}i_1 + R(1+h_{fe})(i_1+i_2) \tag{4.5}$$

and for loop No. 2 we find that

$$\delta V_2 = R(1+h_{fe})(i_1+i_2)+h_{ie}i_2 \tag{4.6}$$

Collecting terms and solving for the current i_2:

$$i_2 = \frac{\delta V_2 . [h_{ie}+R(1+h_{fe})] - \delta V_1 . R(1+h_{fe})}{h_{ie}[h_{ie}+2R(1+h_{fe})]} \tag{4.7}$$

Therefore, the output voltage v_0 is

$$\begin{aligned} v_0 &= -R_L h_{fe} i_2 \\ &= \frac{R_L h_{fe}}{h_{ie}} . \frac{\delta V_1 R(1+h_{fe}) - \delta V_2[h_{ie}+R(1+h_{fe})]}{h_{ie}+2R(1+h_{fe})} \end{aligned} \tag{4.8}$$

The voltages δV_1 and δV_2 can be expressed as follows:

$$\left.\begin{aligned} \delta V_1 &= v_m + v_d \\ \delta V_2 &= v_m - v_d \end{aligned}\right\} \tag{4.9}$$

that is,

$$\left.\begin{aligned} v_m &= \tfrac{1}{2}(\delta V_1 + \delta V_2) \\ v_d &= \tfrac{1}{2}(\delta V_1 - \delta V_2) \end{aligned}\right\} \tag{4.10}$$

Thus v_m is the arithmetic mean of the two residual drift voltages, and v_d is equal to half their difference. When $v_d = 0$, from Equation 4.9 we find that $\delta V_1 = \delta V_2 = v_m$, and so v_m is referred to as the *in-phase component*. When $v_m = 0$, from Equation 4.9 we see that $\delta V_1 = v_d$ and $\delta V_2 = -v_d$, and so v_d is referred to as the *anti-phase component.*

Using Equations 4.8 and 4.9 we can express the output voltage as

$$v_0 = \frac{h_{fe} R_L v_d}{h_{ie}} - \frac{h_{fe} R_L v_m}{h_{ie}+2R(1+h_{fe})} \tag{4.11}$$

When $v_d = 0$ we find from Equation 4.11 that the gain K to the in-phase component v_m is

$$K_m = \frac{v_0}{v_m}$$

$$= \frac{-h_{fe} R_L}{h_{ie}+2R(1+h_{fe})} \tag{4.12}$$

On the other hand, when $v_m = 0$, Equation 4.11 gives the gain K_d to the anti-phase component v_d to be

$$K_d = \frac{v_0}{v_d}$$

$$= \frac{h_{fe} R_L}{h_{ie}} \tag{4.13}$$

The effect cf drift is diminished if $K_d \gg |K_m|$. From Equations 4.12 and 4.13 we observe that this can be achieved if the common resistor R is large enough to satisfy the following inequality:

$$R \gg \frac{h_{ie}}{2(1 + h_{fe})} \tag{4.14}$$

4.1.3. Chopper Amplifier*

Another widely used method of amplifying slowly varying signals is to use a *chopper amplifier*. This involves first converting the input signal into an alternating one, then amplifying it in an a.c. amplifier and finally converting back to d.c.

Fig. 4.6 shows the circuit arrangement of a chopper amplifier. Additional circuitry is provided for opening and closing the switches S_1 and S_2 periodically and synchronously. Transistors can be used for simulating S_1 and S_2. Suppose that a direct voltage V_1 is applied to the input terminal. When S_1 and S_2 are closed the

* It may be found better to read this section after Chapter 6.

voltages at points A and B are both zero. The voltages across the capacitors C_1 and C_2 reach their equilibrium values provided that the closing time of the switches is sufficiently long.

Consider next opening S_1 and S_2. Since the voltage across a capacitor cannot change instantaneously then a voltage step of $(R_{in} V_1)/(R_1+R_{in})$ is applied to the input of the a.c. amplifier. This amplifier is assumed to have an input resistance of R_{in} and a voltage gain of K_v. Therefore, when switch S_2 is opened a voltage step of $(R_{in} K_v V_1)/(R_1+R_{in})$ is developed at point B. This voltage step is

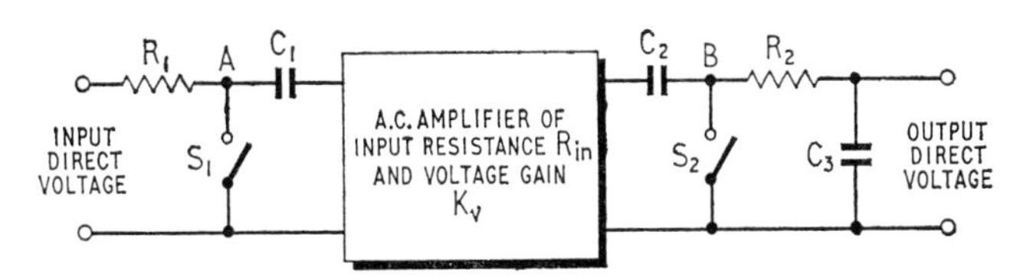

Fig. 4.6. Chopper amplifier

then smoothed by the network of R_2 and C_3 with the result that the direct output voltage developed across the capacitor C_3 is an amplified version of the direct input voltage.

The circuit arrangement of Fig. 4.6 functions with both polarities of input voltage. Further, the effect of drift can be made negligibly small by operating the switches S_1 and S_2 at a sufficiently high rate.

This is in fact a balanced modulator system as described in more detail in section 9.3.3.

4.2. WAVEFORM DISTORTION

When the waveform of the output signal of an amplifier is not a faithful reproduction of the input signal waveform we speak of *waveform distortion.* There are three forms of waveform distortion, which can exist separately or simultaneously; they are

(1) non-linear or harmonic distortion,
(2) amplitude distortion,
(3) phase or delay distortion.

Non-linear distortion was considered earlier in Section 2.3. It is due to the non-linear nature of the transistor characteristic curves. It is of particular concern in the design of power amplifiers.

Amplitude distortion occurs when the different frequency components of an input signal having a complex waveform are not amplified equally. Its presence can be ascertained by examining the *gain-frequency characteristic* whose form is largely a function of the design of the external circuitry associated with the transistor. Fig. 4.7 shows that the gain is constant in the frequency range f_1 to f_2, and so all signal frequencies located inside this frequency range will be amplified equally. Therefore, the amplifier exhibits zero amplitude distortion inside the frequency range f_1 to f_2. However, outside

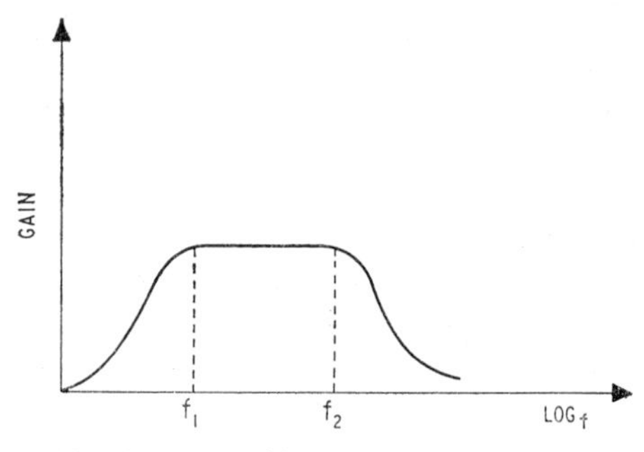

Fig. 4.7. Typical gain-frequency response

this frequency range the amplifier gain begins to fall off and so amplitude distortion is produced.

The third source of waveform distortion, namely, *phase* or *delay distortion*, arises if the various frequency components of a complex input waveform are not delayed equally. Its presence can be established by examining the *phase-frequency characteristic* of the amplifier. If this characteristic is linear as in Fig. 4.8, the resulting output signal is simply delayed with respect to the input signal, and no phase distortion occurs. This form of distortion is of particular concern in the design of wideband amplifiers used in television and radar. Like amplitude distortion, phase distortion is influenced by the external circuitry associated with the transistor.

4.3. WIDEBAND AMPLIFIERS

A *wideband* or *video amplifier* is one capable of providing a uniform amplification of the input signal over a wide frequency range extending from a few hertz to several megahertz. The upper frequency limit of the useful band is determined by the high frequency response of the transistor and the terminating resistors (see Section 3.9). Normally, the various transistor stages are capacitively coupled as in Fig. 4.9, and so the lower frequency limit of the useful band is determined by the coupling capacitor. At the operating frequencies of interest this capacitor presents a negligibly small impedance.

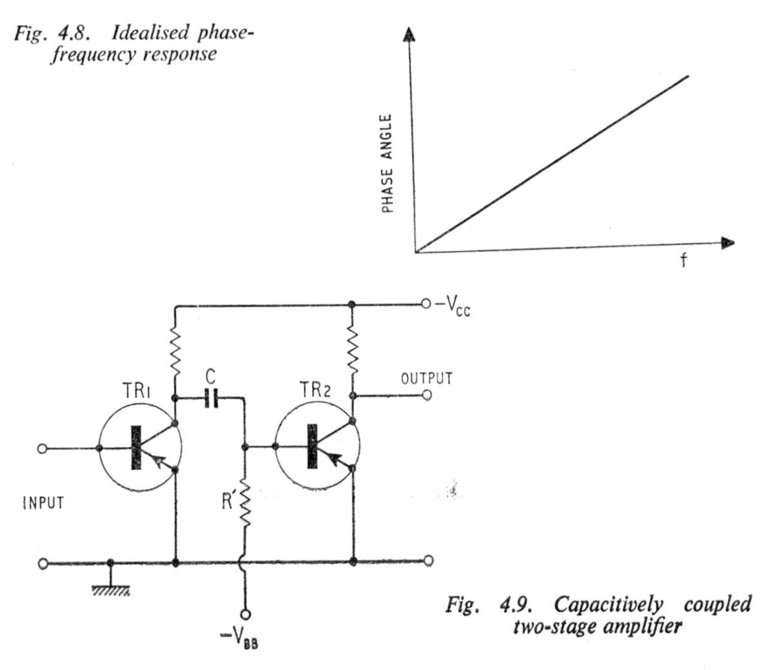

Fig. 4.8. Idealised phase-frequency response

Fig. 4.9. Capacitively coupled two-stage amplifier

Therefore, the complete frequency response of a wideband amplifier can be subdivided into three regions, namely, *high frequency*, *low frequency* and *mid-frequency* regions.

4.3.1. High Frequency Response

In the high frequency region the effect of the coupling capacitor is negligibly small. Therefore, it is only necessary to consider the high frequency effects associated with the transistor. In Section 3.9 we saw that the upper cut-off frequency f_h at which the modulus of the external voltage gain drops to $1/\sqrt{2}$ of its mid-frequency value is

$$f_h = \frac{1}{2\pi C_{eq}}\left(g_{b'e} + \frac{1}{r_{bb'} + R_s}\right) \tag{4.15}$$

where C_{eq} is the equivalent input capacitance defined as

$$C_{eq} \simeq C_{b'e} + g_m R_L C_{b'c} \tag{4.16}$$

Thus f_h can be increased by

(1) Employing a transistor of high common emitter α—cut-off frequency $f_{\alpha e}$, since a high value of $f_{\alpha e}$ corresponds to a low value of $C_{b'e}$ (see Equation 3.106).

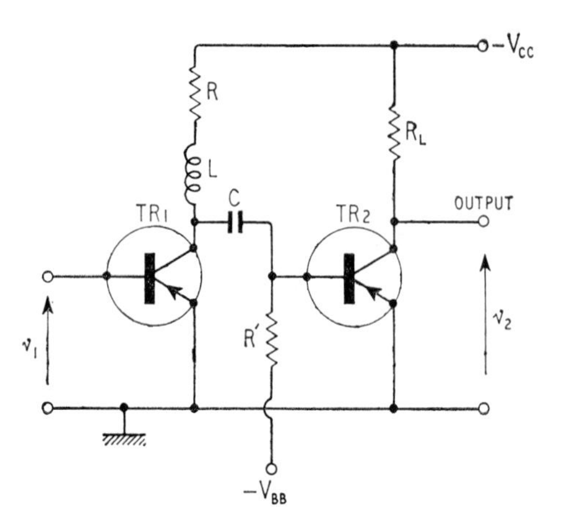

Fig. 4.10. Shunt peaking

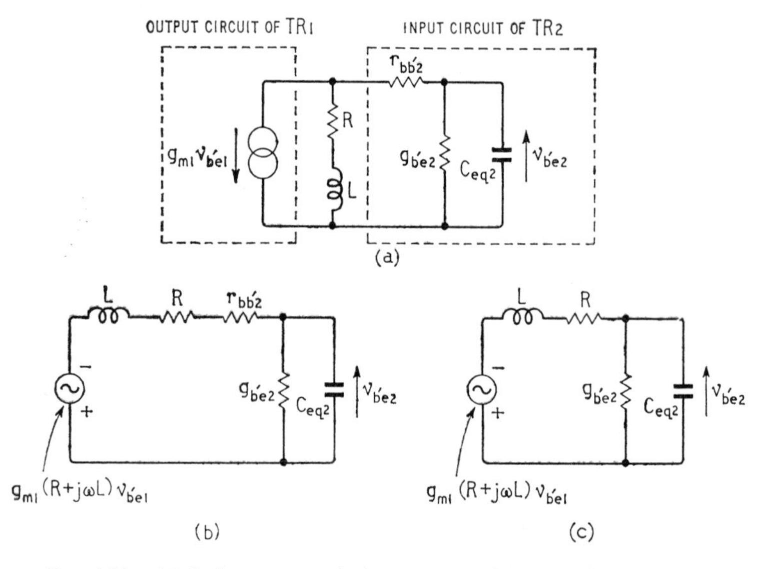

Fig. 4.11. High frequency equivalent circuits: (a) Equivalent circuit at interstage point of Fig. 4.10; (b) Equivalent circuit after replacing current source of (a) with equivalent voltage source; (c) Simplified circuit having neglected $r_{bb'2}$

(2) Reducing source resistance and/or load resistance.

Notice that in the second case a reduction in the value of R_L causes the external voltage gain of the stage to be reduced, and so the frequency response is improved at the expense of useful gain.

If it is found that for a given transistor and for given terminating source and load resistances the upper cut-off frequency is not high enough, then it becomes necessary to add compensating networks so as to neutralise the high frequency effects associated with the transistor.

HIGH-FREQUENCY COMPENSATION

A method widely used for extending the upper cut-off frequency f_h of a transistor stage is to connect an inductor in series with the collector resistance as shown in Fig. 4.10. This method of high frequency compensation is known as *shunt peaking* because the compensating inductor appears in parallel with the input circuit of the second stage as can be seen from the equivalent circuit of Fig. 4.11(a). Here it is assumed that the interstage biasing resistor R' of Fig. 4.10 is large compared to the input impedance of the second stage TR2 so as to justify neglecting its shunting effect. Further, the second stage has been represented by the simplified hybrid equivalent circuit of Section 3.9. Notice that $v_{b'e1}$ and $v_{b'e2}$ are the internal base–emitter voltage drops of stages TR1 and TR2, respectively. The parameter g_{m1} refers to the first stage while $g_{b'e2}$, $r_{bb'2}$ and $C_{eq\,2}$ refer to the second stage. Replacing the controlled current source of Fig. 4.11(a) with an equivalent voltage source we obtain the circuit of Fig. 4.11(b). If R is large compared to the base resistance $r_{bb'2}$ of the second stage, we can neglect the effect of $r_{bb'2}$ and so obtain the simpler circuit of Fig. 4.11(c). Therefore, the high frequency value of the voltage transfer ratio from the internal base of the first stage to that of the second stage is

$$K_{hf} = \frac{v_{b'e2}}{v_{b'e1}}$$

$$\simeq \frac{-g_{m1}(R+j\omega L)}{1+Rg_{b'e2}+j\omega(RC_{eq\,2}+Lg_{b'e2})-\omega^2 LC_{eq\,2}} \tag{4.17}$$

Notice that the overall voltage gain of the circuit of Fig. 4.10 is equal to v_2/v_1, which can be expressed as follows

$$\frac{v_2}{v_1} = \frac{v_{b'e1}}{v_1}\cdot\frac{v_{b'e2}}{v_{b'e1}}\cdot\frac{v_2}{v_{b'e2}} \tag{4.18}$$

However, since we are mainly concerned with evaluating the compensating effect of inductor L then we shall only consider the voltage ratio $v_{b'e2}/v_{b'e1}$.

Let

$$\left.\begin{aligned} K_{mf} &= \frac{-g_{m1}R}{1+g_{b'e2}R} \\ \omega_{h2} &= \frac{1+g_{b'e2}R}{RC_{\mathrm{eq}\,2}} \\ m &= \frac{L(1+g_{b'e2}R)}{R^2 C_{\mathrm{eq}\,2}} \\ n^2 &= 1-\frac{2L}{R^2 C_{\mathrm{eq}\,2}}+\left(\frac{Lg_{b'e2}}{RC_{\mathrm{eq}\,2}}\right)^2 \\ x &= \frac{\omega}{\omega_{h2}} \end{aligned}\right\} \tag{4.19}$$

Thus K_{mf} is the mid-frequency value of the voltage transfer ratio, $f_{h2} = \omega_{h2}/2\pi$ is the upper cut-off frequency of the second stage, and the parameters, m, n and x are all dimensionless. In terms of these parameters we can re-write Equation 4.17 as follows:

$$\frac{K_{hf}}{K_{mf}} \simeq \frac{1+\mathrm{j}xm}{(1-mx^2)+\mathrm{j}x\sqrt{(2m+n^2)}} \tag{4.20}$$

Evaluating the modulus we obtain

$$\begin{aligned} \left|\frac{K_{hf}}{K_{mf}}\right| &\simeq \left[\frac{1+m^2x^2}{(1-mx^2)^2+x^2(2m+n^2)}\right]^{\frac{1}{2}} \\ &= \left[\frac{1+m^2x^2}{1+n^2x^2+m^2x^4}\right]^{\frac{1}{2}} \end{aligned} \tag{4.21}$$

MAXIMALLY FLAT RESPONSE

A frequency response that is of particular interest is the *maximally flat* one which, in the case of Equation 4.21, occurs when the coefficients of the variable x^2 in the numerator and denominator are equal,[1] that is,

$$m^2 = n^2 \tag{4.22}$$

When this relationship is satisfied the first derivative of Equation 4.21

[1] LANDON, V. D., 'Cascade Amplifiers with Maximal Flatness', *R.C.A. Rev.* **5**, *3* and *4* (1941).

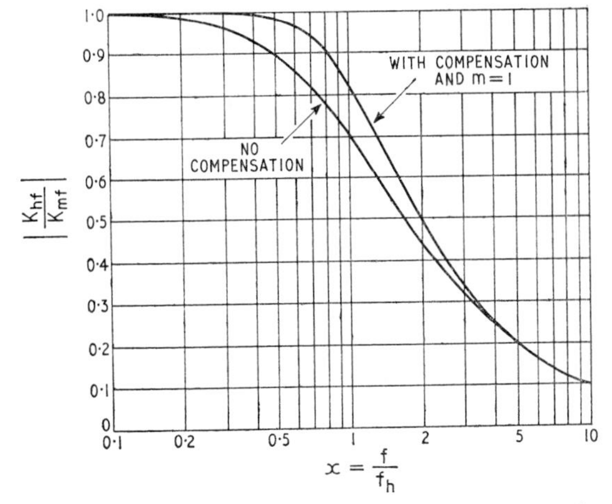

Fig. 4.12. High frequency gain response with and without inductor L

with respect to x is zero at $x = 0$, and the resulting gain–frequency characteristic is maximally flat.

Therefore, for a maximally flat response Equations 4.21 and 4.22 give

$$\left|\frac{K_{hf}}{K_{mf}}\right| = \left[\frac{1+m^2x^2}{1+m^2x^2+m^2x^4}\right]^{\frac{1}{2}} \tag{4.23}$$

Suppose that the circuit conditions are such that $m = 1$. For this value of m Equation 4.23 is shown plotted in Fig. 4.12.

In the uncompensated stage $L = 0$. This corresponds to $m = 0$ and $n = 1$ (see the third and fourth lines of Equation 4.19), and so Equation 4.21 reduces to

$$\left|\frac{K_{hf}}{K_{mf}}\right| = \left[\frac{1}{1+x^2}\right]^{\frac{1}{2}} \tag{4.24}$$

This is also shown plotted in Fig. 4.12, which clearly demonstrates that the addition of the compensating inductor improves the high frequency response of the stage.

PHASE RESPONSE

From Equation 4.20 we find that, at high frequencies, the phase angle ϕ_h is:

$$\phi_h = \tan^{-1}(mx) - \tan^{-1}\left[\frac{x\sqrt{(2m+n^2)}}{(1-mx^2)}\right] \tag{4.25}$$

When the stage is arranged to have a maximally flat gain response, $m = n$ (see Equation 4.22). Hence Equation 4.25 gives the corresponding phase response as

$$\phi_h = \tan^{-1}(mx) - \tan^{-1}\left[\frac{x\sqrt{(2m+m^2)}}{1-mx^2}\right] \tag{4.26}$$

Again suppose that $m = 1$. The phase characteristic as obtained from Equation 4.26 for this value of m is shown plotted in Fig. 4.13.

When the stage is uncompensated $m = 0$, $n = 1$ and Equation 4.25 simplifies to

$$\phi_h = -\tan^{-1} x \tag{4.27}$$

This is also shown plotted in Fig. 4.13. Notice that at $x = \infty$, i.e. $f = \infty$, the phase angle approaches 90° for both compensated and uncompensated circuits.

4.3.2. Low Frequency Response

In the low frequency region we can neglect the high frequency effects associated with the transistor and it is only necessary to account for the effect of the interstage coupling capacitor. Therefore, if in Fig. 4.9 we neglect the shunting effect of resistor R' we obtain the simplified equivalent circuit of Fig. 4.14(a). Replacing the controlled current source with its equivalent voltage source we obtain the circuit of Fig. 4.14(b). Assuming R is large compared to $r_{bb'}$ we find that the low frequency value K_{lf} of the voltage transfer ratio from the internal base of the first stage to that of the second stage is

$$K_{lf} = \frac{v_{b'e2}}{v_{b'e1}}$$

$$\simeq \frac{-g_{m1} r_{b'e2} R}{R + \frac{1}{j\omega C} + r_{b'e2}} \tag{4.28}$$

Let

$$f_l = \frac{\omega_l}{2\pi}$$

$$= \frac{1}{2\pi C(R + r_{b'e2})} \tag{4.29}$$

Then in terms of f_l and K_{mf} we can rewrite Equation 4.28 as

$$\frac{K_{lf}}{K_{mf}} \simeq \frac{x}{x - j} \tag{4.30}$$

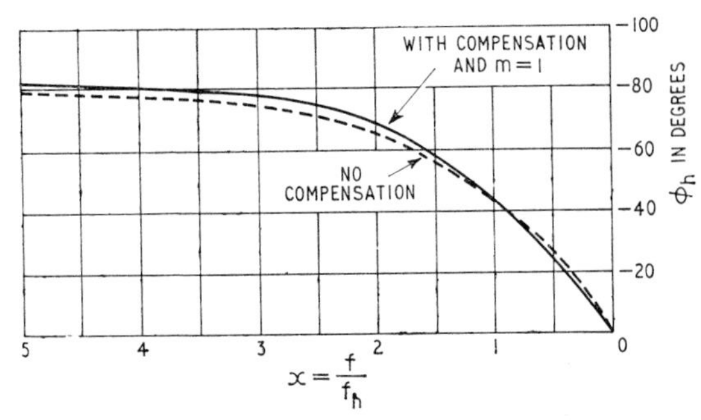

Fig. 4.13. High frequency phase response with and without inductor L

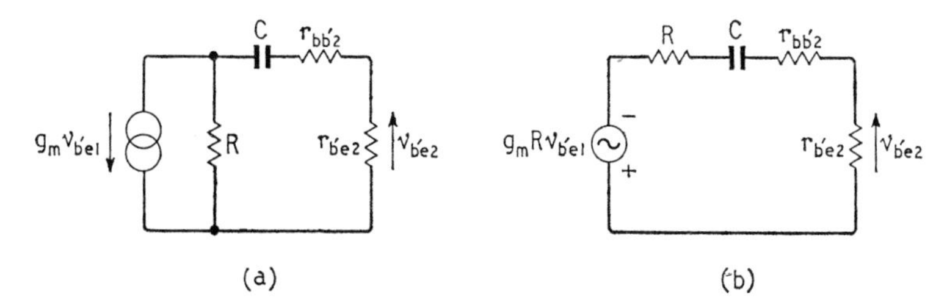

Fig. 4.14. Low frequency equivalent circuits: (a) Equivalent circuit at inter-stage point of Fig. 4.9; (b) Equivalent circuit after replacing current source of (a) with equivalent voltage source

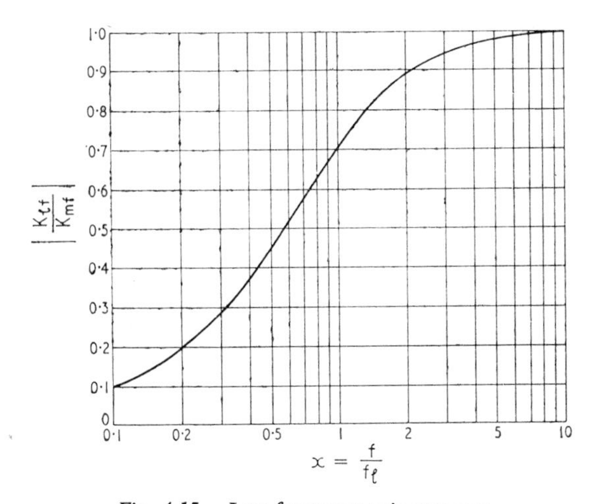

Fig. 4.15. Low frequency gain response

where

$$x = \frac{f}{f_l} \tag{4.31}$$

Evaluating the modulus we find

$$\left|\frac{K_{lf}}{K_{mf}}\right| \simeq \left[\frac{x^2}{x^2+1}\right]^{\frac{1}{2}} \tag{4.32}$$

When $x = 1$,

$$\left|\frac{K_{lf}}{K_{mf}}\right| = \frac{1}{\sqrt{2}}$$

Therefore, f_l is the *lower cut-off frequency* at which the modulus of K_{lf} drops to $1/\sqrt{2}$ of its mid-frequency value. From Equation 4.29 we see that the larger the coupling capacitor C the smaller f_l is, that is the better the low frequency response is.

The gain–frequency characteristic as calculated from Equation 4.32 is shown plotted in Fig. 4.15.

PHASE RESPONSE

From Equation 4.30 we find that, at low frequencies, the phase angle ϕ_l is

$$\phi_l = \tan^{-1}\left(\frac{1}{x}\right) \tag{4.33}$$

This is shown plotted in Fig. 4.16. Notice that at $x = 1$, i.e. $f = f_l$, the phase angle ϕ_l is equal to 45°. When $x = 0$, i.e. $f = 0$, the phase angle ϕ_l reaches its limiting value of 90°.

LOW FREQUENCY COMPENSATION

For a given value of coupling capacitor C the low frequency response can be improved by using the circuit arrangement of Fig. 4.17. At mid-band frequencies the effective load resistance of stage TR1 is equal to R paralleled by the input resistance $R_{\text{in }2}$ of stage TR2. At very low frequencies the current leaving the collector of TR1 sees $R+(1/\text{j}\omega C_1)$ in parallel with $R_{\text{in }2}+(1/\text{j}\omega C)$. This assumes that R' is large and R_1 is much greater than $1/\omega C_1$. We, therefore, find that at low frequencies the current delivered to the input of stage TR2 is practically independent of frequency if

$$\frac{R}{R_{\text{in }2}} = \frac{C}{C_1} \tag{4.34}$$

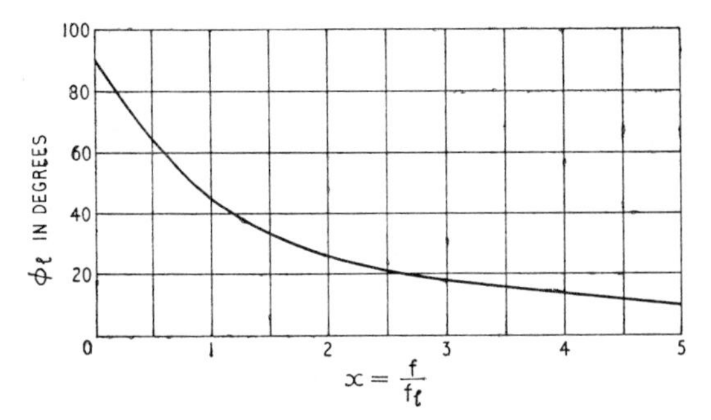

Fig. 4.16. Low frequency phase response

Finally, notice that the application of negative feedback also improves the frequency response of an amplifier. This will be considered further in Chapter 5. However, in this case we shall find that the frequency response is improved at the expense of mid-frequency gain. This is in contrast to the methods of compensation we have considered in the present chapter, the frequency response being improved and the mid-frequency gain left unchanged.

4.3.3. Mid-frequency Response

In the mid-frequency region the high frequency effects of the transistor and the impedance of the coupling capacitor are negligible.

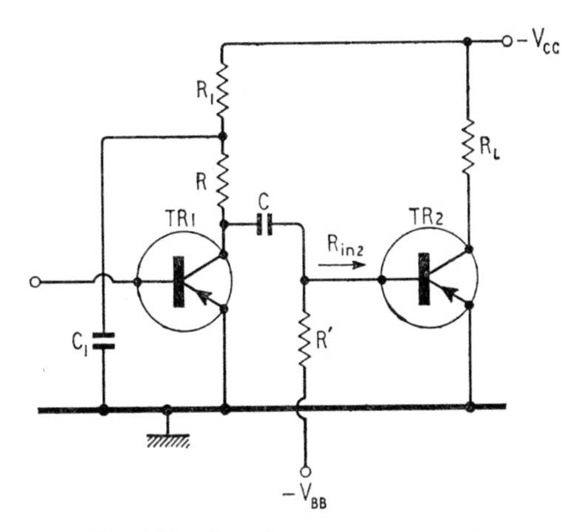

Fig. 4.17. Low frequency compensation

Therefore, the voltage transfer ratio from the internal base of the first stage to that of the second stage becomes purely real and equal to K_{mf} as defined in the first line of Equation 4.19. This expression is approximate in that we have neglected the shunting effects of the output resistance of stage TR1 and the internal feedback associated

Fig. 4.18. Schematic arrangement for tuned amplifier

with the second stage TR2. If a more accurate evaluation of the mid-frequency response is required then the procedure of Section 3.7 can be applied.

4.4. TUNED AMPLIFIERS

The function of a *tuned amplifier* is twofold.

(1) To provide a specified amount of gain from the source to the load over a specified frequency band.

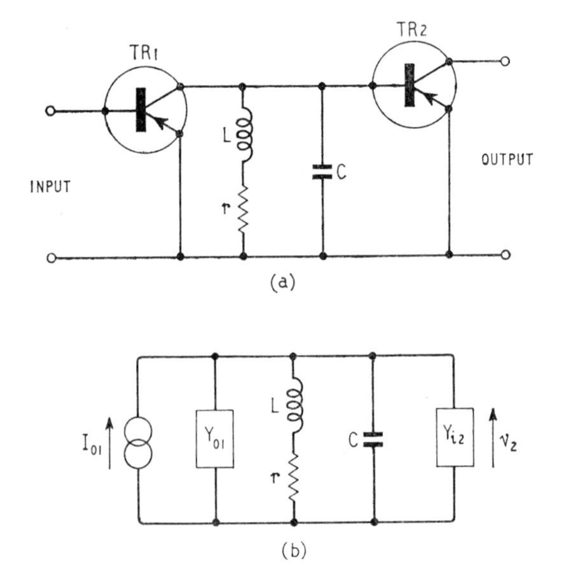

Fig. 4.19. Tuned amplifier: (a) Actual circuit; (b) Equivalent circuit at the interstage point

(2) To attenuate the signal frequencies located outside the required band.

These two major design objectives can be realised using the circuit arrangement of Fig. 4.18. The transistor stages provide the necessary gain while the interstage coupling passive network determines the desired frequency response.

The choice of a transistor is determined by its high frequency response and power handling capacity. In most cases the transistor

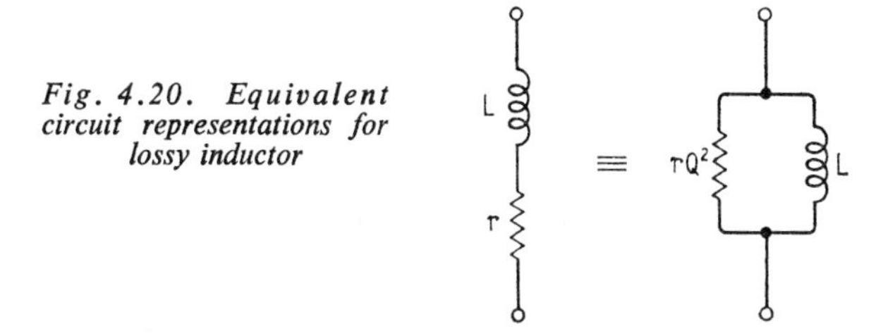

Fig. 4.20. Equivalent circuit representations for lossy inductor

is operated in the common emitter configuration because it provides the highest power gain except at very high frequencies where the common base configuration may be superior. The interstage coupling networks are composed of capacitors and inductors or transformers, the object being to maximise the transfer of power from one transistor stage to the next.

COUPLING NETWORK

A simple coupling network is shown in Fig. 4.19(a). It consists of the parallel combination of an inductor and a capacitor. The resistor r accounts for the unavoidable losses associated with the inductor. The admittance Y of the network is

$$Y = \mathrm{j}\omega C + \frac{1}{\mathrm{j}\omega L + r}$$

$$= \mathrm{j}\omega C + \frac{1}{\mathrm{j}\omega L\left(1 + \frac{r}{\mathrm{j}\omega L}\right)}$$

$$= \mathrm{j}\omega C + \frac{1}{\mathrm{j}\omega L\left(1 + \frac{1}{\mathrm{j}Q}\right)} \tag{4.35}$$

where
$$Q=\frac{\omega L}{r} \tag{4.36}$$

Q is called the *quality* or the Q-*factor* of the coil. The higher its value the smaller is the series resistance r of the coil, that is, the more perfect is the coil.

If Q is very large compared with unity, as is usually the case, the following approximation is quite justified

$$\frac{1}{\mathrm{j}\omega L\left(1+\frac{1}{\mathrm{j}Q}\right)} \simeq \frac{1}{\mathrm{j}\omega L}\left(1-\frac{1}{\mathrm{j}Q}\right)$$
$$=\frac{1}{\mathrm{j}\omega L}+\frac{1}{\omega LQ}$$
$$=\frac{1}{\mathrm{j}\omega L}+\frac{1}{rQ^2} \tag{4.37}$$

where we have made use of Equation 4.36. Therefore, a lossy inductor of high Q can also be represented as an inductance L connected across an equivalent shunt loss conductance of $1/rQ^2$ as in Fig. 4.20. Thus we can rewrite Equation 4.35 as

$$Y\simeq \mathrm{j}\omega C+\frac{1}{\mathrm{j}\omega L}+\frac{1}{rQ^2} \tag{4.38}$$

4.4.1. Selectivity Characteristic

For the purpose of further analysis let us represent the conditions across the output terminals of the stage TR1 by the current source I_{o1} shunted by an admittance Y_{o1} as in Fig. 4.19(b). The current I_{o1} is controlled by the voltage applied to the input terminals of the stage TR1 and Y_{o1} is its output admittance. Y_{i2} of Fig. 4.19(b) is the input admittance of the second stage TR2. Thus the voltage v_2 developed across the input terminals of TR2 is given by

$$v_2=\frac{I_{o1}}{Y_{o1}+Y+Y_{i2}} \tag{4.39}$$

Let Y_{o1} and Y_{i2} be expressed as follows

$$\left.\begin{aligned} Y_{o1} &= G_{o1}+\mathrm{j}\omega C_{o1} \\ Y_{i2} &= G_{i2}+\mathrm{j}\omega C_{i2} \end{aligned}\right\} \tag{4.40}$$

G_{o1} and C_{o1} are the output conductance and capacitance of transistor TR1, respectively; G_{i2} and C_{i2} are the input conductance and capacitance of transistor TR2. In Section 3.9 we found that each of G_{o1}, C_{o1}, G_{i2} and C_{i2} is frequency dependent. We also observed that they have sensibly constant values provided the frequency band of interest is sufficiently narrow. This assumption is perfectly justified in the following analysis.

Therefore, from Equations 4.38, 4.39 and 4.40 we find that the signal transfer ratio from the first to second stage is

$$\frac{I_{o1}}{v_2} \simeq \frac{1}{R_t} + j\omega C_t + \frac{1}{j\omega L} \tag{4.41}$$

where

$$\left.\begin{aligned} \frac{1}{R_t} &= G_{o1} + G_{i2} + \frac{1}{rQ^2} \\ C_t &= C_{o1} + C_{i2} + C \end{aligned}\right\} \tag{4.42}$$

Let $f_o = \omega_o/2\pi$ be the *anti-resonance frequency* of the parallel combination of the tuning inductance L and total capacitance C_t, and Q_e be the effective Q-factor of the loaded tuned circuit, that is

$$\left.\begin{aligned} \omega_0 &= 2\pi f_o = \frac{1}{\sqrt{(LC_t)}} \\ Q_e &= \frac{R_t}{\omega_o L} \end{aligned}\right\} \tag{4.43}$$

In terms of f_o and Q_e we can rewrite Equation 4.41 as follows:

$$\begin{aligned} \frac{I_{o1}}{v_2} &= \frac{1}{R_t}\left[1 + jQ_e\left(\frac{\omega}{\omega_o} - \frac{\omega_o}{\omega}\right)\right] \\ &= \frac{1}{R_t}\left[1 + jQ_e\left(\frac{f}{f_o} - \frac{f_o}{f}\right)\right] \end{aligned} \tag{4.44}$$

If a dimensionless parameter δ is defined as

$$\delta = \frac{f}{f_o} - 1 \tag{4.45}$$

then

$$\begin{aligned} \frac{f}{f_o} - \frac{f_o}{f} &= 1 + \delta - \frac{1}{1+\delta} \\ &= \frac{(1+\delta)^2 - 1}{1+\delta} \\ &= \frac{\delta(2+\delta)}{1+\delta} \end{aligned} \tag{4.46}$$

If the frequency band of interest is narrow then the operating-frequencies will be so close to f_o that δ will be much less than unity. In this case Equation 4.46 approximates to

$$\frac{f}{f_o}-\frac{f_o}{f}\simeq 2\delta \tag{4.47}$$

Therefore, Equation 4.44 becomes

$$\frac{I_{o1}}{v_2}\simeq\frac{1}{R_t}(1+j2\delta Q_e) \tag{4.48}$$

Inverting and evaluating the modulus we obtain

$$\left|\frac{v_2}{I_{o1}}\right|\simeq\frac{R_t}{\sqrt{[1+(2\delta Q_e)^2]}} \tag{4.49}$$

When $\delta = 0$, that is $\omega = \omega_o$ we find that $|v_2/I_{o1}|$ reaches its maximum value of R_t. Equation 4.49 is shown plotted in Fig. 4.21. The

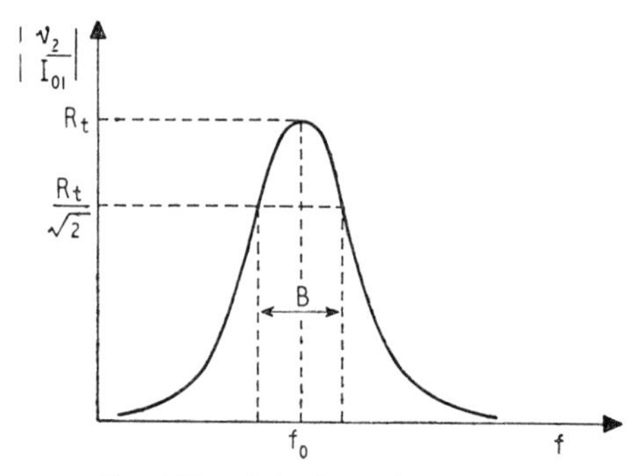

Fig. 4.21. Selectivity characteristic

resulting curve is usually referred to as the *selectivity characteristic* of the tuned amplifier.

4.4.2. Bandwidth

The *bandwidth B* is defined as the difference between the two frequencies at which $|v_2/I_{o1}|$ drops to $1/\sqrt{2}$ of its maximum value R_t. From Equation 4.49 this occurs when

$$\frac{R_t}{\sqrt{2}}=\frac{R_t}{\sqrt{[1+(2\delta Q_e)^2]}}$$

Simplifying and solving for δ we obtain

$$\delta = \pm\frac{1}{2Q_e} \tag{4.50}$$

When $\delta = 1/2Q_e$, Equation 4.45 gives the operating frequency f to be $f_o+(f_o/2Q_e)$ and when $\delta = -1/2Q_e$ the operating frequency is $f_o-(f_o/2Q_e)$. Therefore, the bandwidth B is equal to the difference between these two frequencies, that is

$$B = \left(f_o+\frac{f_o}{2Q_e}\right)-\left(f_o-\frac{f_o}{2Q_e}\right)$$

$$= \frac{f_o}{Q_e} \tag{4.51}$$

For a given value of f_o the larger the effective Q-factor Q_e of the loaded tuned circuit the narrower is the bandwidth B, that is, the more selective is the tuned circuit.

Example 4.2

At the frequency of resonance of 100kHz a transistor has the following parameters (see Figs. 3.59 and 3.60)

$$G_i = 0{\cdot}93\text{mS}$$
$$C_i = 660\text{pF}$$
$$G_o = 136\mu\text{S}$$
$$C_o = 300\text{pF}$$

The tuning coil has an inductance of 1mH and a Q-factor of 100. Determine

(1) The required tuning capacitor
(2) The loaded Q-factor
(3) The bandwidth.

From the first line of Equation 4.43 we find that the total capacitance C_t is

$$C_t = \frac{1}{L(2\pi f_o)^2}$$

$$= \frac{1}{10^{-3}(2\pi\times 10^5)^2}$$

$$= 2540\text{pF}$$

The second line of Equation 4.42 gives the tuning capacitor C to be

$$\begin{aligned} C &= C_t - (C_i + C_0) \\ &= 2540 - (660 + 300) \\ &= 1580\text{pF} \end{aligned}$$

Equation 4.36 gives

$$\begin{aligned} r &= \frac{\omega L}{Q} \\ &= \frac{2\pi \times 10^5 \times 10^{-3}}{100} \\ &= 6{\cdot}28\Omega \end{aligned}$$

Next, the first line of Equation 4.42 gives

$$\begin{aligned} \frac{1}{R_t} &= 136 \times 10^{-6} + 0{\cdot}93 \times 10^{-3} + \frac{1}{6{\cdot}28 \times 100^2} \\ &= 1{\cdot}082 \times 10^{-3}\text{S} \end{aligned}$$

Therefore, $R_t = 920\Omega$ and the second line of Equation 4.43 gives the loaded Q-factor

$$\begin{aligned} Q_e &= \frac{920}{2\pi \times 10^5 \times 10^{-3}} \\ &= 1{\cdot}47 \end{aligned}$$

Equation 4.51 gives the bandwidth

$$\begin{aligned} B &= \frac{100}{1{\cdot}47} \\ &= 68\text{kHz} \end{aligned}$$

Notice that the relatively low value of the loaded Q-factor and the relatively high value of the bandwidth are results of the loading effects of the output impedance of stage TR1 and input impedance of stage TR2.

4.4.3. Transformer Coupling

The tuned amplifier of Fig. 4.19 suffers from the great loss of gain that results from the relatively large impedance mismatch that occurs at the terminals where the two common emitter stages are connected together. This is due to the fact that the input impedance of a common emitter stage is, in general, lower than its output impedance.

The impedance mismatch can be minimised and the gain can, therefore, be increased by using an interstage coupling transformer as in Fig. 4.22. The transformer turns ratio n is chosen such as to match the output conductance G_{o1} of the first stage TR1 to the input conductance G_{i2} of the second stage TR2, that is,

$$n = \sqrt{\frac{G_{i2}}{G_{o1}}} \tag{4.52}$$

The capacitor C is added for tuning the amplifier to the desired frequency, and the shunt inductance L of the primary winding of the transformer serves as the tuning inductor.

4.4.4. Multi-stage Tuned Amplifiers

Most practical tuned amplifiers consist of a number of tuned stages connected in cascade. The overall voltage gain of the complete amplifier is equal to the product of the voltage gains of the individual stages. However, the overall bandwidth becomes narrower with

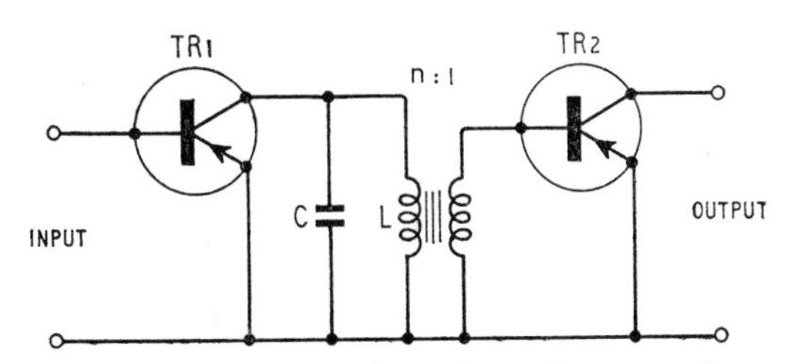

Fig. 4.22. Tuned amplifier with transformer coupling

increasing number of stages. To demonstrate this consider a number n of identical stages tuned to the same centre frequency f_o. The resulting multi-stage amplifier is said to be *synchronously tuned.*

From Equation 4.49 we deduce that the overall voltage gain of the multi-stage amplifier has the following modulus:

$$|K| = \frac{K_{\max}}{[1+(2\delta Q_e)^2]^{n/2}} \tag{4.53}$$

where $K_{\max}$ is the value of $|K|$ at $\delta = 0$, that is, at $f = f_o$. When

$$|K| = \frac{K_{\max}}{\sqrt{2}}$$

Equation 4.53 leads to

$$\sqrt{2} = [1+(2\delta Q_e)^2]^{n/2}$$

Solving for δ we obtain

$$\delta = \pm\frac{1}{2Q_e}\cdot\sqrt{(2^{1/n}-1)} \tag{4.54}$$

Therefore, following an analysis similar to that which led to Equation 4.51 we deduce that the overall bandwidth B_n of the multi-stage tuned amplifier is

$$B_n = \frac{f_o}{Q_e}\cdot\sqrt{(2^{1/n}-1)} \tag{4.55}$$

Equations 4.51 and 4.55 give the ratio of B_n to B as:

$$\frac{B_n}{B} = \sqrt{(2^{1/n}-1)}$$

This ratio has been evaluated for different values of n and the results are given in Table 4.1. This table clearly shows that the overall bandwidth decreases with increasing number of stages.

The process of tuning each inter-stage coupling network of a multi-stage tuned amplifier to the centre frequency f is known as *alignment*. However, as we shall see below, each transistor exhibits internal feedback and, consequently, we find in practice that the tuning of one stage is influenced by the tuning of the neighbouring stages. Therefore, it may be necessary to repeat the alignment process several times until each stage has been tuned to the desired centre frequency.

Table 4.1

n	1	2	3	4	5
$\frac{B_n}{B}$	1	0·643	0·51	0·435	0·387

4.4.5. Neutralisation

The collector depletion layer capacitance $C_{b'c}$ of a transistor is a major source of internal feedback at high frequencies (see the hybrid-π equivalent circuit of Fig. 3.51). This internal feedback effect can be overcome by adding a neutralising capacitor C_N as in Fig. 4.23. The 180° phase reversal provided by the inter-stage

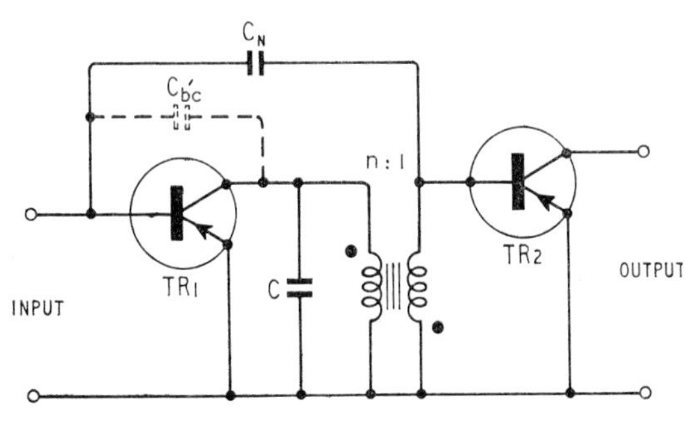

Fig. 4.23. Neutralised tuned amplifier

coupling transformer is necessary so as to ensure that the neutralising capacitor is physically realisable. Thus for neutralisation we have that

$$C_N = nC_{b'c} \tag{4.56}$$

where n is the transformer turns ratio as defined in Fig. 4.23.

Another possible method of neutralisation is shown in Fig. 4.24. The capacitor C_F and the tuning inductor L provide a phase shift

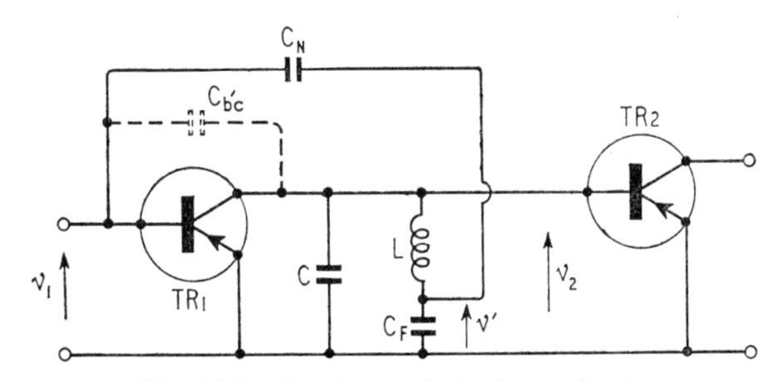

Fig. 4.24. Another method of neutralisation

of nearly 180° between the collector voltage v_2 and the voltage signal v' developed across C_F. Thus at the centre frequency $f_o = \omega_o/2\pi$ we see from Fig. 4.24 that:

$$\frac{v'}{v_2} = \frac{\dfrac{1}{j\omega_o C_F}}{j\omega_o L + \dfrac{1}{j\omega_o C_F}} \tag{4.57}$$

Suppose at $f=f_o$ the reactance of the capacitor C_F is small compared to that of the inductor L, that is

$$\omega_o L \gg \frac{1}{\omega_o C_F} \tag{4.58}$$

Then, Equation 4.57 simplifies to

$$\frac{v'}{v_2} \simeq -\frac{1}{\omega_o^2 L C_F} \tag{4.59}$$

Eliminating ω_o between this and the first line of Equation 4.43 gives

$$\frac{v'}{v_2} \simeq \frac{C_t}{-C_F} \tag{4.60}$$

Therefore, in Fig. 4.24 the capacitor C_N provides a feedback signal equal to and 180° out of phase with respect to the internal feedback signal produced by $C_{b'c}$, that is, the stage is neutralised if

$$C_N = \frac{C_F}{C_t} . C_{b'c} \tag{4.61}$$

Notice that $C_t \gg C_{b'c}$; therefore, $C_F \gg C_N$.

4.5. POWER AMPLIFIERS

The function of a power amplifier is to deliver a specified amount of power to its load. For satisfactory operation it is necessary to ensure that the following transistor ratings are not exceeded:

(1) Maximum collector voltage $V_{C\,max}$

(2) Maximum collector power dissipation $P_{C\,max}$

(3) Maximum collector current $I_{C\,max}$.

The various factors which determine the magnitude of $V_{C\,max}$, $P_{C\,max}$ and $I_{C\,max}$ were considered in Section 1.5.

The power level which an output stage handles can be relatively large, of the order of 1 watt or even more. Therefore, the power stage is essentially a large signal amplifier and it is not justifiable to apply the small signal equivalent circuits of Chapter 3. Their circuit response is best determined by graphical means as in Chapter 2. When studying power amplifiers it is necessary to investigate the effects of non-linearities associated with the input and output characteristic curves of the transistor. A consequence

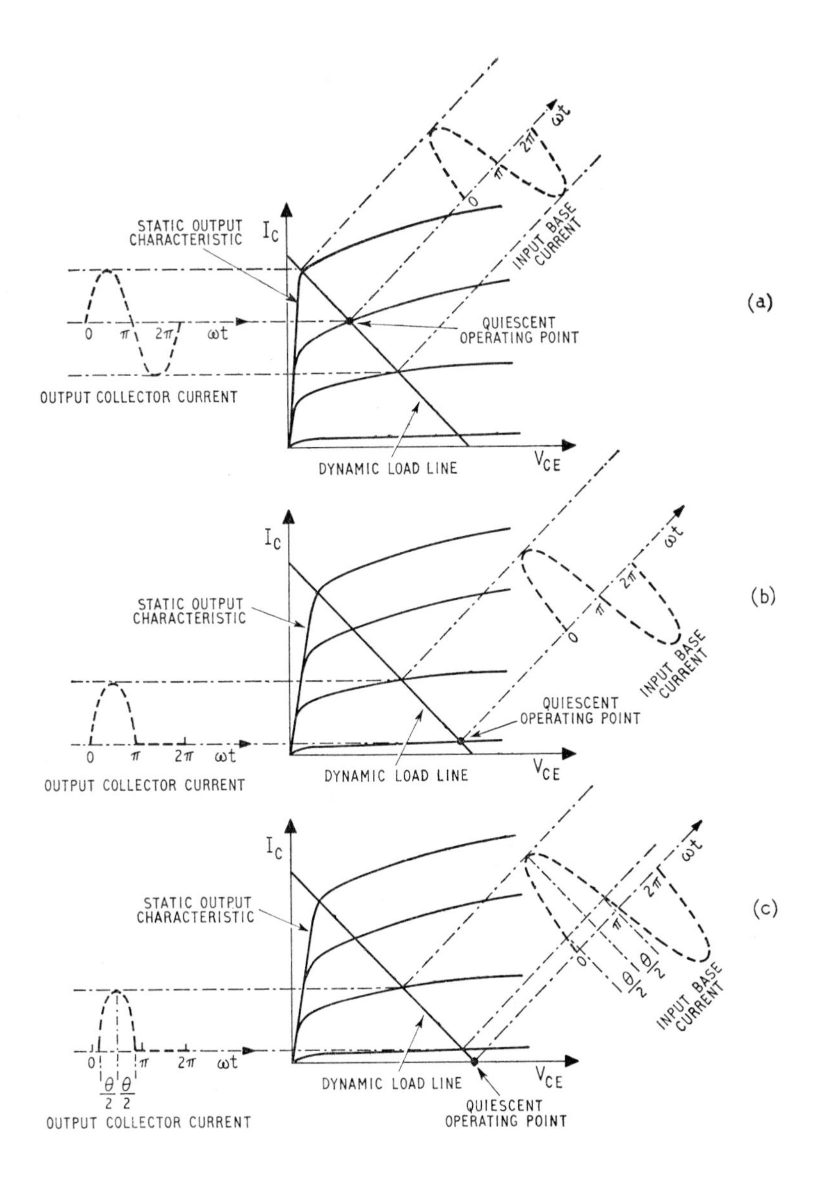

Fig. 4.25. The three different classes of operating power amplifiers: (a) Class A; (b) Class B; (c) Class C

of these non-linear characteristics is that if the input signal is sinusoidal of frequency f and relatively large amplitude, it is found that the resulting output signal contains sinusoidal components of frequencies equal to $2f$, $3f$, etc. in addition to the amplified fundamental component of frequency f. The component of frequency $2f$ is referred to as the *second harmonic*, that of frequency $3f$ is the *third harmonic* and so on. The presence of these harmonic components gives rise to the non-linear distortion which was evaluated in Section 2.3. Consequently, the output signal is no longer a truly faithful reproduction of the input signal. For satisfactory operation it is necessary to keep the levels of the undesirable harmonic components below prescribed values. In this respect, non-linear distortion is reduced by applying negative feedback as we shall see in Chapter 5.

Another factor that is of particular concern in the study of a power amplifier is its *conversion efficiency* η. This is defined as the ratio of the a.c. power output (delivered to the load) to the power drawn from the d.c. supply. The closer η is to unity the more useful is the transistor stage as a power amplifier.

The non-linear distortion and conversion efficiency of a power amplifier depend on the class in which it is operated. When the stage is biased such that the output current is finite during quiescent conditions and it flows continuously during one cycle of the input signal, it is said to be operating in *Class A* as in Fig. 4.25(a).

If the stage is biased such that the output current is zero in the quiescent condition and it flows for only one-half cycle of the input signal waveform, then *Class-B* operation results as in Fig. 4.25(b). The stage is said to be operating in *Class C* if the output current is again zero in the quiescent condition but it flows for less than one-half cycle of the input signal waveform. This is illustrated in Fig. 4.25(c) where θ is termed the *angle of flow*. We shall only consider Class-A and Class-B power amplifiers.

4.5.1. Class-A Power Amplifiers

A transistor output stage can be coupled to its load either directly or by means of a transformer. In practice, however, transformer coupling is usually preferred because the small d.c. losses in the transformer enable the realisation of a relatively high conversion efficiency.

Consider the transformer coupled stage of Fig. 4.26. The d.c. resistance of the primary winding is normally quite small. Therefore, the static load-line is practically a vertical line passing through the point $V_{CE} = -V_{CC}$ on the collector voltage axis as indicated in Fig. 4.27.

If n is the transformer turns ratio as defined in Fig. 4.26 then under a.c. conditions, the load resistance R_L' measured looking into the primary winding is

$$R_L' = n^2 R_L \tag{4.62}$$

Here it is assumed that the shunt inductance of the primary winding is such that, at the operating-frequency, its impedance is large

Fig. 4.26 (right). Transformer coupled output stage

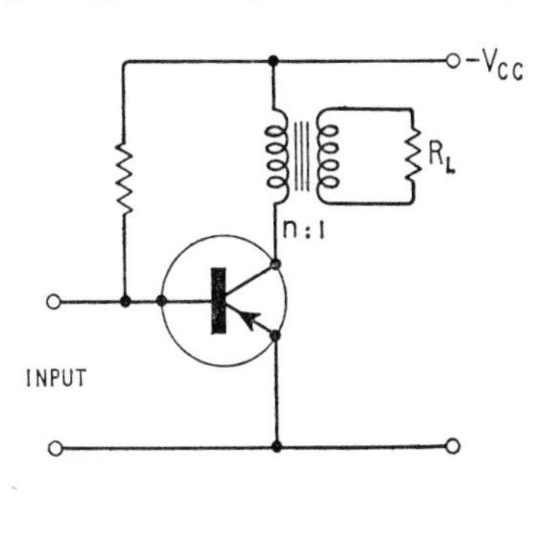

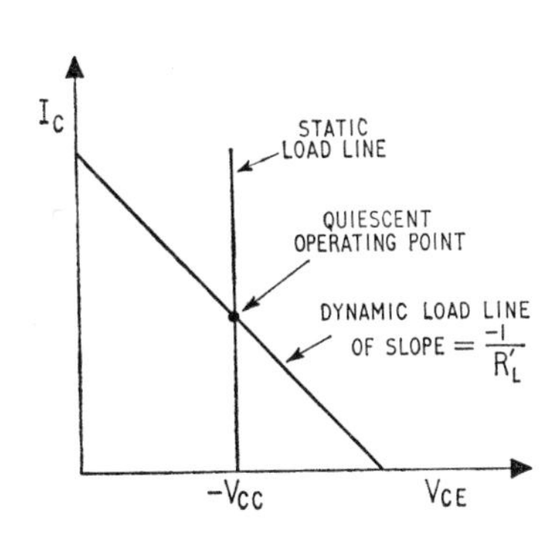

Fig. 4.27 (left). Static and dynamic load-line of transformer coupled stage

compared with R_L'. Thus the dynamic load-line will have a slope equal to $-1/R_L'$, and it intersects the static load-line at the quiescent operating-point as shown in Fig. 4.27. Notice that the instantaneous collector voltage of a transformer coupled stage can swing above the d.c. supply voltage V_{CC}.

The a.c. power output P_{ac} of the stage is

$$P_{ac} = \frac{V_m^2}{2R_L'} \tag{4.63}$$

where V_m is the peak value of sinusoidal voltage developed at the collector. Hence, maximum a.c. power is delivered to the load when the reflected load resistance R_L' is as small as possible and V_m is as large as is allowed by the maximum collector voltage rating $V_{C\,max}$.

The smallest possible reflected load resistance R_L' can be determined from the graphical construction of Fig. 4.28. The dynamic load-line intersects the voltage axis at $V_{CE} = -V_{C\,max}$, and it is tangential to the maximum collector power dissipation hyperbola at the quiescent operating-point. Thus

$$\left.\begin{aligned} V_{CC} &= \frac{V_{C\,\max}}{2} \\ I_Q &= \frac{2P_{C\,\max}}{V_{C\,\max}} \\ R_L' &= \frac{V_C^2{}_{\max}}{4P_{C\,\max}} \end{aligned}\right\} \qquad (4.64)$$

These equations, therefore, enable the evaluation of the quiescent operating-point and the reflected load resistance R_L' for maximum

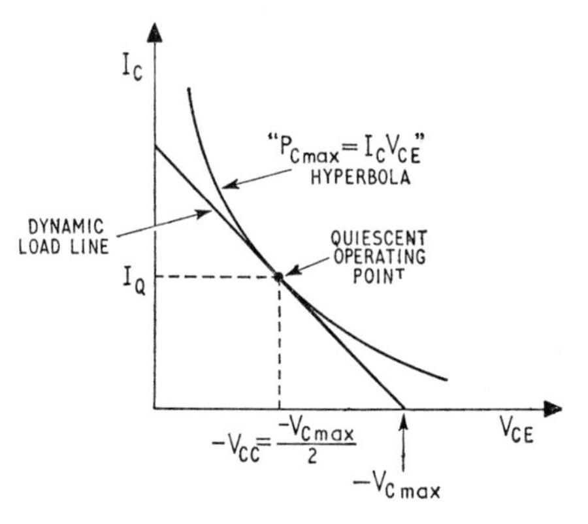

Fig. 4.28. Graphical construction for evaluating minimum possible value of R_L'

a.c. power output. Notice that the operating-point is located midway along the dynamic load-line, and the collector voltage can swing between a minimum value of zero and a maximum value of $2V_{CC}$. Thus $V_m = V_{CC}$ and Equation 4.63 gives the maximum a.c. power output to be

$$P_{ac} = \frac{V_{CC}^2}{2R_L'}$$

Combining this with the first and third lines of Equation 4.64 we obtain

$$P_{ac} = \tfrac{1}{2}P_{C\,\max} \qquad (4.65)$$

However, the power P_{dc} drawn from the d.c. supply is

$$P_{dc} = V_{CC} I_Q$$
$$= P_{C\,\max} \tag{4.66}$$

where we have made use of the first and second lines of Equation 4.64. Therefore, the maximum conversion efficiency is

$$\eta = \frac{P_{ac}}{P_{dc}}$$
$$= \tfrac{1}{2} \tag{4.67}$$

Thus the transformer coupled stage of Fig. 4.26 has a maximum possible conversion efficiency of 50%.

4.5.2. Class-A Push–Pull Amplifiers

By connecting two transistors in Class-A push–pull as in Fig. 4.29 it is possible to increase the power output above that obtainable from a single stage. The two input signals applied to the base

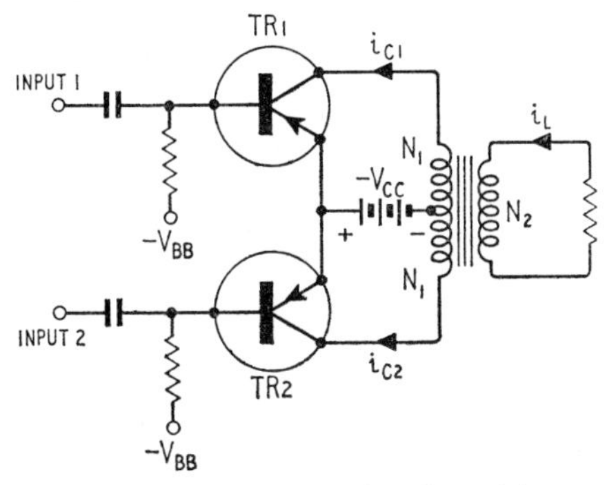

Fig. 4.29. Class-A push–pull amplifier

terminals of the two transistors must be equal in magnitude but of opposite phase. A centre-tapped transformer or one of the driver circuits to be discussed later can be used for generating the required input signals.

In a Class-A push–pull amplifier each transistor conducts at all times and, as a result of the input signals, the collector current in one transistor increases when that in the other decreases. Each transistor supplies one half of the power output delivered to the load.

A Class-A push–pull amplifier has the added advantage of reduced non-linear distortion. To illustrate this suppose that in Fig. 4.29

the two transistors have identical characteristics and that the signal v_1 applied to transistor TR1 is

$$v_1 = V_1 \cos \omega t \tag{4.68}$$

Owing to the non-linear characteristic curves it follows that the resulting collector current i_{C1} of TR1 can, in general, be expressed as

$$i_{C1} = I_o + I_1 \cos \omega t + I_2 \cos 2\omega t + I_3 \cos 3\omega t + \ldots \tag{4.69}$$

With the applied input signals equal in magnitude and 180° out of phase we see that the signal v_2 applied to transistor TR2 is

$$v_2 = V_2 \cos (\omega t + 180°) \tag{4.70}$$

Therefore, the collector current i_{C2} of transistor TR2 is similar to Equation 4.69 but with ωt replaced by $\omega t + 180°$. Thus

$$\begin{aligned} i_{C2} &= I_o + I_1 \cos (\omega t + 180°) + I_2 \cos (2\omega t + 360°) \\ &\quad + I_3 \cos (3\omega t + 540°) + \ldots \\ &= I_o - I_1 \cos \omega t + I_2 \cos 2\omega t - I_3 \cos 3\omega t + \ldots \end{aligned} \tag{4.71}$$

Assuming the output transformer is ideal, the load current i_L is

$$i_L = \frac{N_1}{N_2}(i_{C1} - i_{C2}) \tag{4.72}$$

where $2N_1$ = total number of turns of primary winding

N_2 = number of turns of secondary winding.

Equations 4.69, 4.71 and 4.72 give

$$i_L = \frac{2N_1}{N_2}(I_1 \cos \omega t + I_3 \cos 3\omega t + \ldots) \tag{4.73}$$

We thus deduce that if the two transistors of a Class-A push–pull power amplifier are identical, even order harmonics in the output cancel.

The conversion efficiency of a Class-A push–pull amplifier like a simple stage power amplifier, cannot exceed 50%. Therefore, if a higher efficiency is required then Class-B amplification must be used.

4.5.3. Class-B Power Amplifiers

In Class-B operation the transistor is so biased that the output current flows for one half-cycle only. If the transistor is operated in the common emitter mode and it is biased as above, it will cut off the positive half-cycle and amplify the negative half-cycle. Therefore, in order to obtain an amplified output signal that is sinusoidal for a sinusoidal input it is necessary to employ two transistors operated in push–pull.

In the push–pull circuit of Fig. 4.30 the quiescent collector current of each transistor is approximately zero. The base terminals of the two transistors are fed from an input transformer with a centre-

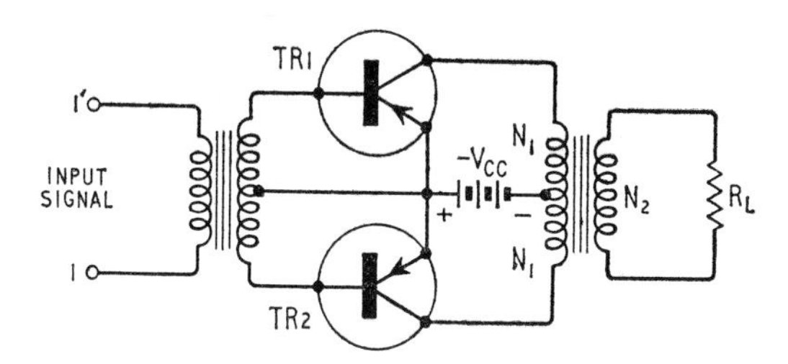

Fig. 4.30. Class-B push–pull amplifier

tapped secondary winding. The two halves of this winding are series aiding such that the voltage signals developed at the base terminals are equal in magnitude and opposite in phase. Consequently, during one half-cycle transistor TR1 is cut off and TR2 is conducting, and during the next half-cycle the states of the two transistors are interchanged.

The load resistance R_L' reflected to either half of the primary winding of the output transformer is

$$R_L' = \left(\frac{N_1}{N_2}\right)^2 R_L \qquad (4.74)$$

where $2N_1$ = total number of turns of primary winding

N_2 = number of turns of secondary winding.

Let us next evaluate the peak collector current and collector voltage that occur in a Class-B push–pull amplifier. Suppose that transistor TR1 is conducting and TR2 is cut off. The peak collector current I_m of TR1 occurs when the instantaneous operating-point lies on the current axis, that is, when the instantaneous collector–emitter voltage is zero. Thus

$$I_m = \frac{V_{CC}}{R_L'} \qquad (4.75)$$

The peak collector voltage V_m of the cut-off transistor TR2 occurs when the collector current of the conducting transistor TR1 has the value I_m of Equation 4.75. Then

$$V_m = V_{CC} + I_m R_L'$$
$$= 2V_{CC} \tag{4.76}$$

Similarly, when the states of the two transistors are interchanged, transistor TR1 experiences a peak collector voltage of $2V_{CC}$ and TR2 draws a peak collector current of V_{CC}/R_L'.

Therefore, if the maximum permissible collector voltage $V_{C\,max}$ is not to be exceeded we deduce from Equation 4.76 that, for optimum operating conditions, the d.c. supply voltage V_{CC} is given by

$$V_{CC} = \frac{V_{C\,max}}{2} \tag{4.77}$$

Provided that V_{CC} does not exceed the value defined in Equation 4.77 the instantaneous collector voltage will not exceed $V_{C\,max}$.

From Equation 4.76 and Fig. 4.30 we deduce that the total a.c. power P_{ac} delivered to the load has a maximum value of

$$P_{ac} = \left(\frac{1}{\sqrt{2}}\cdot\frac{N_2}{N_1}\cdot 2V_{CC}\right)^2 \cdot R_L^{-1}$$
$$= \frac{2V_{CC}^2}{R_L}\cdot\left(\frac{N_2}{N_1}\right)^2 \tag{4.78}$$

Theoretical considerations[1] show that the conversion efficiency of a Class-B push–pull amplifier can have a maximum possible value of 78·5%. Compare this with the maximum conversion of 50% that is achievable with a transformer-coupled Class-A power amplifier.

If the 2 transistors of a Class-B push–pull amplifier are not identical then the 2 half-cycles of the input waveform will not be amplified equally resulting in harmonic distortion. This can be overcome by the use of matched transistors; in particular it is necessary to ensure that the common emitter current amplification factors α_e of the 2 transistors are nearly equal and their α_e versus I_C characteristics (see Fig. 1.23) are as nearly identical as possible.

Another form of distortion that is experienced in Class-B transistor push–pull amplifiers is *cross-over distortion* illustrated in Fig. 4.31. This occurs when the collector currents of the 2 transistors are arranged to be zero (or nearly so) in the quiescent

[1] See HAYKIN, S. S., *Junction Transistor Circuit Analysis*, Iliffe Books Ltd. (1962).

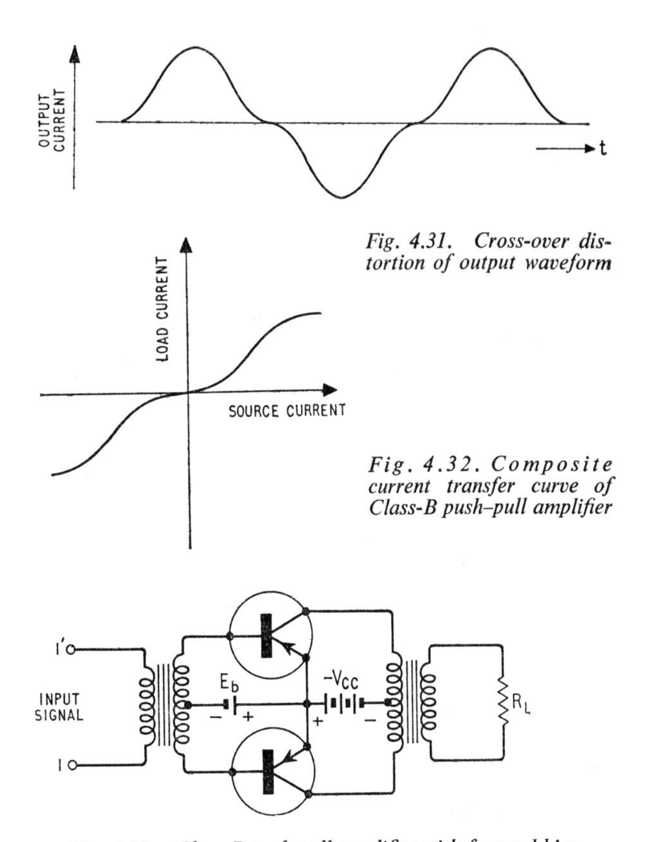

Fig. 4.31. Cross-over distortion of output waveform

Fig. 4.32. Composite current transfer curve of Class-B push–pull amplifier

Fig. 4.33. Class-B push-pull amplifier with forward bias

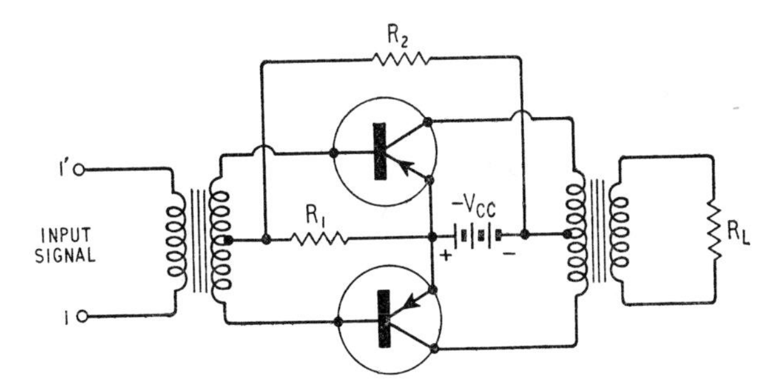

Fig. 4.34. Alternative arrangement for applying forward bias to push–pull amplifier

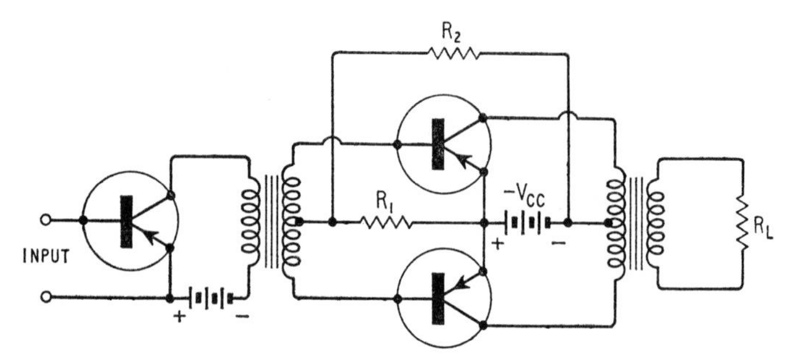

Fig. 4.35. Push–pull amplifier with its driver stage

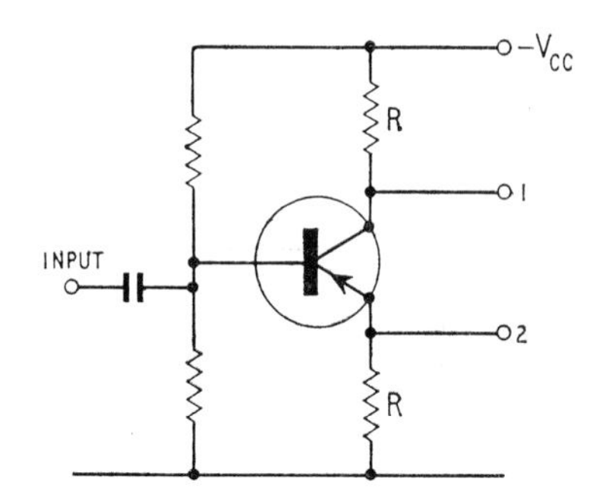

Fig. 4.36. Single-stage phase splitter

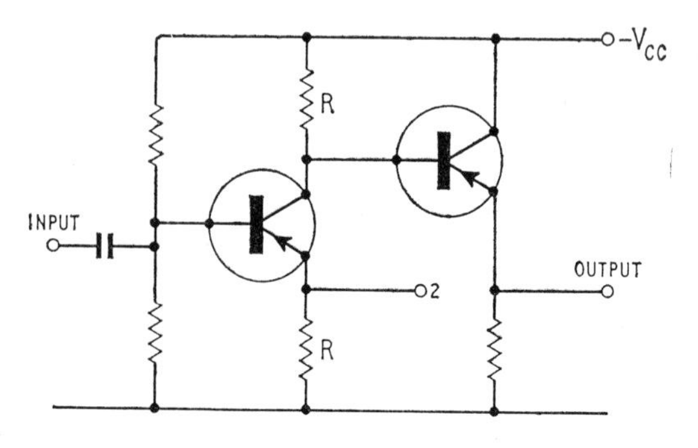

Fig. 4.37. Two-stage phase splitter

condition, giving rise to a dynamic current transfer curve of the form shown in Fig. 4.32. Thus the lower input signal levels are amplified to a lesser extent than the higher levels causing harmonic distortion of the output signal. Cross-over distortion can be reduced by increasing the series resistance of the driver stage and/or by forward-biasing of the emitter–base junction of each transistor as in Fig. 4.33. Alternatively, the circuit arrangement of Fig. 4.34 can be used. The resistors R_1 and R_2 have been included to apply the desired forward bias to each transistor. Normally the required forward bias voltage need only be of the order of a few tenths of a volt.

4.5.4. Driver Circuits

Push–pull operation of a pair of transistors (either both pnp or npn) requires 2 input signals that are 180° out of phase relative to each other. This can be achieved by using a centre-tapped transformer as in Fig. 4.30. It has been reproduced in Fig. 4.35 together

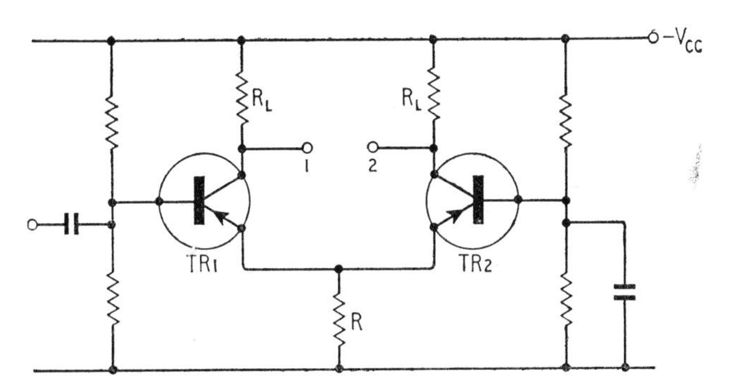

Fig. 4.38. Emitter coupled pair as phase splitter

with the driver stage, which is simply a transformer-coupled stage capable of supplying sufficient power to drive the output push–pull stage.

Alternatively, the phase-splitting circuits of Figs. 4.36, 4.37 and 4.38 can be used. The single stage of Fig. 4.36 has a fairly high input resistance because of the unbypassed emitter resistor. The voltage signal developed at the output emitter terminal 2 is in phase with the input signal whereas that developed at the output collector terminal 1 is 180° out of phase. Therefore, the two output signals are 180° out of phase as required. The circuit of Fig. 4.36 suffers from the following major source of unbalance. The load connected to terminal 1 is fed from a common emitter type of output resistance whereas the load connected to output terminal 2 is fed from a common collector type of output resistance. Therefore, from Section 3.4 we deduce that the load connected to terminal 2 is driven from a much lower source resistance than the load connected to terminal 1.

The above source of unbalance can be overcome by the use of an additional transistor operated in the common collector mode and connected into circuit as shown in Fig. 4.37. Alternatively we can use the emitter-coupled pair circuit of Fig. 4.38. A variation of this circuit was considered in Section 4.1. The input signal is applied to the base terminal of transistor TR1, and the base terminal of TR2 is bypassed for a.c. signals by the use of a suitably large capacitor. Therefore, TR2 is operated as a common base stage. The voltage signal developed at the collector of TR1 is 180° out of phase with the input. The voltage developed across the common resistor R and, therefore, the collector of TR2 is in phase with the input. Thus the 2 output signals are 180° out of phase. Further, the amplitudes of the output signal are nearly equal if the resistor R is large compared to the resistance measured looking into the emitter terminal of the common base stage TR2. Hence, from Table 3.2 we deduce that for satisfactory operation

$$R \gg r_e + r_b(1-\alpha_b) \tag{4.79}$$

Other applications of the emitter-coupled pair will be considered in Chapters 6, 9 and 10.

CHAPTER 5

FEEDBACK AMPLIFIER AND OSCILLATOR CIRCUITS

5.1. EFFECTS OF FEEDBACK

A *feedback amplifier* is one in which a signal proportional to the output is fed back to the input circuit. The application of this feedback signal results in a number of interesting and useful circuit properties. Thus the gain, input resistance, output resistance, non-linear distortion, noise output, frequency response and stability of the amplifier become affected by the applied feedback. These various effects will be demonstrated later. However, we shall first discuss the different ways in which feedback can be applied to an amplifier. The feedback signal can be proportional to the output current or to the output voltage. It can be applied in series or in shunt with the input circuit. Therefore, there are four basic types of feedback circuits and they are:

(1) The *voltage output–series input* type where the feedback signal is proportional to the output voltage and it is applied in series with the input circuit.

(2) The *voltage output–shunt input* type where the feedback signal is proportional to the output voltage and it is applied in shunt with the input circuit.

(3) The *current output–series input* type where the feedback signal is proportional to the output current and it is applied in series with the input circuit.

(4) The *current output–shunt input* type where the feedback signal is proportional to the output current and it is applied in shunt with the input circuit.

Examples of these four basic feedback circuits will be considered later.

The input and output resistances of these four feedback circuits are affected differently by the applied feedback, but their individual external voltage gains, harmonic distortion and noise output are affected similarly.

5.1.1. Effect on Gain

Consider the feedback circuit of Fig. 5.1. The basic amplifier provides forward signal transmission from the source to the load, while the feedback network (which is normally a passive one) extracts a signal proportional to the output voltage and applies it in series with the input circuit of the basic amplifier. Therefore, the

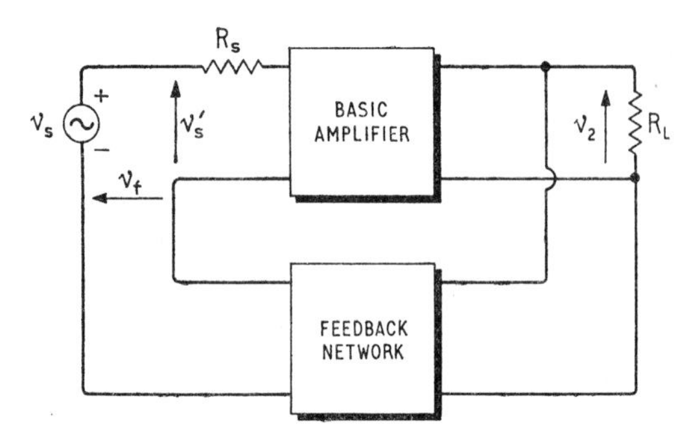

Fig. 5.1. Voltage output–series input feedback amplifier

feedback amplifier of Fig. 5.1 is of the voltage output–series input type.

The complete amplifier of Fig. 5.1 is fed from a voltage source v_s of series resistance R_s and it is coupled to a load R_L. Assuming that K is the external voltage gain of the basic amplifier as defined in Section 3.2, then from Fig. 5.1 it follows that

$$v_2 = Kv_s' \tag{5.1}$$

Further, the voltage signal v_f fed back to the input circuit is given by

$$v_f = \beta v_2 \tag{5.2}$$

where β is the reverse voltage transmission ratio of the feedback network. However, the feedback voltage v_f and source voltage v_s are related as follows

$$v_s = v_s' - v_f \tag{5.3}$$

Eliminating v_s' and v_f between Equations 5.1 to 5.3 we obtain

$$v_2 = K(v_s+\beta v_2)$$

Solving this equation for the output voltage v_2 and then evaluating the voltage ratio v_2/v_s we find that

$$K_c = \frac{v_2}{v_s}$$

$$= \frac{K}{1-\beta K} \tag{5.4}$$

K_c is the external voltage gain of the complete feedback amplifier and it is, therefore, termed the *closed loop gain.*

The product term βK is termed the *open loop gain* of the system and it is usually denoted by T; thus

$$T = \beta K \tag{5.5}$$

It can be evaluated by, first, reducing the voltage source v_s of Fig. 5.1 to zero and then breaking the feedback loop at a convenient point

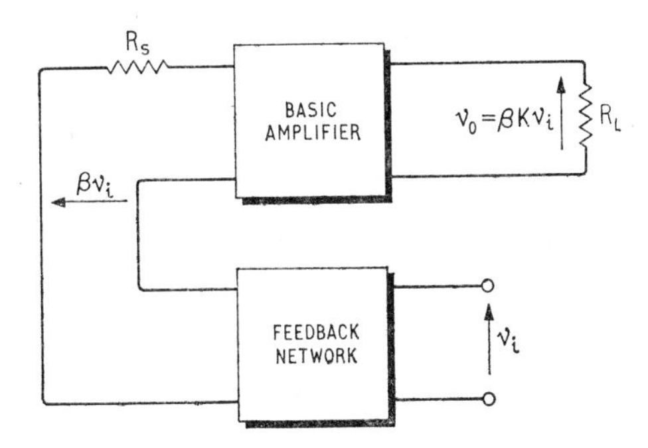

Fig. 5.2. Circuit arrangement for measuring open loop gain

as shown in Fig. 5.2. Here it is assumed that the loading effect of the feedback network on the output circuit of the basic amplifier is sufficiently small to ensure that the total impedance across the output terminals of the basic amplifier is practically the same before and after breaking the feedback loop. If in Fig. 5.2 a voltage v_i is applied to the feedback network then the voltage v_o developed across the load is equal to $\beta K v_i$. Therefore, the voltage ratio v_o/v_i of Fig. 5.2 is equal to the open loop gain of the amplifier.

If the open loop gain T of any feedback amplifier is negative then the resulting amplifier is said to be a *negative* or *degenerative* feedback one. On the other hand, *positive* or *regenerative* feedback results if the open loop gain of the circuit is positive. Therefore, from Equation 5.4 the closed loop gain K_c of a negative feedback amplifier is always less than the gain K of the basic amplifier, whereas in a positive feedback amplifier K_c is greater than K.

The term $(1-\beta K)$ is a measure of the amount of feedback applied to the circuit, and it is thus termed the *feedback factor F*. However,

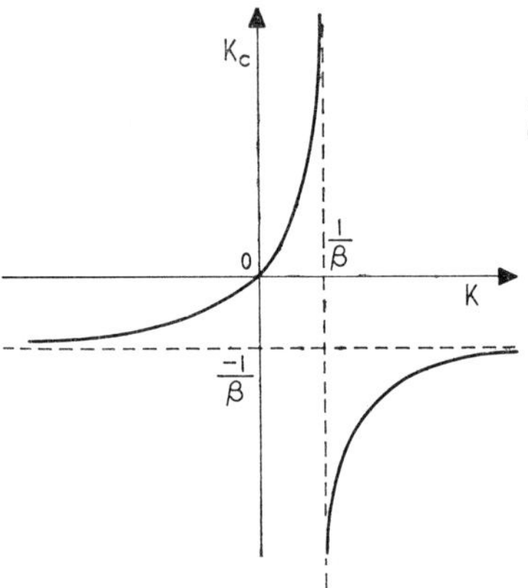

Fig. 5.3. Effect of varying basic amplifier gain on closed loop gain

since βK is equal to T (see Equation 5.5) it follows that the feedback factor F and the open loop gain T are related as

$$F = 1 - T \tag{5.6}$$

in view of which we can rewrite Equation 5.4 as

$$K_c = \frac{K}{F} \tag{5.7}$$

Normally, the feedback factor is expressed in decibels. Thus, the amount of applied feedback, in decibels, is equal to $20 \log_{10} F$.

It is of interest to plot the closed loop gain K_c as a function of the gain K without feedback. Using Equation 5.4 we obtain the characteristic of Fig. 5.3. Notice that when $\beta K = 1$ the closed loop gain K_c is infinite. In this case the feedback amplifier becomes unstable, and this effect will be considered in greater detail in Section 5.5. Further, from Fig. 5.3 we see that when the applied feedback is of the degenerative type and the open loop gain is very large compared to unity in magnitude, then the closed loop gain approaches the limiting value of $-1/\beta$. Therefore, the closed loop gain becomes determined by the passive feedback network and so practically independent of the variations in the parameters of the transistors which constitute the basic amplifier. This is a very attractive feature of negative feedback amplifiers. In fact, even when the open loop gain is not large, the *sensitivity* of a negative feedback amplifier to variations in transistor parameters is reduced as a result of the feedback application. Sensitivity is defined as the relative change in the closed loop gain produced by a change in the gain without feedback. Differentiating Equation 5.4 with respect to K we obtain

$$\frac{\mathrm{d}K_c}{\mathrm{d}K} = \frac{1}{(1-\beta K)^2}$$

Therefore,

$$\mathrm{d}K_c = \frac{\mathrm{d}K}{(1-\beta K)^2} \tag{5.8}$$

Dividing Equation 5.8 by Equation 5.4 gives the sensitivity of a feedback amplifier to be:

$$\frac{\mathrm{d}K_c}{K_c} = \frac{1}{1-\beta K} \cdot \frac{\mathrm{d}K}{K}$$

$$= \frac{1}{F} \cdot \frac{\mathrm{d}K}{K} \tag{5.9}$$

However, for a negative feedback amplifier F is greater than unity, and so Equation 5.9 shows that the relative change in the closed loop gain is reduced with increasing feedback factor.

Example 5.1

The basic amplifier of a negative feedback circuit has an external voltage gain K of -1000. Transistor parameter variations cause K to change by $\pm 10\%$. If the reverse transmission ratio β is equal to 0·01, evaluate the percentage change in the closed loop gain K_c.

$$\text{Open loop gain } T = \beta K$$

$$= (0{\cdot}01)(-1{,}000) = -10$$

$$\text{Feedback factor } F = 1 - T = 11$$

$$\therefore \text{ Percentage change in } K_c = \frac{\pm 10\%}{11}$$

$$= \pm 0{\cdot}91\%$$

5.1.2. Effect on Distortion

We shall next investigate the effect of feedback on non-linear distortion and noise output of an amplifier. For this purpose consider Fig. 5.4, where the voltage source D accounts for any non-linear distortion or noise present at the output of the basic amplifier in the absence of any feedback. From Fig. 5.4 we see that:

$$v_2 = Kv_1' + D \tag{5.10}$$

and

$$v_1 = v_1' - \beta v_2 \tag{5.11}$$

From Equations 5.10 and 5.11 we find that

$$v_2 = \frac{K}{1-\beta K} v_1 + \frac{D}{1-\beta K} \tag{5.12}$$

Hence, the output signal-to-noise ratio with feedback is equal to Kv_1/D. On the other hand, in the absence of feedback, that is, when $\beta = 0$ we find from Equation 5.11 that $v_1 = v_1'$. Thus

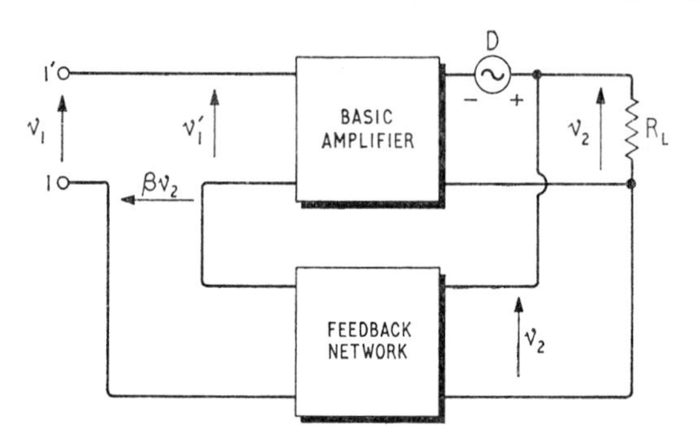

Fig. 5.4. Effect of extraneous signal

Equation 5.10 gives the output signal-to-noise ratio without feedback to be Kv_1/D. In both cases the signal-to-noise ratio has been defined as the ratio of the output signal voltage to the output noise voltage. Therefore, if the input voltage v_1 is maintained constant with and without feedback, the output signal-to-noise ratio is unaffected by the applied feedback. If, however, negative feedback is applied and at the same time the input voltage is increased to such an extent that the output signal remains at the same level it had before the application of negative feedback, then from Equations 5.10 and 5.12 we find that the net effect at the output terminal is to reduce the output noise and non-linear distortion component by the factor of $(1-\beta K)^{-1}$ that is, F^{-1}. In other words, provided that the output power is maintained the same, with and without feedback, then the application of negative feedback results in a reduction in the non-linear distortion and noise components present at the output.

5.1.3. Effect on Input and Output Resistance

The applied feedback also affects the input and output resistances of the amplifier. However, the resulting effect not only depends on the amount of applied feedback but also on the nature of the feedback signal and the manner in which it is applied at the input circuit.

Table 5.1. EFFECTS OF FEEDBACK ON INPUT AND OUTPUT RESISTANCE

Type of feedback	*Input resistance*	*Output resistance*
Feedback signal proportional to output voltage	Unaffected	Decreases
Feedback signal proportional to output current	Unaffected	Increases
Feedback signal applied in shunt with input circuit	Decreases	Unaffected
Feedback signal applied in series with input circuit	Increases	Unaffected

We shall find that the output resistance of the amplifier is increased if the applied feedback is degenerative and if the feedback signal is proportional to the output current. However, if the feedback signal is proportional to the output voltage then the output resistance is reduced as a result of the negative feedback application. Next, if the feedback signal is applied in shunt with the input circuit of the basic amplifier, the input resistance is reduced, while it is increased if the feedback signal is applied in series. However, the manner in which the feedback signal is applied at the input circuit does not affect the output resistance, and similarly the nature of the feedback signal does not influence the input resistance. Table 5.1 summarises the effects of negative feedback on the input and output resistance of an amplifier.

The above effects of feedback on the input resistance and output resistance of the amplifier will be demonstrated in Section 5.2 when specific feedback circuits are considered.

5.1.4. Effect on Frequency Response

In a practical transistor amplifier the gain decreases at high frequencies mainly due to the frequency dependence of α_b and to the collector depletion layer capacitance $C_{b'c}$ (see Section 3.9). The gain also decreases at low frequencies because of the interstage coupling capacitor (see Section 4.3). We shall now demonstrate that the application of negative feedback improves the high as well as low frequency response of the amplifier.

Consider first the effect on high frequency response. In Section 3.9 we showed that at high frequencies the external voltage gain can be expressed as

$$K_{hf} = \frac{K_{mf}}{1+\dfrac{jf}{f_h}} \tag{5.13}$$

where K_{mf} is the mid-frequency gain and f_h is the upper cut-off frequency. Suppose that the feedback network is purely resistive and, therefore, β is purely real. Equations 5.4 and 5.13 give the corresponding closed loop gain to be

$$K_c = \frac{K_{hf}}{1-\beta K_{hf}}$$

$$= \frac{\dfrac{K_{mf}}{1+\dfrac{jf}{f_h}}}{1-\dfrac{\beta K_{mf}}{1+\dfrac{jf}{f_h}}}$$

$$= \frac{K_{mf}}{1-\beta K_{mf}+\dfrac{jf}{f_h}} \tag{5.14}$$

Dividing the numerator and denominator by the factor $(1-\beta K_{mf})$ we obtain

$$K_c = \frac{K_{mf}'}{1+\dfrac{jf}{f_h'}} \tag{5.15}$$

where K_{mf}' = mid-frequency value of closed loop gain

$$= \frac{K_{mf}}{1-\beta K_{mf}} \tag{5.16}$$

f_h' = upper cut off frequency with feedback

$$= f_h(1-\beta K_{mf}) \tag{5.17}$$

If the applied feedback is negative, $1-\beta K_{mf}>1$, $K_{mf}'<K_{mf}$ and $f_h'>f_h$. Therefore, the application of negative feedback improves the high frequency response at the expense of mid-frequency gain.

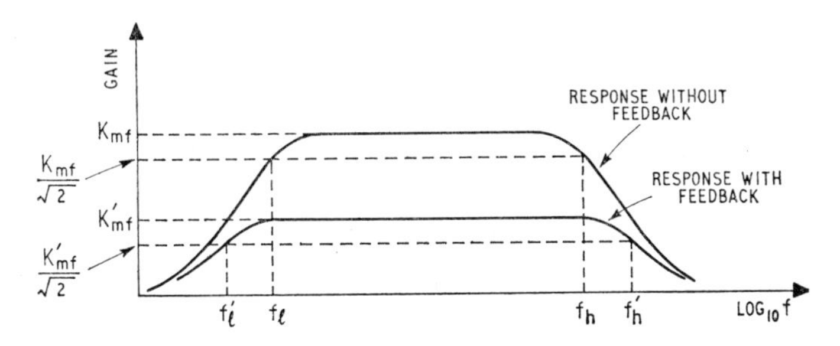

Fig. 5.5. Effect of feedback on frequency response

Consider next the effect on low frequency response. From Section 4.3, the gain at low frequencies is

$$K_{lf} = \frac{K_{mf}}{1+\dfrac{f_l}{jf}} \tag{5.18}$$

where f_l is the lower cut-off frequency. Assuming β to be real and using Equation 5.4 we find that the closed loop gain at low frequencies is

$$K_c = \frac{K_{lf}}{1-\beta K_{lf}}$$

$$= \frac{\dfrac{K_{mf}}{1+\dfrac{f_l}{jf}}}{1-\dfrac{\beta K_{mf}}{1+\dfrac{f_l}{jf}}}$$

$$= \frac{K_{mf}}{1-\beta K_{mf}+\dfrac{f_l}{jf}} \tag{5.19}$$

Dividing the numerator and denominator by $1-\beta K_{mf}$ we obtain

$$K_c = \frac{K_{mf}'}{1+\dfrac{f_l'}{jf}} \tag{5.20}$$

where K_{mf}' is as defined in Equation 5.16 and

$$f_l' = \text{lower cut off frequency with feedback}$$

$$= \frac{f_l}{1-\beta K_{mf}} \tag{5.21}$$

Therefore with negative feedback $f_l > f_l'$, that is, the low frequency response is improved.

Fig. 5.5 shows the complete frequency response of a capacitively coupled single-stage amplifier with and without feedback.

5.2. SINGLE-STAGE FEEDBACK CIRCUITS

Negative feedback can be applied to a single common emitter stage in one of two basic ways: *series* or *shunt* feedback. In each case the applied feedback is usually referred to as *local feedback.*

5.2.1. Local Series Feedback

The circuit of Fig. 5.6 illustrates the application of this type of local feedback to a common emitter stage. The resistor R_e, which is

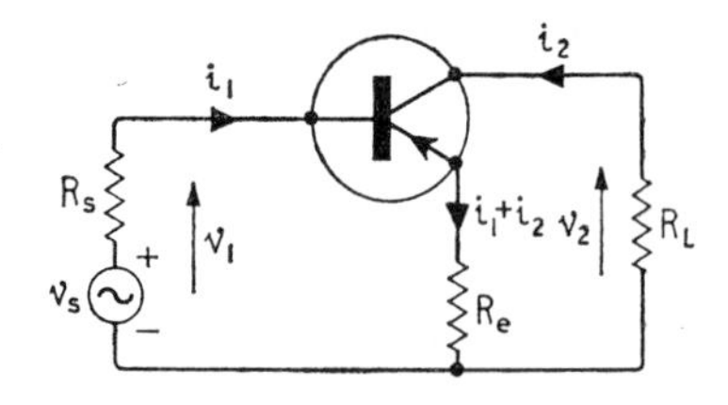

Fig. 5.6. Local series feedback

connected in series with the emitter terminal, is responsible for the applied feedback. It enables a signal proportional to the output current i_2 to be fed back and applied in series with the input circuit. Therefore, the feedback amplifier of Fig. 5.6 is of the current output–series input type.

It is assumed that the stage is driven from a voltage source of resistance R_s and that it is connected to a load resistance R_L. We shall develop expressions for the closed loop gain, open loop gain, input and output resistance of the stage.

Normally the feedback resistor R_e is small compared to the load resistor R_L. The applied feedback will affect the voltage gain of the stage while its current gain remains practically unchanged. In other words, the applied negative feedback will only stabilise the voltage gain of the stage. Therefore,

$$K_{if} \simeq K_i \tag{5.22}$$

where K_{if} is the current gain of the stage with feedback, and K_i is its current gain without feedback. Thus the output current i_2 of Fig. 5.6 is related to the input current i_1 as follows:

$$\begin{aligned} i_2 &= K_{if}\, i_1 \\ &= K_i\, i_1 \end{aligned} \tag{5.23}$$

However, from Fig. 5.6 we see that the input voltage v_1 is

$$v_1 \simeq i_1 R_{\text{in}} + R_e(i_1 + i_2) \tag{5.24}$$

where R_{in} is the input resistance of the common emitter stage without feedback. Therefore Equations 5.23 and 5.24 give

$$\begin{aligned} v_1 &\simeq i_1 R_{\text{in}} + R_e(i_1 + K_i\, i_1) \\ &= i_1[R_{\text{in}} + R_e(1 + K_i)] \end{aligned} \tag{5.25}$$

Hence, the input resistance $R_{\text{in}\,f}$ of the stage of Fig. 5.6 with feedback is

$$\begin{aligned} R_{\text{in}\,f} &= \frac{v_1}{i_1} \\ &\simeq R_{\text{in}} + R_e(1 + K_i) \end{aligned} \tag{5.26}$$

However, in Section 3.4 we saw that the common emitter stage has a large positive current gain, that is, $K_i \gg 1$. Hence, we may simplify Equation 5.26 as follows:

$$R_{\text{in}f} \simeq R_{\text{in}} + K_i R_e \tag{5.27}$$

which shows that the input resistance of the stage of Fig. 5.6 has been increased by an amount approximately equal to $K_i R_e$ as a result of the applied negative feedback due to R_e.

Next, from Equation 3.37 we find that if we replace K_v with K_{vf}, K_i with K_{if} and R_{in} with $R_{\text{in}\,f}$ we obtain

$$K_{vf} = -\frac{K_{if}\, R_L}{R_{\text{in}\,f}} \tag{5.28}$$

which, together with Equations 5.22 and 5.27 gives

$$K_{vf} \simeq \frac{-K_i\, R_L}{R_{\text{in}} + K_i\, R_e} \tag{5.29}$$

To evaluate the closed loop gain K_c of the stage, we replace K_i with K_{if} and R_{in} with $R_{in\,f}$ in Equation 3.44 obtaining

$$K_c = -\frac{R_L K_{if}}{R_{in\,f}+R_s}$$

$$\simeq \frac{-R_L K_i}{R_{in}+R_s+K_i R_e} \tag{5.30}$$

Dividing the numerator and denominator of this equation by the term $R_{in}+R_s$ we can rewrite it as follows

$$K_c \simeq \frac{-\dfrac{K_i R_L}{R_{in}+R_s}}{1+\dfrac{K_i R_e}{R_{in}+R_s}} \tag{5.31}$$

However, the term $-K_i R_L/(R_{in}+R_s)$ is the external voltage gain K of the stage without feedback. Hence, comparing Equations 5.4 and 5.31 we deduce the following expression for the open loop gain T of the feedback circuit of Fig. 5.6:

$$T \simeq -\frac{K_i R_e}{R_{in}+R_s} \tag{5.32}$$

where the minus sign accounts for the negative feedback existing in the circuit.

Having determined the open loop gain it is instructive to relate it to the input resistance $R_{in\,f}$ of the stage with feedback. For this purpose let us eliminate the term $K_i R_e$ between Equations 5.27 and 5.32 and so obtain

$$R_{in\,f} \simeq R_{in} - T(R_{in}+R_s) \tag{5.33}$$

Adding R_s to both sides of this equation and then dividing by $R_{in}+R_s$ we find that:

$$\frac{R_{in\,f}+R_s}{R_{in}+R_s} = 1-T$$

$$= F \tag{5.34}$$

Similarly, the output resistance $R_{out\,f}$ of the stage of Fig. 5.6 with feedback is related to the feedback factor F and the output resistance R_{out} without feedback as follows:

$$\frac{R_{out\,f}+R_L}{R_{out}+R_L} = F \tag{5.35}$$

From this equation we deduce that the output resistance of the stage of Fig. 5.6 is increased as a result of the applied local series feedback. Since the feedback circuit of Fig. 5.6 is of the current output–series input type, we conclude that current output feedback increases the output resistance while series input feedback increases the input resistance of the stage. This confirms the respective results of Table 5.1.

Example 5.2

A transistor has the following common emitter h-parameters

$$h_{ie} = 1400\Omega$$

$$h_{re} = 5 \times 10^{-4}$$

$$h_{fe} = 49$$

$$h_{oe} = 25\mu\text{S}$$

This transistor is used in the stage of Fig. 5.6 with

$$R_s = R_L = 1000\Omega$$

Find the value of R_e to apply 6dB of feedback. Determine the corresponding values of closed loop gain, input resistance and output resistance of the stage.

$$\text{6dB of feedback} = 20\log_{10} F$$

Therefore,

$$F = 2$$

and

$$T = -1$$

Next, from Equations 3.46 and 3.47 we find

$$K_i = \frac{h_{fe}}{1 + h_{oe} R_L}$$

$$= \frac{49}{1 + 25 \times 10^{-6} \times 10^3}$$

$$= 48$$

$$R_{\text{in}} = h_{ie} - \frac{h_{re} h_{fe} R_L}{1 + h_{oe} R_L}$$

$$= 1400 - \frac{5 \times 10^{-4} \times 49 \times 10^3}{1 + 25 \times 10^{-6} \times 10^3}$$

$$= 1380\Omega$$

Therefore, from Equation 5.32 we find that the feedback resistor R_e is given by:

$$R_e \simeq \frac{-T(R_{in}+R_s)}{K_i}$$

$$= \frac{1(1380+1000)}{48}$$

$$= 50\Omega$$

From Equation 5.30 we find that the closed loop gain is

$$K_c \simeq \frac{-10^3 \times 48}{1380+1000+48\times 50}$$

$$= -10$$

From Equation 5.27 we find that the input resistance $R_{in\,f}$ with feedback is

$$R_{in\,f} = 1380+48\times 50$$

$$= 3780\Omega$$

Next, from Equation 3.49

$$\frac{1}{R_{out}} = h_{oe} - \frac{h_{re}h_{fe}}{h_{ie}+R_s}$$

$$= 25\times 10^{-6} - \frac{5\times 10^{-4}\times 49}{1400+1000}$$

$$= 14{\cdot}8\times 10^{-6}\text{S}.$$

Therefore,

$$R_{out} = \frac{1}{14{\cdot}8\times 10^{-6}}$$

$$= 68\times 10^3\Omega$$

Applying Equation 5.35 we obtain:

$$\frac{R_{out\,f}+10^3}{68\times 10^3+10^3} = 2$$

Solving for $R_{out\,f}$ we find that

$$R_{out\,f} = 2\times 69\times 10^3 - 10^3$$

$$= 137\times 10^3\Omega$$

5.2.2. Local Shunt Feedback

Fig. 5.7 shows another way of applying local feedback to a common emitter stage. The resistor R_c connected between the collector and base terminals enables a voltage signal proportional to the output voltage to be fed back and applied across the input circuit. Therefore, the feedback circuit of Fig. 5.7 is of the voltage output–shunt input type.

Normally the shunt feedback resistor is large compared with the load resistance, and the applied feedback does not appreciably affect

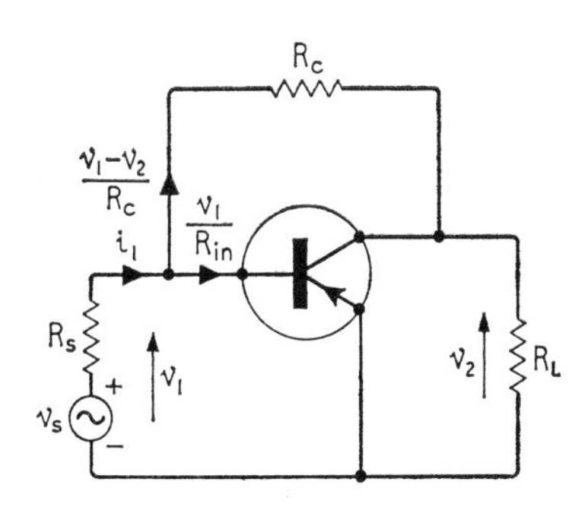

Fig. 5.7. Local shunt feedback

the voltage gain of the stage while it reduces its current gain appreciably. In other words, the local shunt feedback stabilises the current gain of the common emitter stage. This is in direct contrast to the local series feedback of Fig. 5.6.

Therefore, for the feedback circuit of Fig. 5.7 we have that

$$K_{vf} \simeq K_v \tag{5.36}$$

where K_{vf} is the voltage gain of the stage with feedback and K_v is its voltage gain without feedback.

Thus the output voltage v_2 of Fig. 5.7 is related to the input voltage v_1 as follows:

$$\begin{aligned} v_2 &= K_{vf} v_1 \\ &\simeq K_v v_1 \end{aligned} \tag{5.37}$$

Next, from Fig. 5.7 we deduce that the input current i_1 is

$$i_1 = \frac{v_1}{R_{\text{in}}} + \frac{v_1 - v_2}{R_c} \tag{5.38}$$

where, as before, R_{in} is the input resistance of the common emitter stage without feedback. From Equations 5.37 and 5.38 we find

$$i_1 \simeq \frac{v_1}{R_{\text{in}}} + \frac{v_1 - K_v v_1}{R_c}$$

$$= v_1 \left[\frac{R_c + R_{\text{in}}(1 - K_v)}{R_{\text{in}} R_c} \right] \tag{5.39}$$

Therefore, the input resistance $R_{\text{in}\,f}$ of the stage of Fig. 5.7 with feedback is given by

$$R_{\text{in}\,f} = \frac{v_1}{i_1}$$

$$= \frac{R_{\text{in}} R_c}{R_c + R_{\text{in}}(1 - K_v)} \tag{5.40}$$

In Section 3.4 we saw that the common emitter stage has a large negative voltage gain, that is, $-K_v \gg 1$. Hence, we may simplify Equation 5.40 as follows:

$$R_{\text{in}f} \simeq \frac{R_{\text{in}} R_c}{R_c - K_v R_{\text{in}}} \tag{5.41}$$

or equivalently,

$$\frac{1}{R_{\text{in}f}} \simeq \frac{1}{R_{\text{in}}} + \frac{-K_v}{R_c} \tag{5.42}$$

Equation 5.42 shows that the input conductance of the stage of Fig. 5.7 has been increased by an amount approximately equal to $|K_v|/R_c$ as a result of the applied negative feedback due to R_c. In other words, shunt feedback reduces the input resistance of the stage.

Next, from Equation 5.28

$$K_{if} = -\frac{R_{\text{in}\,f} K_{vf}}{R_L} \tag{5.43}$$

Therefore, using Equations 5.36, 5.40 and 5.43 we obtain

$$K_{if} \simeq \frac{-R_{\text{in}} R_c K_v}{R_L(R_c - K_v R_{\text{in}})} \tag{5.44}$$

To determine the closed loop gain K_c we use the first line of Equation 5.30 and so obtain

$$K_c = -\frac{R_L K_{if}}{R_{\text{in}\,f} + R_s}$$

$$\simeq \frac{K_v R_{\text{in}} R_c}{R_c(R_{\text{in}} + R_s) - K_v R_{\text{in}} R_s} \tag{5.45}$$

where we have made use of Equations 5.40 and 5.44. Dividing the numerator and denominator of Equation 5.45 by the term $R_c(R_{\text{in}}+R_s)$ we find that:

$$K_c \simeq \frac{\dfrac{K_v R_{\text{in}}}{R_{\text{in}}+R_s}}{1-\dfrac{K_v R_{\text{in}} R_s}{R_c(R_{\text{in}}+R_s)}} \tag{5.46}$$

Therefore, on comparing Equations 5.4 and 5.46 we deduce that the open loop gain T of the feedback circuit of Fig. 5.7 is

$$T \simeq \frac{K_v R_{\text{in}} R_s}{R_c(R_{\text{in}}+R_s)} \tag{5.47}$$

Here again it will prove useful to relate the open loop gain of the stage of Fig. 5.7 to its input resistance with feedback. For this purpose let us eliminate the term K_v/R_c between Equations 5.42 and 5.47; we thus obtain

$$\frac{1}{R_{\text{in}\,f}} \simeq \frac{1}{R_{\text{in}}} - T\left(\frac{1}{R_{\text{in}}}+\frac{1}{R_s}\right) \tag{5.48}$$

Adding $1/R_s$ to both sides of this equation, and then dividing throughout by the term

$$\frac{1}{R_{\text{in}}}+\frac{1}{R_s}$$

we find that:

$$\frac{\dfrac{1}{R_{\text{in}\,f}}+\dfrac{1}{R_s}}{\dfrac{1}{R_{\text{in}}}+\dfrac{1}{R_s}} = 1-T$$
$$= F \tag{5.49}$$

Similarly, we deduce that the output resistance $R_{\text{out}\,f}$ of the stage of Fig. 5.7 is related to its output resistance R_{out} without feedback as follows:

$$\frac{\dfrac{1}{R_{\text{out}\,f}}+\dfrac{1}{R_L}}{\dfrac{1}{R_{\text{out}}}+\dfrac{1}{R_L}} = F \tag{5.50}$$

From this expression we deduce that $1/R_{\text{out}\,f}$ is greater than $1/R_{\text{out}}$, that is, the output resistance with feedback is less than the output resistance without feedback.

We therefore conclude that since the local shunt feedback stage of Fig. 5.7 is of the voltage output–shunt input type, then voltage output feedback reduces the output resistance while shunt input feedback reduces the input resistance. This confirms the results of Table 5.1.

Example 5.3

The transistor whose parameters were given in Example 5.2 is used in the local shunt feedback stage of Fig. 5.7 with

$$R_s = R_L = 1000\Omega$$

Evaluate the resistor R_c so as to apply 6dB of feedback, and determine the corresponding values of closed loop gain, input resistance and output resistance of the stage.

As in Example 5.2, we find that 6dB of feedback corresponds to $F = 2$ and $T = -1$. Further, $R_{\text{in}} = 1380\Omega$ and $K_i = 48$. Therefore, from Equation 3.37, the voltage gain without feedback is:

$$K_v = -\frac{K_i R_L}{R_{\text{in}}}$$

$$= -\frac{48 \times 1000}{1380}$$

$$= -35$$

Solving Equation 5.47 for the feedback resistor we find

$$R_c \simeq \frac{K_v R_{\text{in}} R_s}{T (R_{\text{in}} + R_s)}$$

$$= \frac{35 \times 1380 \times 1000}{1380 + 1000}$$

$$= 20 \times 10^3\Omega$$

Next, from Equation 5.45 the closed loop gain is

$$K_c \simeq \frac{-35 \times 1380 \times 20 \times 10^3}{20 \times 10^3(1380 + 1000) + 35 \times 1380 \times 1000}$$

$$= -10$$

Equation 5.41. gives the input resistance with feedback to be:

$$R_{\text{in}\,f} \simeq \frac{1380 \times 20 \times 10^3}{20 \times 10^3 + 35 \times 1380}$$

$$= 400\Omega$$

From Example 5.2 we have $R_{\text{out}} = 68 \times 10^3\Omega$. Hence Equation 5.50 gives

$$\frac{\dfrac{1}{R_{\text{out}\,f}} + \dfrac{1}{10^3}}{\dfrac{1}{68 \times 10^3} + \dfrac{1}{10^3}} = 2$$

Solving for $1/R_{\text{out}\,f}$ we find that

$$\frac{1}{R_{\text{out}f}} = 1{\cdot}03 \times 10^{-3}\text{S}$$

Therefore, $R_{\text{out}\,f} = 970\Omega$.

5.3. TWO-STAGE AND THREE-STAGE FEEDBACK AMPLIFIERS

We shall now consider the case of feedback amplifiers involving two and three transistor stages.

The circuits of Figs. 5.8 and 5.9 show two feedback amplifiers, each involving two transistor stages. In the circuit of Fig. 5.8 the resistor R_1

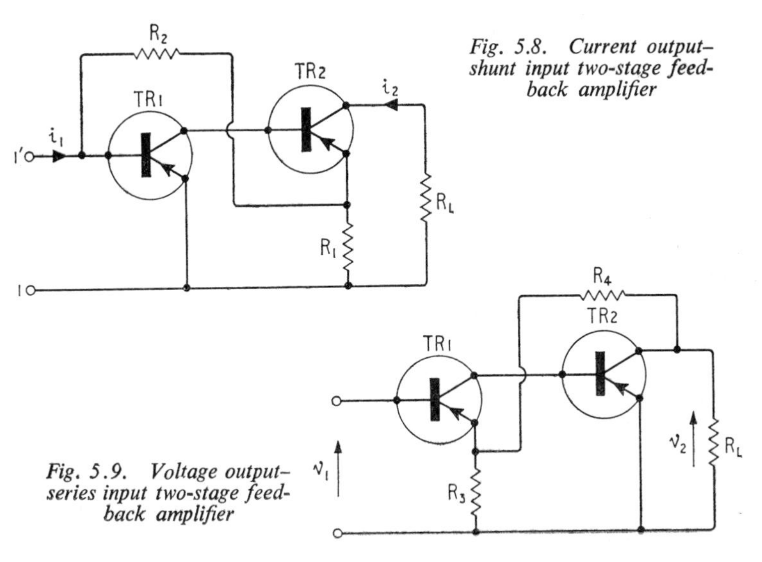

Fig. 5.8. Current output–shunt input two-stage feedback amplifier

Fig. 5.9. Voltage output–series input two-stage feedback amplifier

extracts a voltage signal proportional to the output current and the resistor R_2 applies it across the input circuit. Therefore, this feedback amplifier is of the current output–shunt input type. On the other hand, the resistor R_4 of the amplifier arrangement of Fig. 5.9 extracts a feedback signal proportional to the output voltage, and the resistor R_3 applies it in series with the input circuit, and so this feedback amplifier is of the voltage output–series input type. However, in both amplifiers the applied feedback is negative because if, in either case, a voltage signal is applied at any point inside the feedback loop then it will circulate around the loop and return to the same point with its phase angle shifted by 180°.

From Table 5.1 we deduce that the applied feedback will reduce the input resistance of the current output–shunt input feedback

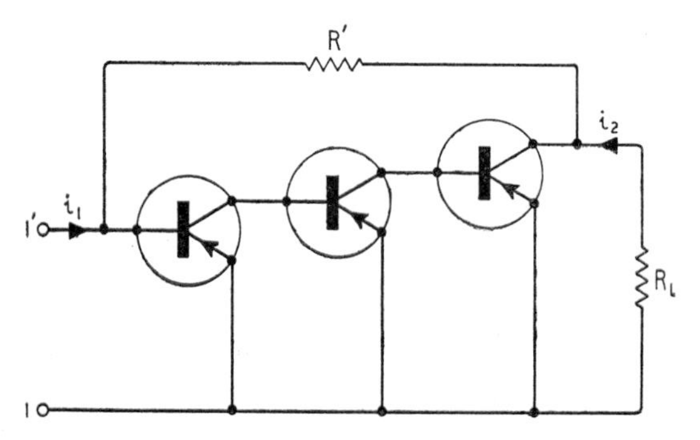

Fig. 5.10. Voltage output–shunt input three-stage feedback amplifier

amplifier of Fig. 5.8 and increase its output resistance. Therefore, it is best driven from a current source, and is most useful as a current amplifier. In fact, if the amount of applied feedback is large then the current gain of the circuit (that is, i_2/i_1) will approximate to the limiting value of

$$\left(-\frac{R_1+R_2}{R_1}\right)$$

Therefore, it becomes practically independent of transistor parameters.

On the other hand, the applied feedback will increase the input resistance and reduce the output resistance of the voltage output–series input feedback amplifier of Fig. 5.9. Therefore, this amplifier is best driven from a voltage source and is most useful as a voltage amplifier. Again if the amount of applied feedback is large then the voltage gain of the circuit (that is, v_2/v_1) approaches the limiting value of $(R_3+R_4)/R_3$.

Consider next the feedback amplifiers of Figs. 5.10 and 5.11, both of which involve three common emitter stages. In the circuit of

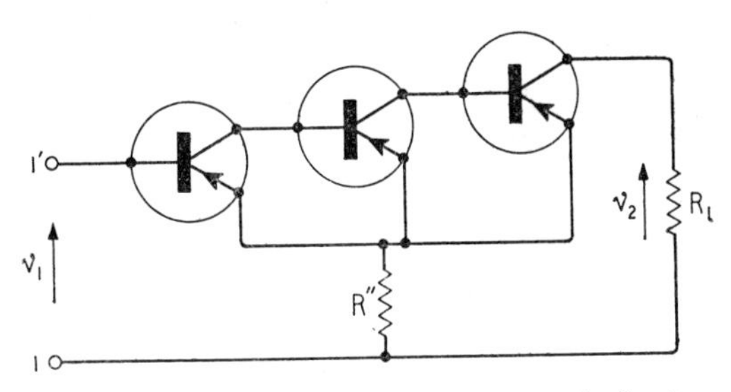

Fig. 5.11. Current output–series input three-stage feedback amplifier

Fig. 5.10 the shunt feedback resistor R' extracts a voltage signal proportional to the output voltage and applies it across the input circuit. This amplifier is, therefore, of the voltage output–shunt input type, and its input resistance as well as output resistance are reduced by the applied feedback. If the amount of feedback is large then the current gain closely approaches the limiting value of R'/R_L.

In the circuit of Fig. 5.11 the resistor R'' extracts a voltage signal proportional to the output current and applies it in series with the input circuit. Therefore, the feedback amplifier of Fig. 5.11 is of the current output–series input type and, consequently, the applied feedback will increase its input resistance as well as the output resistance. If the amount of applied feedback is large, then the voltage gain will closely approach the limiting value of $-R_L/R''$.

The amplifier arrangement of Figs. 5.8 to 5.11 are only 4 examples of the many different ways in which negative feedback can be applied to two and three transistor stages. A much more detailed treatment is covered elsewhere.[1]

5.4. FEEDBACK AND STABILITY

In the basic feedback equation, Equation 5.4, each of the terms K and β may be a function of the complex frequency $j\omega$. The frequency dependence of the gain K without feedback can be due to the high frequency effects associated with the transistor (see

[1] HAYKIN, S. S., *Junction Transistor Circuit Analysis*, Iliffe Books Ltd. (1962), Chapter 7.

Section 3.9) or due to the influence of interstage coupling capacitors (see Section 4.3). On the other hand, the frequency dependence of the reverse transmission ratio β can be produced by the presence of capacitors or inductors in the feedback network.

Thus the open loop gain T which, by definition, is equal to $K\beta$ can be expressed as follows:

$$T = |T| . e^{j\phi}$$
$$= |T| . \cos\phi + j|T| . \sin\phi \qquad (5.51)$$

where T is the modulus of the open loop gain and ϕ is its phase angle. Equation 5.51 can be represented by the vector **OP** in the complex plane of Fig. 5.12. The frequency dependence of the open loop gain can be represented graphically, in the complex plane, by the curve traced out by the tip of the vector **OP** as the frequency is varied from zero to infinity. The resulting curve may be as shown in Fig. 5.12.

The locus of Fig. 5.12 can be used for evaluating the feedback factor F. From Equation 5.6

$$F = 1 - T$$

Therefore, the vector corresponding to the feedback factor can be obtained by joining the tip of the vector **OP** and the point (1, 0) as

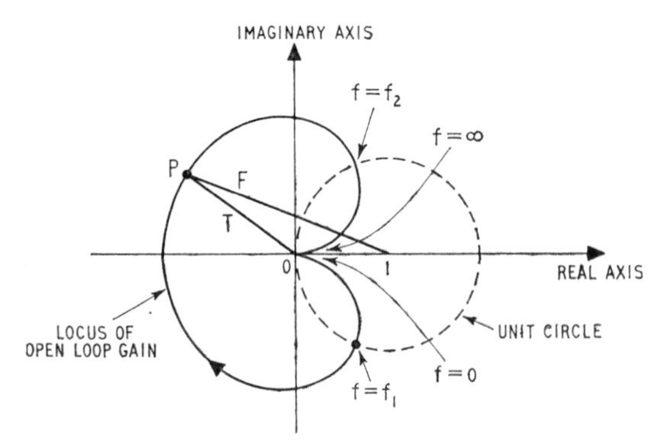

Fig. 5.12. Relationship between open loop gain and feedback factor vectors

indicated in Fig. 5.12. If, at any operating frequency f, the point P lies inside the unit circle of Fig. 5.12, then the modulus of the feedback factor will be less than unity and, consequently, the modulus of the closed loop gain will be greater than that of the gain without feedback. Therefore, the whole area inside the unit circle of Fig. 5.12 corresponds to positive feedback. If, on the other hand, the point P of the vector **OP** lies outside the unit circle then the modulus of the feedback factor will be greater than unity and so the resulting feedback is negative. Thus, in Fig. 5.12 the applied feedback is positive for the frequency range of zero to f_1 as well as the frequency range of f_2 to infinity, while it is negative for the frequency range f_1 to f_2. The frequency response of the gains with and without feedback can therefore be as illustrated in Fig. 5.13.

If, at the operating-frequency f_0 (say), the locus of the open loop gain passes through the point (1, 0) then from Equation 5.6 it follows that at this frequency the feedback factor is zero and, therefore, the closed loop gain is infinite. Thus, oscillation occurs and the amplifier becomes unstable. The point (1, 0) is referred to as the *critical point* and the frequency at which the locus passes through it is termed the *oscillation frequency*. A more general criterion for

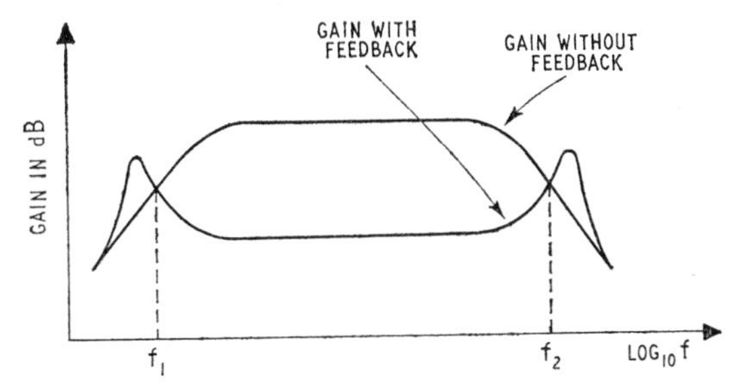

Fig. 5.13. Gain-frequency response with and without feedback

stability has been derived by Nyquist and it can be stated as follows:[1]

'If the locus traced out by the tip of the open loop gain vector passes through or encloses the critical point (1, 0) then the feedback circuit is unstable and it will, therefore, oscillate. On the other hand, the circuit is stable if the locus does not pass through or enclose the critical point (1, 0).'

Therefore, in view of Nyquist's criterion we can state that the locus of Fig. 5.14(a) corresponds to an unstable feedback circuit, while the locus of Fig. 5.14(b) applies to a stable circuit.

5.5. FEEDBACK OSCILLATORS

When designing a feedback amplifier one of the principal objectives is to ensure that it will remain stable under all possible operating

[1] NYQUIST, H., 'Regeneration Theory', *Bell System Tech. J.* **11**, 1 (1932).

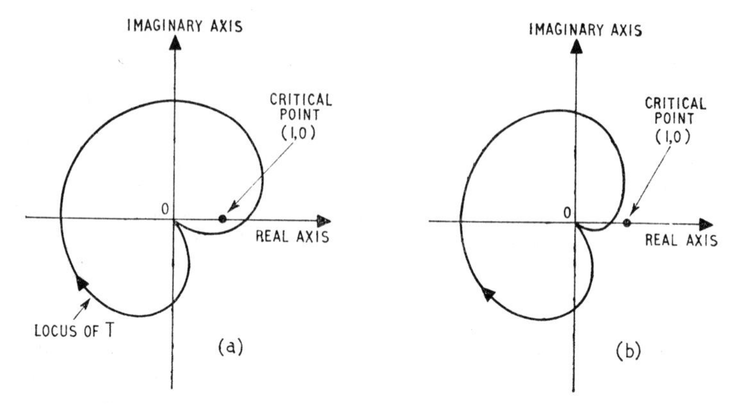

Fig. 5.14. Nyquist loci. (a) Unstable circuit; (b) Stable circuit

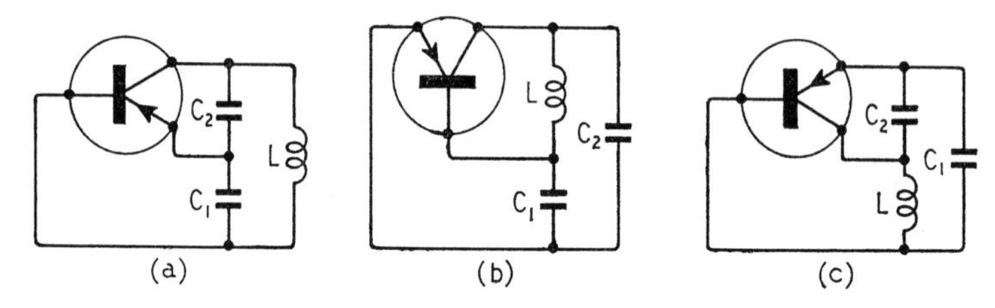

Fig. 5.15. Copitts oscillator. (a) Common emitter orientation; (b) Common base orientation; (c) Common collector orientation

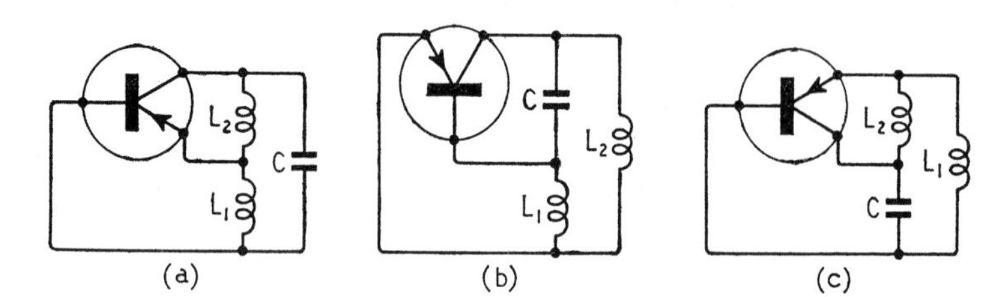

Fig. 5.16. Hartley oscillator. (a) Common emitter orientation; (b) Common base orientation; (c) Common collector orientation

conditions. On the other hand, when designing an oscillator the circuit is deliberately arranged to be unstable, that is, the locus of the open loop gain is made to enclose the critical point (1, 0).

There are many different oscillator circuits that are widely used in practice. We shall consider only a few representative types.

5.5.1. Inductance–Capacitance Oscillators

Figs. 5.15(a) and 5.16(a) show the a.c. circuits of the well-known Colpitts and Hartley oscillators, respectively; for simplicity we have omitted the biasing arrangements. In each case, the feedback network consists of inductors and capacitors, and it is arranged to introduce, at the desired oscillation frequency, a phase shift of 180° between the voltage signal developed at the collector of the transistor and the voltage signal fed back to the base. However, the common emitter stage produces a phase shift of 180° between its input and output voltage signals. Therefore, at the desired oscillation frequency there is effectively zero phase shift around the feedback loop. If at this frequency the modulus of the open loop gain is equal to or greater than unity, the resulting locus of the open loop gain will enclose the critical point, thereby resulting in oscillation as desired.

Notice that both the Colpitts and Hartley oscillator circuits can be re-arranged in such a way that the transistor appears operated in any one of the 3 basic configurations, namely, the common emitter, common base and common collector modes. Thus in Figs. 5.15(a) and 5.16(a) the transistor is operated in the common emitter mode. In Figs. 5.15(b) and 5.16(b) the transistor is operated in the common base mode, and in Figs. 5.15(c) and 5.16(c) the transistor is operated in the common collector mode. However, the analysis of any of the 3 representations of Figs. 5.15 will lead to the same results. This equally applies to the representations of Figs. 5.16. We shall only consider the common emitter versions.

For the purpose of analysis we shall represent the oscillator circuits of Figs. 5.15(a) and 5.16(a) by the more general form shown in Fig. 5.17. Thus comparing Figs. 5.15(a) and 5.17 we observe that for the Colpitts oscillator the impedances Z_1, Z_2 and Z_3 are:

$$\left.\begin{aligned} Z_1 &= \frac{1}{\mathrm{j}\omega C_1} \\ Z_2 &= \mathrm{j}\omega L \\ Z_3 &= \frac{1}{\mathrm{j}\omega C_2} \end{aligned}\right\} \qquad (5.52)$$

On the other hand, for the Hartley oscillator of Fig. 5.16(a) we see that the impedances Z_1, Z_2 and Z_3 of Fig. 5.17 are given by

$$\left.\begin{aligned} Z_1 &= j\omega L_1 \\ Z_2 &= \frac{1}{j\omega C} \\ Z_3 &= j\omega L_2 \end{aligned}\right\} \tag{5.53}$$

Here it is assumed that there exists no mutual coupling between the inductors L_1 and L_2 of Fig. 5.16(a).

Returning to the general circuit of Fig. 5.17, if we assume that at the desired oscillation frequency the impedance Z_3 is sufficiently

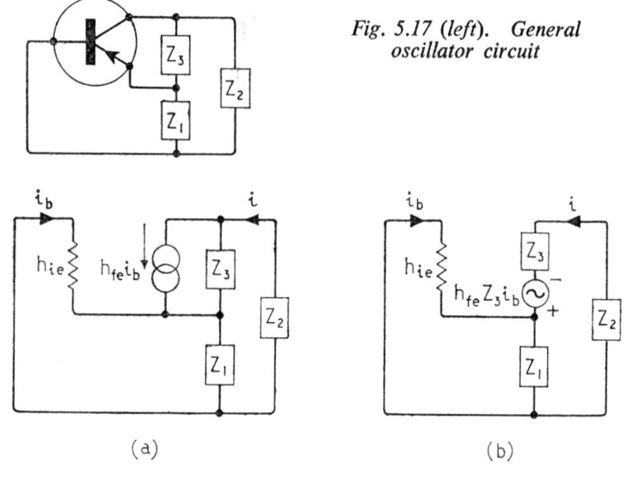

Fig. 5.17 (left). General oscillator circuit

Fig. 5.18. Equivalent circuits for general oscillator of Fig. 5.17. (a) Simplified equivalent circuit; (b) Equivalent circuit after replacing the current source of (a) with equivalent voltage source

small, then to a first order of approximation we can represent the transistor by the simplified common emitter h-parameters' equivalent circuit of Fig. 5.18(a). Converting from a current source to an equivalent voltage source we obtain the modified circuit of Fig. 5.18(b). Thus:

$$i = \frac{Z_3 h_{fe} i_b}{Z_2 + Z_3 + \dfrac{Z_1 h_{ie}}{Z_1 + h_{ie}}} \tag{5.54}$$

However, the base current i_b is given by

$$i_b = \frac{-Z_1 i}{Z_1 + h_{ie}} \tag{5.55}$$

Therefore, from Equations 5.54 and 5.55 it follows that the open loop gain T of the general feedback oscillator circuit of Fig. 5.17 is:

$$T = \frac{-Z_1}{Z_1 + h_{ie}} \cdot \frac{Z_3 h_{fe}}{Z_2 + Z_3 + \dfrac{Z_1 h_{ie}}{Z_1 + h_{ie}}}$$

$$= \frac{-Z_1 Z_3 h_{fe}}{(Z_2 + Z_3)(Z_1 + h_{ie}) + Z_1 h_{ie}} \tag{5.56}$$

We shall next use this equation to study the oscillator circuits of Figs. 5.15(a) and 5.16(a). For the Hartley oscillator of Fig. 5.16(a) we can combine Equations 5.53 and 5.56 to give:

$$T = \frac{\omega^2 L_1 L_2 h_{fe}}{\left(\dfrac{L_1}{C} - \omega^2 L_1 L_2\right) + jh_{ie}\left(\omega L_1 + \omega L_2 - \dfrac{1}{\omega C}\right)} \tag{5.57}$$

Assuming that f_o is equal to the frequency of oscillation, then at $\omega = \omega_o = 2\pi f_o$ we see from Equation 5.57 that the phase shift of the open loop gain T is zero if the imaginary term in the denominator is zero, that is

$$h_{ie}\left(\omega_o L_1 + \omega_o L_2 - \frac{1}{\omega_o C}\right) = 0$$

Solving this equation for f_o we find that

$$f_o = \frac{\omega_o}{2\pi}$$

$$= \frac{1}{2\pi} \cdot \frac{1}{\sqrt{[C(L_1 + L_2)]}} \tag{5.58}$$

This equation represents the resonant frequency of a tuned circuit consisting of a capacitor C and two inductors L_1 and L_2, all connected in series. When $\omega = \omega_o$, we find that Equation 5.57 reduces as:

$$T = \frac{\omega_o{}^2 L_1 L_2 h_{fe}}{\dfrac{L_1}{C} - \omega_o{}^2 L_1 L_2} \tag{5.59}$$

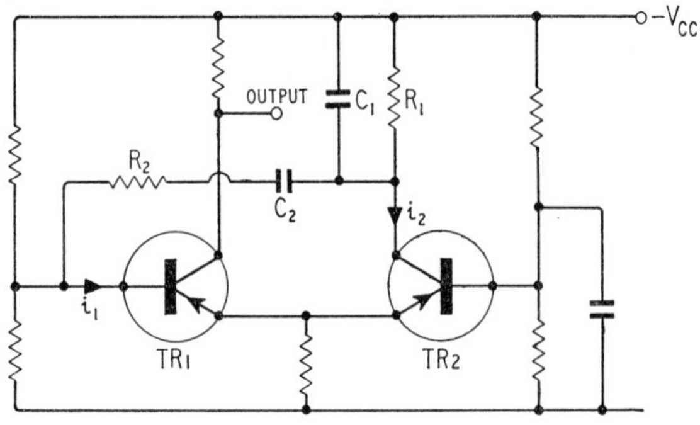

Fig. 5.19. Wien-bridge oscillator

Eliminating ω_o between Equations 5.58 and 5.59 we obtain that the value of the open loop gain at $\omega = \omega_o$ is given by

$$T = \frac{h_{fe} L_2}{L_1} \tag{5.60}$$

However, from Nyquist's criterion we know that oscillations can occur if, at $\omega = \omega_o$, the modulus of the open loop gain is equal to

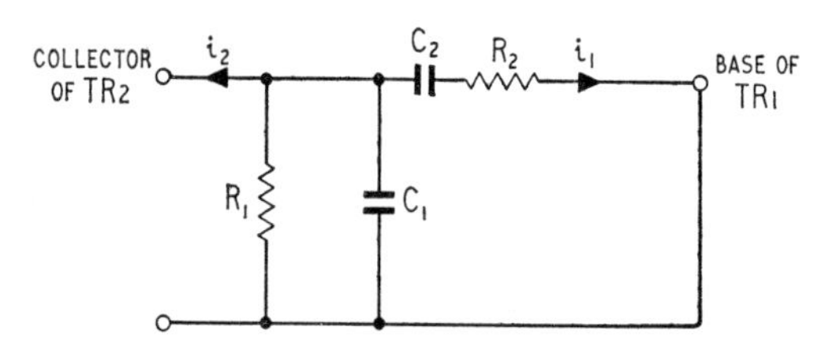

Fig. 5.20. Simplified equivalent circuit for Wien-bridge oscillator

or greater than unity. Therefore, from Equation 5.60 the necessary condition for oscillation is:

$$1 \leqslant \frac{h_{fe} L_2}{L_1}$$

that is, the common emitter short circuit current gain of the transistor satisfies the following condition:

$$h_{fe} \geqslant \frac{L_1}{L_2} \tag{5.61}$$

Similarly, using Equations 5.52 and 5.56 the frequency of oscillation f_o of the Colpitts oscillator of Fig. 5.15(a) is given by

$$f_o = \frac{1}{2\pi}\sqrt{\left[\frac{1}{L}\left(\frac{1}{C_1}+\frac{1}{C_2}\right)\right]} \tag{5.62}$$

and the necessary condition for oscillation is

$$h_{fe} \geqslant \frac{C_2}{C_1} \tag{5.63}$$

5.5.2. Resistance–Capacitance Oscillators

If the desired oscillation frequency is low (of the order of 100Hz and less) then it is preferable to use capacitors and resistors for the feedback network as shown in the Wien-bridge oscillator of Fig. 5.19. Here the basic amplifier is the emitter coupled pair which we considered earlier (see Section 4.1). Provided that

(1) the resistor R_1 is low compared with the output resistance of the transistor stage TR2, and

(2) the resistor R_2 is large compared with the input resistance of TR1,

then the current ratio i_1/i_2 of the feedback network can be obtained from the simplified equivalent circuit of Fig. 5.20. Notice that i_1 and i_2 are the input and output currents, respectively, of the emitter coupled pair.

From this diagram we obtain

$$\frac{i_1}{i_2} = \frac{-Z_1}{Z_1+Z_2} \tag{5.64}$$

where

$$Z_1 = \frac{R_1}{1+j\omega C_1 R_1}$$

and

$$Z_2 = \frac{1+j\omega C_2 R_2}{j\omega C_2} \tag{5.65}$$

Denoting the current gain of the emitter coupled pair by K_i we find that the open loop gain T of the oscillator circuit of Fig. 5.19 is

$$T = K_i . \frac{i_1}{i_2}$$

$$= \frac{-Z_1 K_i}{Z_1+Z_2} \tag{5.66}$$

where use has been made of Equation 5.64. Combining Equations 5.65 and 5.66 we find that

$$T = \frac{-j\omega C_2 R_1 K_i}{j\omega(C_2 R_1 + C_1 R_1 + C_2 R_2) + (1 - \omega^2 C_1 C_2 R_1 R_2)} \tag{5.67}$$

The phase shift of the open loop gain will be zero if the current gain K_i is negative and, at the oscillation frequency f_o, the real term of the denominator of Equation 5.67 vanishes, that is, if

$$1 - \omega_o^2 C_1 C_2 R_1 R_2 = 0$$

Hence,

$$f_o = \frac{\omega_o}{2\pi} \tag{5.68}$$

$$= \frac{1}{2\pi \sqrt{(C_1 C_2 R_1 R_2)}}$$

When $\omega = \omega_o$ we find that Equation 5.67 reduces to

$$T = \frac{-C_2 R_1 K_i}{C_2 R_1 + C_1 R_1 + C_2 R_2} \tag{5.69}$$

Therefore, from Nyquist's criterion we deduce that oscillations will occur if

$$1 \leqslant \frac{-C_2 R_1 K_i}{C_2 R_1 + C_1 R_1 + C_2 R_2}$$

that is, if the current gain K_i satisfies the following necessary condition:

$$-K_i \geqslant 1 + \frac{C_1}{C_2} + \frac{R_2}{R_1} \tag{5.70}$$

For the special case of $C_1 = C_2 = C$ and $R_1 = R_2 = R$ we see that Equations 5.68 and 5.70 simplify as follows, respectively

$$f_o = \frac{1}{2\pi CR} \tag{5.71}$$

$$-K_i \geqslant 3 \tag{5.72}$$

Normally, there is no difficulty in satisfying the necessary condition of Equation 5.72. Notice that the current gain K_i of the emitter coupled pair is negative.

The resistance–capacitance oscillator of Fig. 5.19 employs two transistors. It is possible to use one transistor only and yet still use resistors and capacitors for the feedback network as shown by the

RC phase shift oscillator circuit of Fig. 5.21 where R_{in} is the input resistance of the common emitter stage. The elements R and C of the feedback network are chosen so as to introduce a phase shift of 180° into the open loop gain response of the circuit at the desired oscillation frequency. It can be shown that the frequency of oscillation has the following approximate value

$$f_o \simeq \frac{1}{2\pi} \cdot \sqrt{\left(\frac{1}{2C^2 RR_L(2R_L+3R)}\right)} \tag{5.73}$$

The necessary condition for oscillation is

$$h_{fe} \geqslant 23 + \frac{4R_L}{R} + \frac{29R}{R_L} \tag{5.74}$$

5.5.3. Frequency Stability

The above analysis of oscillator circuits is very approximate in that it ignored effects of the parameters h_{re}, h_{oe} and the effects of external loading conditions such as biasing elements, load resistance and parisitic capacitances. All these effects cause the frequency of oscillation to depend on transistor parameters and external loading

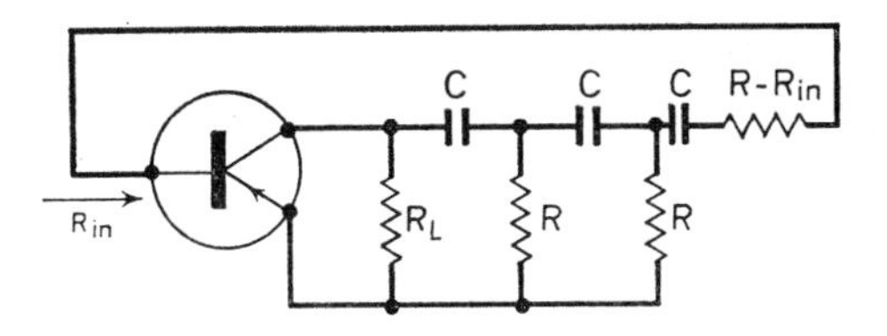

Fig. 5.21. RC phase shift oscillator

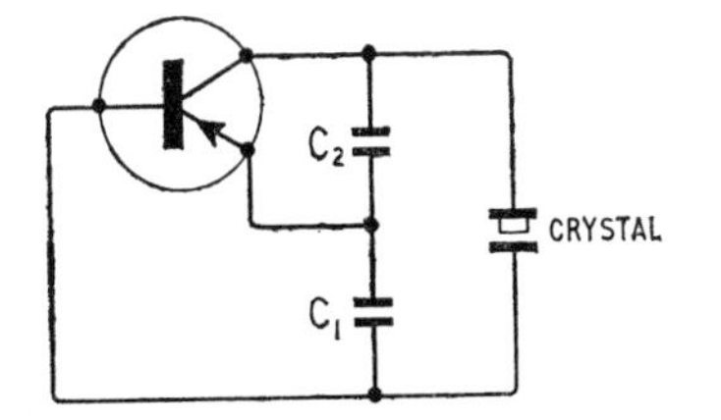

Fig. 5.22. Crystal-controlled oscillator

conditions. Consequently, the frequency of oscillation invariably drifts from the desired value as the ambient temperature and the supply voltage are changed, and as different transistor samples are used.

The frequency dependence on transistor parameters can be minimised by the proper choice of the circuit elements of the feedback network. Thus in the Colpitts oscillator, for example, it can be shown that[1] the frequency of oscillation is very nearly equal to the value given by Equation 5.62 provided that

$$\frac{C_1+C_2}{L} \gg \frac{h_{oe}}{h_{ie}} \tag{5.75}$$

The frequency stability can be improved further by using a

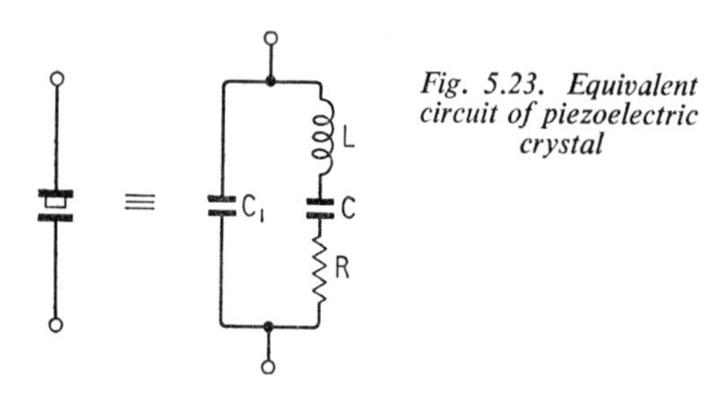

Fig. 5.23. Equivalent circuit of piezoelectric crystal

piezoelectric crystal as the frequency controlling element as in Fig. 5.22. The *piezoelectric effect* is encountered in certain crystals such as quartz. If such crystals are deformed mechanically, a distribution of electric charges appears on their surfaces. Conversely, if they are subjected to electric fields, they become mechanically deformed. Therefore, the piezoelectric effect can provide

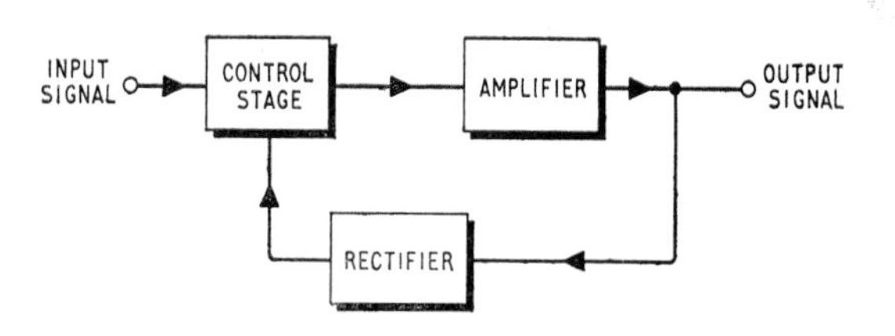

Fig. 5.24. Automatic gain control circuit

electromechanical coupling between a transistor and a mechanical resonator consisting of a block of the crystal. The electrical behaviour of the crystal can be represented by the equivalent circuit of Fig. 5.23. It consists basically of a series resonant circuit shunted by a capacitor C_1 which represents the electrostatic capacitance of the electrodes when the crystal is not vibrating. The equivalent

[1] HAYKIN, S. S., *Junction Transistor Circuit Analysis*, Iliffe Books Ltd. (1962), Chapter 9.

resonant circuit of the vibrating crystal is characterised by a remarkably high Q-factor, a typical value being 10^5. Hence, the oscillation frequency of a crystal controlled oscillator is highly stable.

5.6. AUTOMATIC GAIN CONTROL

The object of *automatic gain control* (AGC) is to maintain the output signal level of an amplifier constant despite variations in the input signal level. Normally this has to be achieved without causing appreciable harmonic distortion.

Fig. 5.24 shows the main features of an automatic gain control circuit. The presence of feedback in the circuit is clearly observed.

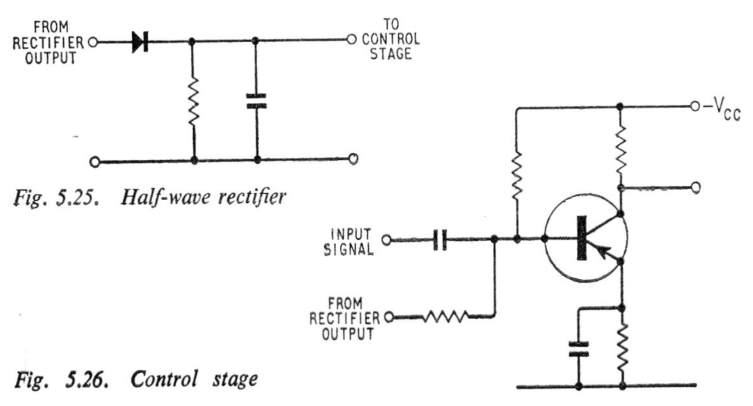

Fig. 5.25. Half-wave rectifier

Fig. 5.26. Control stage

The *rectifier* component can consist of a half-wave rectifier (see Section 9.3) with a smoothing capacitor as shown in Fig. 5.25. The d.c. control voltage developed across the smoothing capacitor is, therefore, directly proportional to the amplitude of the amplifier sinusoidal output.

This control voltage determines the forward gain of the *control stage*. This can be accomplished by making use of the dependence of the gain of a transistor stage on the quiescent operating-point (see Section 3.8). Thus the gain can be varied appreciably by varying the emitter current or collector voltage. In Fig. 5.26 the d.c. control voltage is applied to the base of a common emitter stage, thereby influencing the emitter current.

Therefore, by deriving the gain-control voltage from the amplifier output it is possible to ensure that any increase (or decrease) in the output signal level will automatically reduce (or increase) the overall amplifier gain and so maintain the amplifier output signal at a nearly constant level.

CHAPTER 6

SWITCHING CIRCUITS

6.1. SWITCHING PERFORMANCE

In considering semiconductor devices as switches we are not concerned with the linearity of their characteristics. In fact, as we shall see, we are only interested in the non-linear extremities.

6.1.1. The Ideal Switch*

In order to understand the action and limitations of a semiconductor switch, consider first a circuit consisting of a battery V, load resistor R and an ideal switch S, in the form of a pair of contacts which have zero internal (or *switch closed*) resistance and infinite leakage (or *switch open*) resistance, as shown in Fig. 6.1(a). If we plot switch current I_S against switch voltage V_S we obtain the two points A and B shown in Fig. 6.1(b).

When the switch is open, $V_S = V$ and $I_S = 0$, giving point A.

When the switch is closed, $V_S = 0$ and $I_S = V/R$, giving point B.

The two states of the switch thus correspond to points A and B, and the switch transfers instantaneously from one state to the other along the load-line, which has a slope of $-1/R$ as indicated in Fig. 6.1(b). Clearly there is no power dissipated in the switch.

Notice that the two states of the switch could also be described by the conditions:

$$\left.\begin{aligned} &V_S = 0 \quad \text{or} \quad V_S = V \\ \text{or alternatively} \quad &I_S = V/R \quad \text{or} \quad I_S = 0 \end{aligned}\right\} \tag{6.1}$$

6.1.2. The Effect of Internal Resistance

Now consider the effect of including an internal resistance, i.e. the switch is no longer a short circuit when closed but presents a small resistance R_c to the circuit, as in Fig. 6.2(a).

* This treatment follows that by Neeteson (see p. 360).

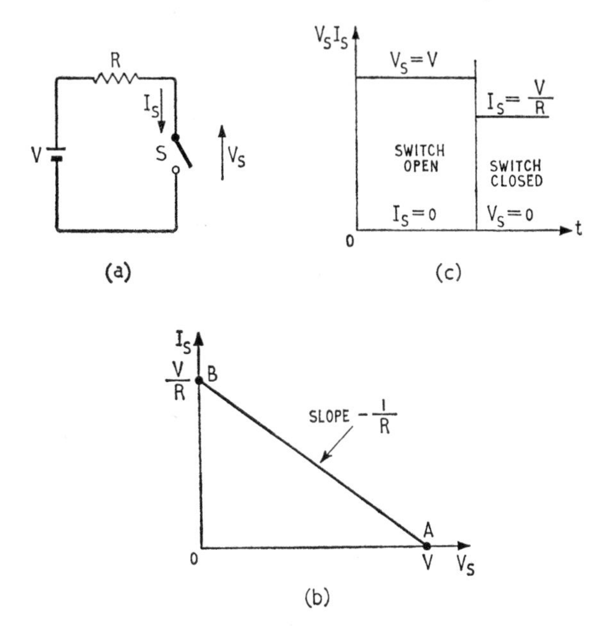

Fig. 6.1. The ideal switch. (a) Circuit diagram; (b) Switching characteristic; (c) The states of the switch

When the switch is open, $V_S = V$ and $I_S = 0$, giving point A as before.

When the switch is closed, however, V_S is no longer zero but equal to $V.R_c/(R+R_c)$ and I_S is reduced to $V/(R+R_c)$, giving point C in Fig. 6.2(b).

The change in voltage between the two states is thus reduced to

$$V - \frac{R_c}{R+R_c}.V = \frac{R}{R+R_c}.V$$

If in Fig. 6.2(b) a straight line is drawn through the origin with a slope equal to $1/R_c$, then its intersection with the load-line at point C gives the characteristic of the closed switch. There will be a power loss P_S in the closed switch given by

$$P_S = I_S^2 . R_c$$

$$= \frac{V^2 . R_c}{(R+R_c)^2} \qquad (6.2)$$

6.1.3. The Effect of Leakage Resistance

Taking the discussion a stage further we shall now consider the effects of finite leakage resistance, i.e. the presence of a resistance R_o across the terminals of the switch as in Fig. 6.3(a). With the switch in the open condition we now have

$$V_S = \frac{R_o}{R+R_o}V; \quad I_S = \frac{V}{R+R_o}$$

This condition gives point D on the load-line in Fig. 6.3(b) corresponding to the intersection of a straight line through the origin with

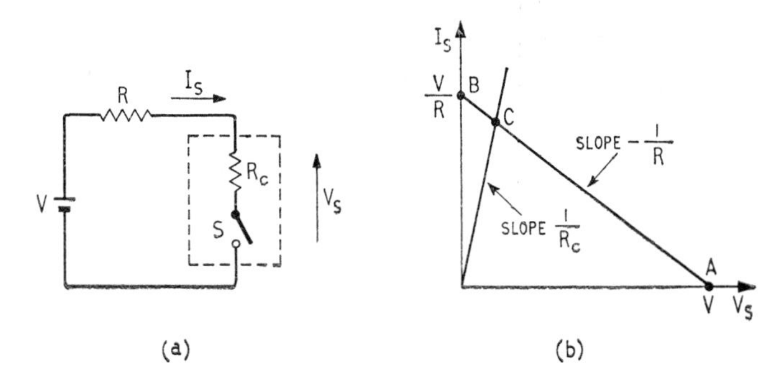

Fig. 6.2. The effect of internal resistance. (a) Circuit diagram; (b) Switching characteristic

a slope of $1/R_o$ and the load-line. If R_o is very large compared to R_c, as is usually the case, then the presence of R_o will not appreciably alter the switch-closed condition from point C. Thus the two states of the switch now correspond to point C and point D and the distinction between the two states has been reduced as illustrated in Fig. 6.3(c). In addition some power is dissipated in the open switch. Since the efficiency of the switch is impaired by the reduction of R_o or the increase of R_c, the ratio R_o/R_c may be used as a figure of merit to describe any given switch.

6.1.4. The Effect of Inertia

So far we have discussed only the reduction in amplitude of the transition, which can still take place instantaneously. Now consider the effect of inertia in the switch, which may be represented in our equivalent circuit by a capacitor C across the terminals as in Fig. 6.4(a).

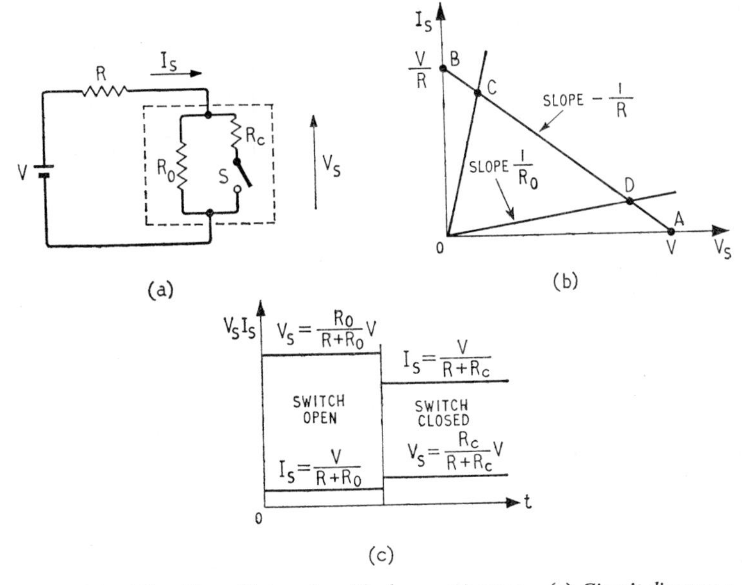

Fig. 6.3. The effect of internal and leakage resistance. (a) Circuit diagram; (b) Switching characteristic; (c) The states of the switch

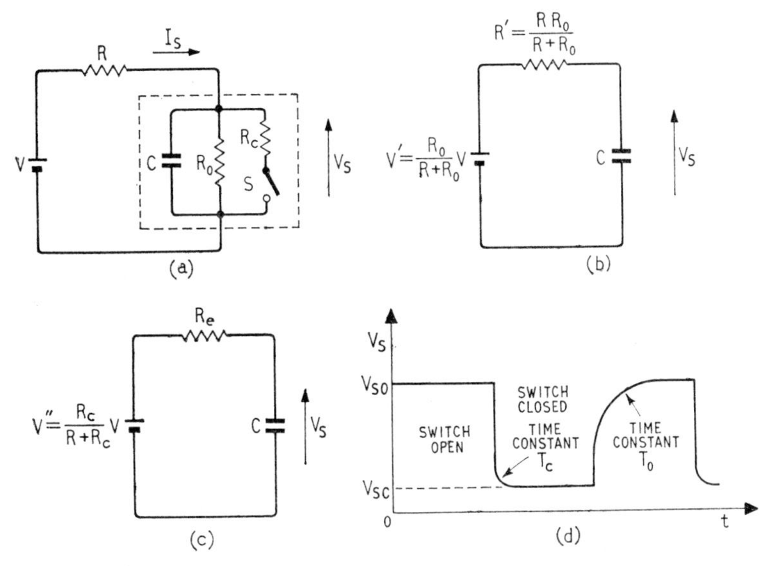

Fig. 6.4. A practical switch. (a) Circuit diagram; (b) Equivalent circuit with switch open; (c) Equivalent circuit with switch closed; (d) Output waveform switch

If the switch is initially closed and the voltage across it is $V_S = V.R_c/(R+R_c)$ then at the instant the switch is opened the equivalent circuit may be simplified by the application of Thevenin's Theorem to that of Fig. 6.4(b). Thus the capacitor will charge exponentially from the initial value given above to the final value $V_{SO} = V.R_o/(R+R_o)$ with a time constant $T_o = C(R.R_o)/(R+R_o)$. This process will be considered in detail in Section 6.5.

Now assuming that the capacitor has charged to V_{SO}, when the switch is closed the equivalent circuit can again be simplified by use of Thevenin's Theorem to that of Fig. 6.4(c). The capacitor will now discharge exponentially from V_{SO}, to V_{SC} with a time constant

$$T_c = C_1 R_e$$

where

$$\frac{1}{R_e} = \frac{1}{R} + \frac{1}{R_o} + \frac{1}{R_c}.$$

If these time constants are both relatively small compared to the switching period, then when the switch is successively opened and closed the voltage waveform obtained across the switch is as in Fig. 6.4(d). Notice the very important fact that since R_e is less than $R.R_o/(R+R_o)$, then when capacitance is present the switch can be closed faster than it can be opened.

6.2. THE JUNCTION DIODE AS A SWITCH

The switching characteristic of an ideal diode is illustrated in Fig. 6.5(a). For all negative applied voltages it is open circuit and for all positive voltages it is short circuit. The break point need not necessarily be at the origin but may be displaced as in Fig. 6.5(b). Notice that the diode is a controlled switch, i.e. the state of the switch depends on the polarity of the applied voltage.

Fig. 6.6 shows the characteristic of a typical practical germanium junction diode. In order to determine the switching action of such a diode, consider the circuit of Fig. 6.7(a). When V is positive the operating-point can be determined by superimposing a load-line corresponding to R on the diode forward characteristic. Thus the voltage drop across the forward-biased diode is not zero but V_D. For many purposes the forward characteristic may be approximated by a straight line through the origin, intersecting the characteristic at the operating point as in Fig. 6.7(b). Similarly the operating point for negative applied voltage may be determined. An

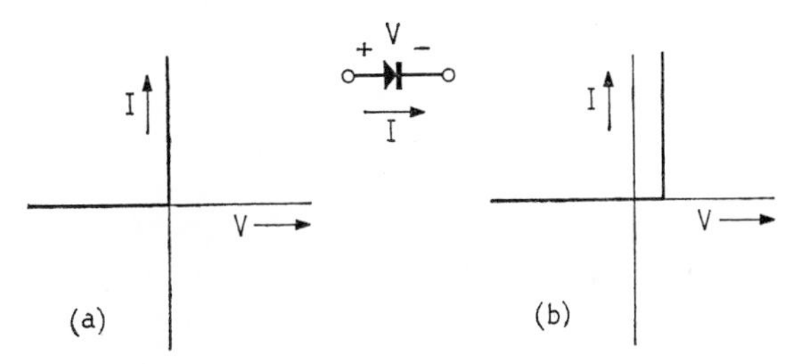

Fig. 6.5. Ideal diode characteristics. (a) Break point at zero voltage; (b) Break point at small positive voltage

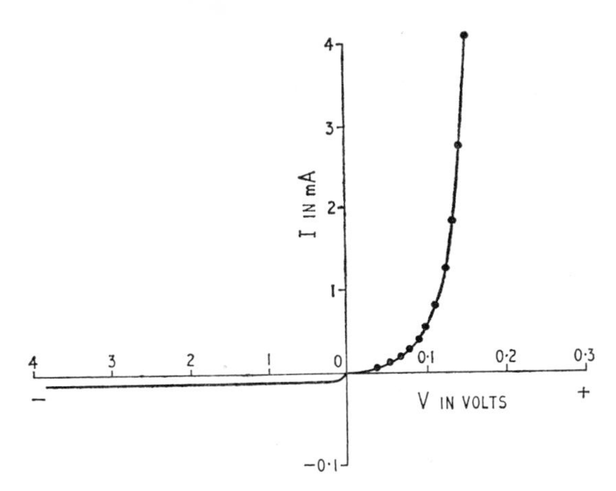

Fig. 6.6. V/I characteristic of a typical germanium junction diode. Note change of scales for forward and reverse characteristics

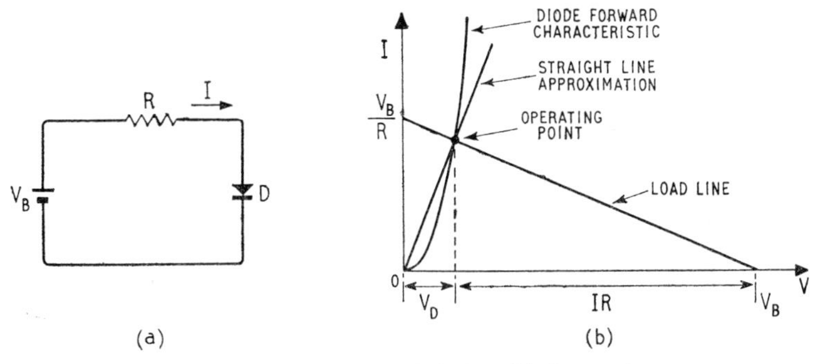

Fig. 6.7. Biasing the practical diode. (a) Circuit diagram; (b) Operating characteristic

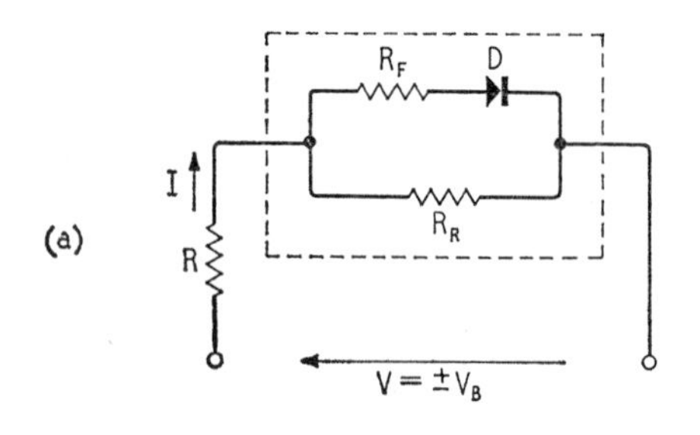

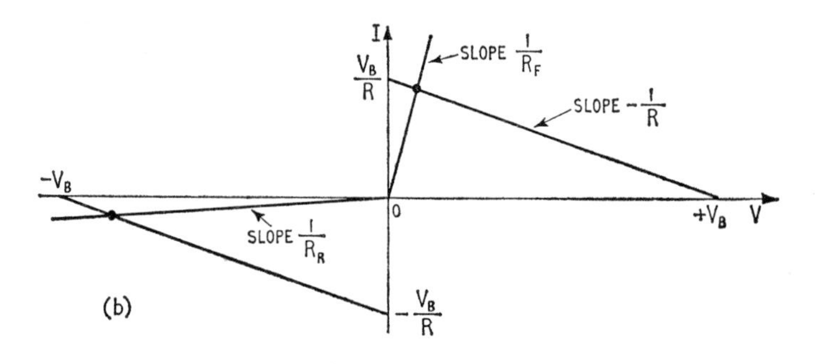

Fig. 6.8 (above). Approximating diode characteristic. (a) The approximate equivalent circuit; (b) V/I characteristics

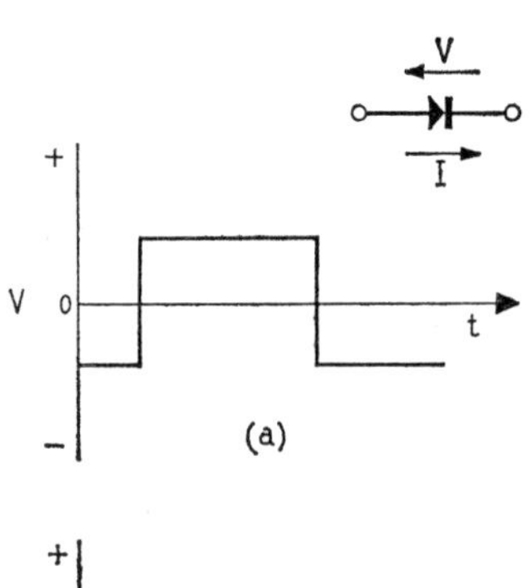

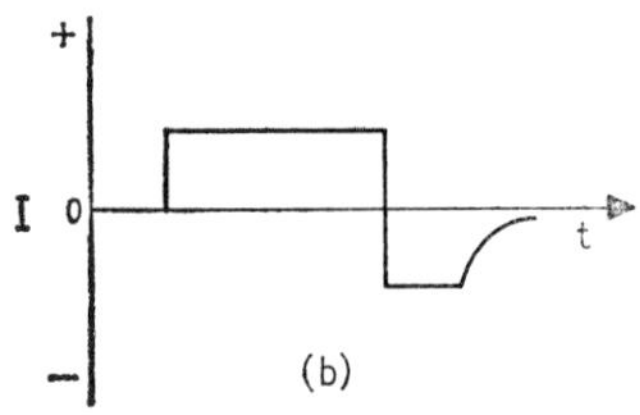

Fig. 6.9 (left). The effect of charge storage on diode current. (a) Voltage waveform applied to diode; (b) Resulting curren waveform

approximate equivalent circuit to represent the practical diode may thus be deduced as in Fig. 6.8(a), where D is an ideal diode.

The related $V-I$ characteristic is shown in Fig. 6.8(b), assuming $R_R \gg R_F$. In discussing various circuit functions in subsequent work we shall often achieve sufficient accuracy by assuming the ideal characteristic of Fig. 6.5(a).

In addition to the inertia due to shunt capacitance in the junction diode, charge storage effects also limit the speed of switching. When the diode is forward biased there is a high charge density built up at the junction compared to the reverse condition. When the applied voltage is suddenly reversed current continues to flow until the charge distribution has readjusted as indicated in Fig. 6.9. Thus for a period after reversal of the applied voltage the diode still presents a low resistance, before reverting to the high reverse resistance. This obviously sets a limit on switching speed since, if the reversals are made too fast, the diode will in effect always be forward-biased.

6.3. THE JUNCTION TRANSISTOR AS A SWITCH

Fig. 6.10(b) shows the common emitter output characteristics of a typical pnp junction transistor with a load-line superimposed, corresponding to the circuit of Fig. 6.10(a). Fig. 6.10(c) shows the low collector voltage region of Fig. 6.10(b) and Fig. 6.10(d) shows a typical input characteristic. If a small positive bias is applied to the base, $I_B = -I_{CBO}$ resulting in the usual collector leakage current I_{CBO}, point A (Fig. 6.10(b)). This is the extreme *OFF* condition of the transistor when used as a switch, corresponding to point D on the switching characteristic of Fig. 6.3(b). In this state the transistor is said to be *cut-off.* If now a negative bias is applied, base current flows, resulting in collector current I_C. The operating point traverses the load line to point B in Fig. 6.10(b) which represents the extreme *ON* condition corresponding to point C of Fig. 6.3(b). The transistor is thus acting as a separately controlled switch, the condition in the input base–emitter circuit determining the condition in the output collector–emitter circuit. If the base current is greater than the minimum value required to cause the transistor to operate at point B in Fig. 6.10(b) the transistor is said to be *saturated.* Under these conditions the collector–emitter voltage V_{CE} is very small, of the order of $-0{\cdot}1$V, Fig. 6.10(c), representing from 5 to 50Ω internal resistance for a low power transistor. Thus the transistor forms a very efficient switch. The base–emitter diode is now forward-biased and when the transistor is saturated the base–emitter voltage V_{BE} is of the order of $-0{\cdot}2$V, Fig. 6.10(d). Thus the collector is more positive than the base and emits holes into the base region in addition to those from the emitter.

6.3.1. Transistor Switching Times

Charge storage effects limit the speed of switching of the transistor even more so than was the case for the semiconductor diode. In the normal active region of the transistor the collector is negatively biased with respect to the base and holes injected into the base region from the emitter diffuse across the base region and then cross the collector–base junction. If, however, the collector–base junction potential is not sufficiently large to remove the holes as they are injected the phenomenon of *hole storage* occurs and charge builds

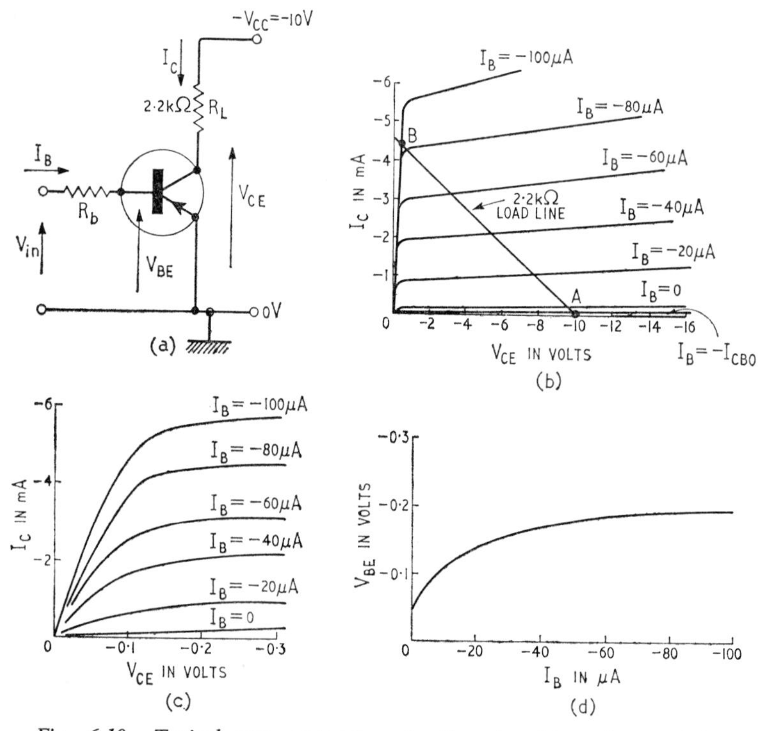

Fig. 6.10. Typical pnp germanium junction transistor characteristics. (a) Circuit diagram; (b) Common emitter output characteristics; (c) Enlarged section of (b) for low V_{CE}; (d) Common emitter input characteristic

up in the base region. In the saturated condition the collector itself emits holes into the base and so the build up of charge in the base region can be very large. When it is required to cut off the transistor, therefore, it is first necessary to remove the excess charge from the base region before collector current can begin to decrease. There is

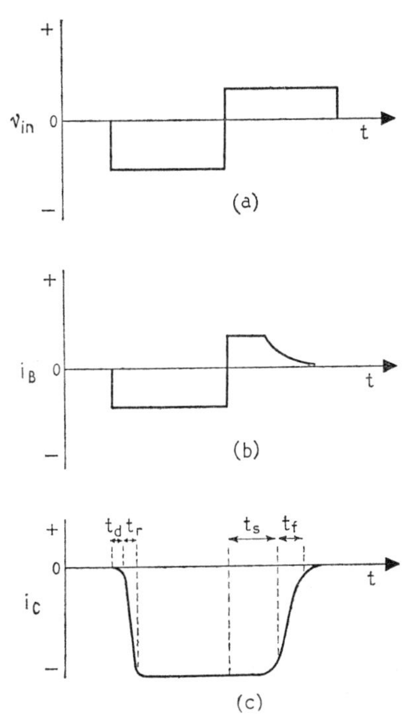

Fig. 6.11. Typical transistor pulse response. (a) Input voltage waveform; (b) Base current waveform; (c) Collector current waveform

thus a time delay in switching to the *OFF* condition after removal of the drive to the transistor. If the drive, instead of being removed, is reversed in polarity a reverse base current will flow initially, thus speeding up the removal of excess charge and reducing the *storage time*, t_s, of the transistor.

In addition to the hole-storage effect there are other inertia effects to be considered when attempting to operate the transistor at high switching-rates. Since there is a finite time associated with the diffusion of minority carriers across the base region there will be a small delay termed the *delay time*, t_d, before any appreciable rise in collector current occurs after applying forward drive to the base. Following this stage a finite time termed the *rise time*, t_r, is required for the collector current to reach the saturation value due to the combined effect of the frequency dependence of the α_B and depletion layer capacitances (see Section 3.9). For the same reasons a finite *fall time*, t_f, will follow the storage time when the drive voltage is either removed or reversed. The typical response of a transistor to a voltage pulse input will therefore be as illustrated in Fig. 6.11.

6.3.2. Methods of Reducing Switching Time

The hole-storage effect is the greatest failing of the transistor when used in very high speed switching-circuits and special precautions must be taken to minimise its effect. There are two alternative ways in which this may be done:

(1) By using a transistor with a high cut-off frequency. Normally such transistors have a narrow base-width, thus reducing the volume available for charge storage. Also charge movement is necessarily speedier.

(2) By preventing the transistor from saturating in the *ON* condition, i.e. providing just sufficient base drive to set the transistor operating-point at B on Fig. 6.10(b), thus preventing the collector from acting as an emitter. Though this gives a lead to a method of solution of the problem it is not in itself satisfactory, since spread of characteristics between samples of transistors especially with regard to α_E will invariably cause point B to be non-constant.

Consider now the circuit of Fig. 6.12. When the input voltage is positive the transistor is cut-off, the collector voltage is very nearly $-V_{CC}$ and the diode D is reverse-biased. As the input voltage becomes more negative the transistor begins to conduct and the collector voltage rises towards zero, according to the load-line. When the base current I_B reaches such a value that the collector–emitter voltage V_{CE} is equal to $(V_{BE}+I_B R_1)$, where V_{BE} is the base–emitter voltage, then diode D conducts and any further base current is bypassed through the low forward resistance of D. The collector voltage at which D conducts can be controlled by adjustment of R_1 and may be chosen to be just within the linear operating region. The only hole-storage effect is now in the forward-biased diode and, since the current it carries is only the excess drive current,

Fig. 6.12. A method of preventing the transistor from saturating

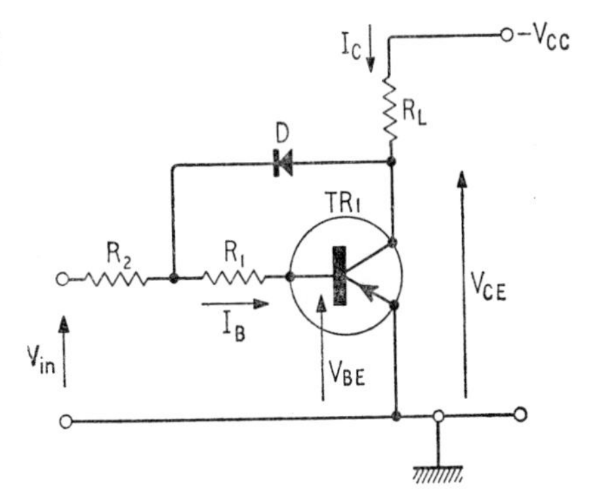

much improved performance over the saturated transistor may be obtained. There is obviously a penalty to pay in that the *ON* resistance is higher than for the saturated case, as also is the collector dissipation. The *clamping* action of the diode in Fig. 6.12 will be discussed in more detail in Section 6.4.

6.3.3. Simplified representation of the transistor switch

Referring again to the common emitter output characteristics we may now identify three distinctly separate regions as indicated in Fig. 6.13:

(1) The cut-off region where both the emitter–base and collector–base junctions are reverse-biased and which is located below the characteristic curve representing $I_B = -I_{CBO}$.

(2) The saturation region where both the emitter–base and collector–base junctions are forward-biased and which is located to the left of the line where the characteristic curves come together.

(3) The active or linear operating region where the emitter–base junction is forward-biased and the collector–base junction is reverse-biased and which is located between the cut-off and saturation regions.

For many purposes of analysis in switching-circuits we may simplify the characteristics of Fig. 6.13 to the idealised set shown in Fig. 6.14. These assume that the transistor is open circuited between collector and emitter, when cut off, and short circuited when the base current is sufficient to cause saturation. Thus an elementary equivalent circuit which accounts for the action of the idealised transistor in

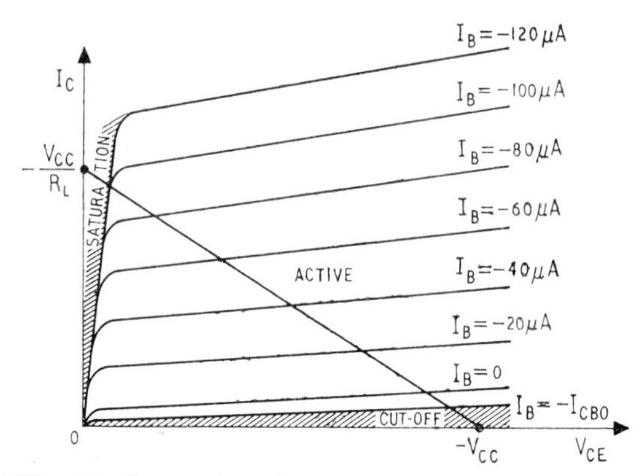

Fig. 6.13. The three regions of operation of the common emitter stage

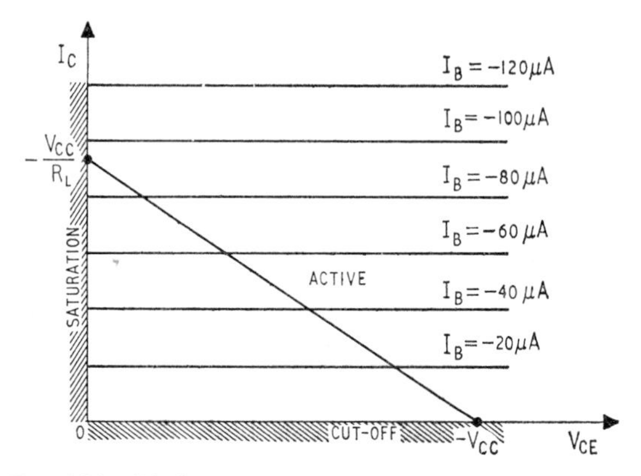

Fig. 6.14. Idealised common emitter output characteristics

the three regions is shown in Fig. 6.15, where diodes D_e and D_c are ideal and represent the emitter–base and collector–base junctions respectively.

In the cut-off region, both D_e and D_c are reverse-biased, and I_B and I_C are zero. In the active region, D_e is forward-biased, D_c is reverse-biased and $I_C = \alpha_E I_B$. In the saturation region, both D_e and D_c are forward-biased, $I_C = -V_{CC}/R_L$ and $I_B > I_C/\alpha_E$.

For true cut off of the practical transistor, positive base bias must be applied so that $I_B = -I_{CBO}$ and collector current is I_{CBO}. If now the base is open circuited the collector current will be the familiar

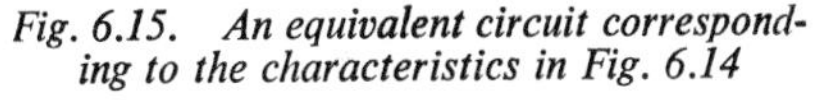

Fig. 6.15. An equivalent circuit corresponding to the characteristics in Fig. 6.14

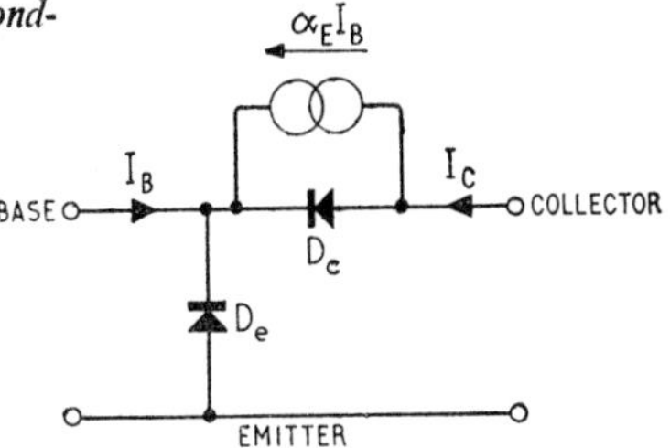

I_{CEO}. Connecting the base to the emitter directly or through a resistance results in a collector current somewhere between I_{CBO} and I_{CEO}. We shall make the approximation in many instances of assuming that for any of these conditions in the base circuit the collector circuit is open circuited. Therefore we may summarise the actual switching states for a typical junction transistor as follows:

(1) In the *OFF* condition the input and output resistances are high, consistent with low leakage currents, and the collector voltage is almost equal to the supply voltage $-V_{CC}$.

(2) In the saturated *ON* condition the input and output resistances are low

$$I_C \simeq -\frac{V_{CC}}{R_L}, \quad V_{BE} \simeq -0{\cdot}2\text{V}, \quad V_{CE} \simeq -0{\cdot}1\text{V} \quad \text{and} \quad -I_B > 100\mu\text{A}$$

The simplified representation of the two states is shown in Fig. 6.16.

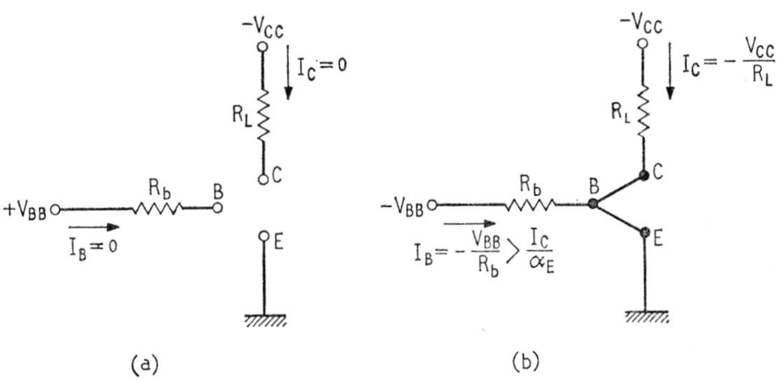

Fig. 6.16. The two states of the ideal transistor. (a) OFF; (b) ON

6.4. CLIPPING AND SQUARING CIRCUITS

6.4.1. Diode Squaring Circuits

Consider the diode rectifier circuit of Fig. 6.17(a) and assume initially that the diode has an ideal characteristic. If a sine wave of voltage is applied to the input terminals, varying from +10V to −10V, then during the positive excursion the diode is forward biased, connecting the output and input terminals together, with the result that the output voltage will be identical to the input. During the negative excursion of the input signal the diode is reverse-biased, resulting in zero current flow in *R* and hence zero output signal. Reversal of the diode connections obviously changes the output waveshape as in Fig. 6.17(b). If the circuits of Fig. 6.17(a) and (b) are rearranged as in Fig. 6.17(c) and (d) respectively, the previous results will be unaffected. These circuits constitute particular forms of waveform clipping circuits in that a complete half-cycle of the input waveform is removed. For example, the output of Fig. 6.17(b) is limited to all values less than zero volts. It often happens that it is

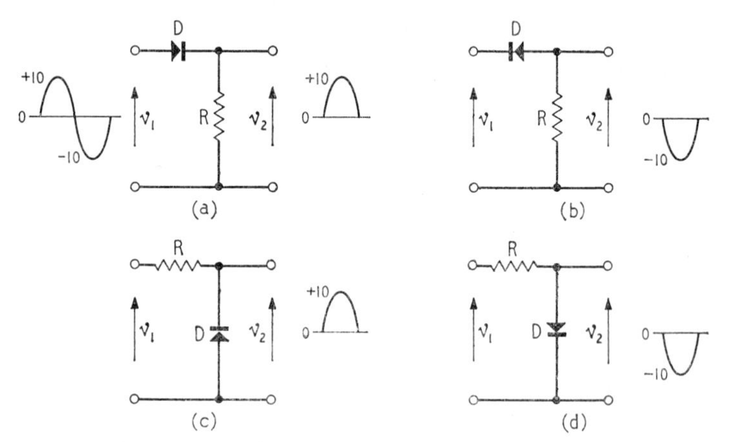

Fig. 6.17. The half-wave rectifier as a waveform clipper

required to clip part of the input waveform only, i.e. at some value other than zero. If a battery is included in the circuit of Fig. 6.17(d) with an e.m.f. $E < 10$V and polarity as shown in Fig. 6.18(a), then for all values of input voltage less than *E* the diode is reverse-biased and the output follows the input. For all input levels greater than *E* the diode conducts and the output is *clamped* to *E*. The input voltage in excess of *E* produces a voltage drop in the resistor *R*. A similar result can be obtained by inclusion of a battery in Fig. 6.17(c) as shown in Fig. 6.18(b).

Now consider the circuit of Fig. 6.19 where two diodes are connected in parallel but with opposite polarities. The excursions of the output signal are now limited to $+E$ and $-E$. If the input signal amplitude is very much greater than E then the excursions of the output signal are limited to those parts of the input signal which

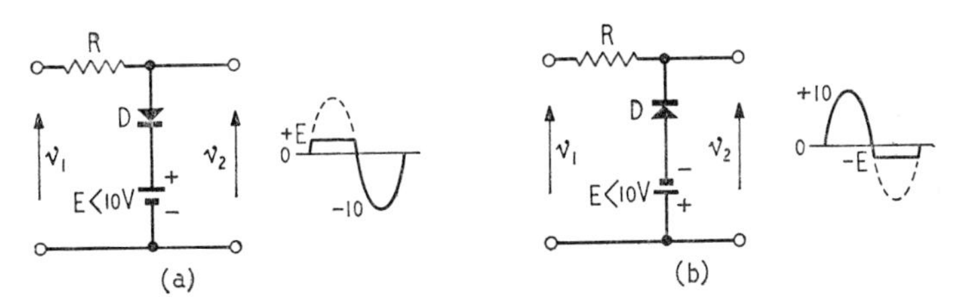

Fig. 6.18. Moving the clipping level by means of a battery

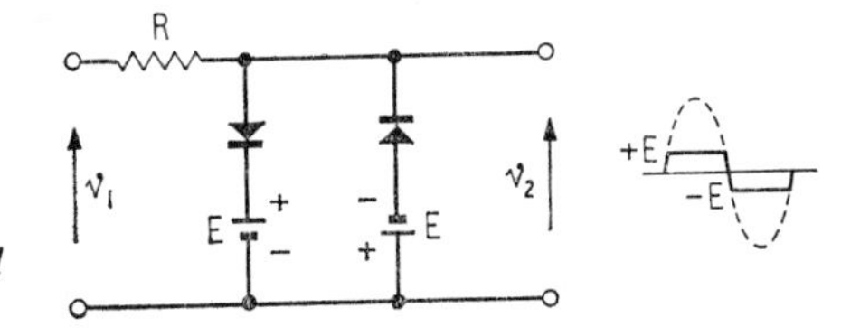

Fig. 6.19. A symmetrical clipping circuit

approximate to vertical characteristics. Therefore the larger the input signal is compared with E, the squarer will be the output waveform.

6.4.2. The Effect of Diode Voltage Drop

In the previous discussion the diodes have been assumed to be ideal. Now consider the effect of using practical diodes. In the reverse-biased state the diode was assumed to be an open circuit, but in fact, it has a finite reverse resistance R_R which will vary with diode samples. It is therefore necessary for satisfactory operation of the above circuits that $R \ll R_R$. Similarly, in the forward-biased state there will be a voltage drop across the diode so that in Fig. 6.18(a) the output will not be clamped to the battery voltage, but to some other value. To determine this value we must construct the load-line corresponding to R on the diode forward characteristic, as in Fig. 6.7(b). The intersection of the two characteristics then gives the operating-point for a given supply voltage. The actual output of the clipping-circuit of Fig. 6.18(a) will thus be as in Fig. 6.20, where it will be seen that the flat top of the waveform has been lost. To reduce this effect it is necessary to make the forward resistance $R_F \ll R$. We thus have two conflicting requirements on the

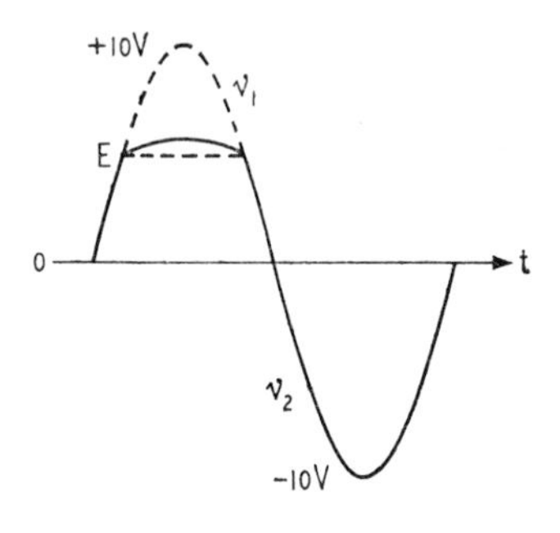

Fig. 6.20. The effect of the diode forward voltage drop in the circuit of Fig. 6.18(a)

value of R, and the magnitude of the ratio of R_R to R_F is thus an important quantity. Typical figures lie between 10^4 and 10^5.

Example 6.1

Determine the output waveform of the circuit of Fig. 6.21(a) when the input waveform is as shown and the diode is (a) ideal, (b) has infinite reverse resistance and a forward characteristic as shown in Fig. 6.6.

(a) During the positive half-cycle of the input wave the diode is reverse-biased and is open circuit, thus the output is given by

$$v_2 = +\frac{R_L}{R+R_L} \cdot v_1$$

$$= +\frac{10}{11} \cdot 10$$

$$= +9{\cdot}1\text{V}$$

When the input wave is negative the diode is forward-biased and the output is clamped to the battery voltage, i.e. -6V.
(b) The output for the positive half-cycle will be as for case (a) but for the negative half-cycle we must make use of the forward characteristic of the diode reproduced in Fig. 6.21(d) as follows. To determine the diode operating-point in this state apply Thevenin's Theorem to the circuit of Fig. 6.21(a). The steps in the procedure are illustrated in

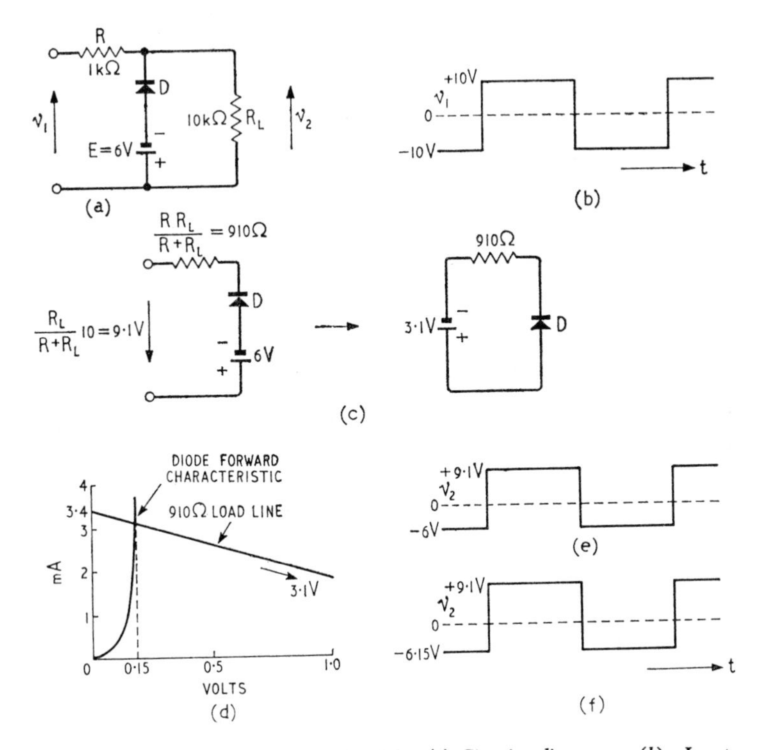

Fig. 6.21. Solution of Example 6.1. (a) Circuit diagram; (b) Input waveform; (c) Equivalent circuit for part (b); (d) Determination of diode voltage drop; (e) Output waveform part (a); (f) Output waveform part (b)

the associated diagrams. Thus the output is now clamped to the battery voltage plus the diode voltage drop, i.e. −6·15V.

The output waveform for the two cases are thus as shown in Fig. 6.21(e) and (f).

6.4.3. Transistor Squaring-circuits

We have shown a method of squaring a sine wave using diodes, but it is inefficient since a large input voltage is required compared to the squared output amplitude. Also two reference batteries are required. A more efficient circuit can be obtained using the transistor as a switch.

The application is best demonstrated by an example.

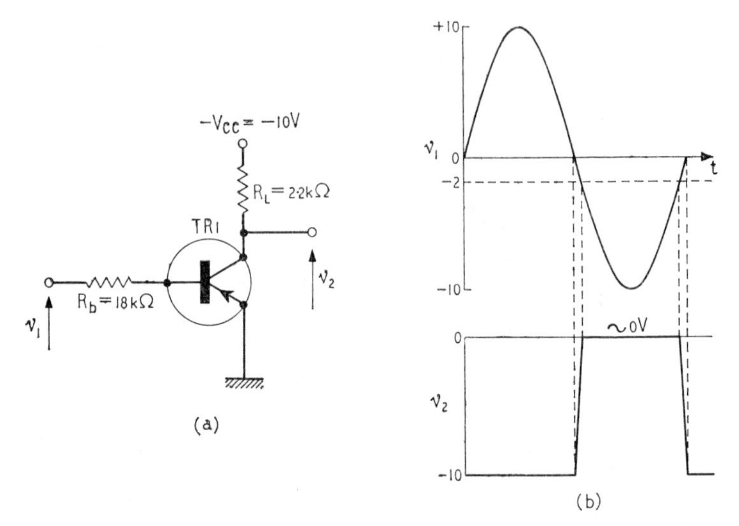

Fig. 6.22. A transistor squaring circuit. (a) Circuit diagram; (b) Input and output waveforms

Example 6.2

Determine the output waveform if a sine wave v_1 of amplitude ± 10V is applied to the circuit of Fig. 6.22(a).

For all values of v_1 more than a few millivolts positive the emitter–base junction is reverse-biased and the collector current is effectively I_{CBO}, say -10μA. The voltage drop in the load resistor R_L due to this current is

$$2{\cdot}2 \times 10^3 \times 10^{-5} = 22\text{mV},$$

thus the collector voltage is effectively equal to the supply voltage of -10V. As v_1 goes negative the emitter–base junction is forward-biased and base current flows, resulting in collector current according to the load-line construction. Thus the collector voltage will rise towards zero and the transistor will saturate with a 2·2kΩ load when the base current exceeds -100μA, from Fig. 6.10(b). The collector voltage will then be approximately $-0{\cdot}1$V and the base voltage approximately $-0{\cdot}2$V. Thus the value of v_1 when the transistor saturates is

$$-0{\cdot}2-(18 \times 10^3 \times 10^{-4}) = -2\text{V}.$$

The output voltage has therefore changed effectively from -10V to

0V as the input changed from 0V to -2V. For all values of input voltage more negative than -2V the output is 0V. The resulting output waveform is a very good approximation to a square wave as is shown in Fig. 6.22(b).

Notice that although the input wave in the previous example was symmetrical about 0V the circuit only responds to the negative half-cycle giving an asymmetric squaring action. This is often a disadvantage in circuit application, especially in amplitude limiters used in

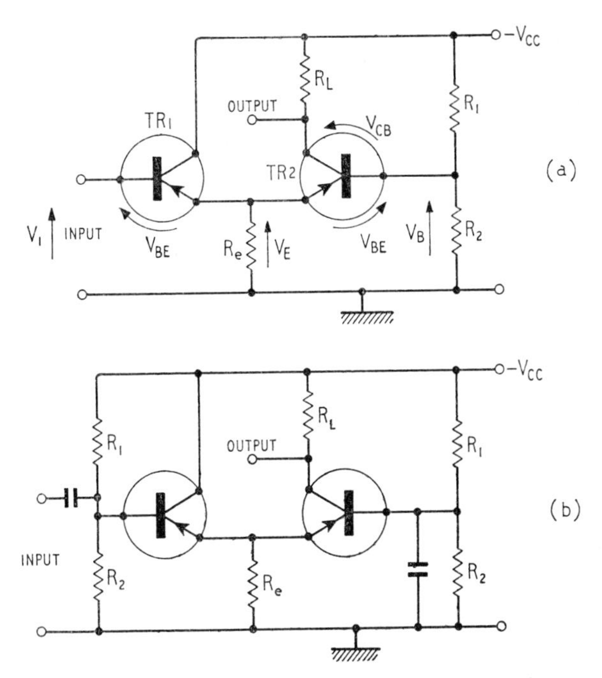

Fig. 6.23. The emitter coupled pair as a symmetrical clipper. (a) Basic circuit; (b) Circuit modified for a.c. input

communications systems (see Chapter 9). A circuit which avoids this difficulty uses an emitter-coupled pair of transistors assumed identical, as in Fig. 6.23(a). With zero input signal the transistor TR2 is biased into the active region and the output voltage is $V_B + V_{CB}$. This condition holds as the input is made more negative until it exceeds V_E when TR1 begins to conduct. Since $V_E = V_B - V_{BE}$ then for all values of V_1 less than $V_B - V_{BE}$ TR1 is cut off and TR2 is conducting. Now as V_1 is further increased V_E will also increase, thus reducing the positive emitter–base voltage on TR2 and tending to cut it off. When $V_E = V_B$, TR2 will cut off and since now $V_1 = V_E + V_{BE}$, then for all values of V_1 greater than $V_B + V_{BE}$ transistor TR2 is cut off and the output will be $-V_{CC}$. Now let us modify the circuit as in Fig. 6.23(b) and arrange that transistors TR1 and TR2 have the same standing bias. If the input is a sine-wave with peak amplitude less than $\pm V_{BE}$ the circuit will act as a linear amplifier, but for all levels above this symmetrical limiting will occur.

6.5. TRANSIENT RESPONSE OF SOME LINEAR NETWORKS

6.5.1. Methods of Circuit Analysis

As has been shown, the basic repetitive waveform associated with switching circuits is the square wave, or more generally the rectangular wave, and we shall be interested in the response of various circuits to

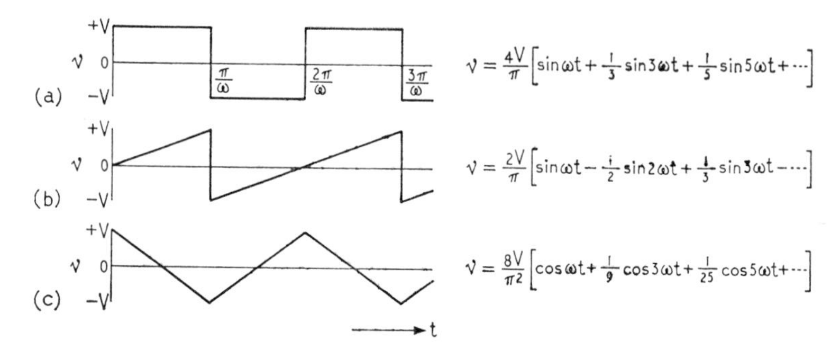

Fig. 6.24. Analysis of typical waveforms. (a) Square wave; (b) Sawtooth wave; (c) Triangular wave

rectangular wave excitation. How shall we determine this response? So far we have dealt with circuit response to sinusoidal excitation, but now we are concerned with instantaneous values of non-sinusoidal waveforms. Any periodic non-sinusoidal wave can be shown to be made up of numbers of harmonically related sinewaves. Fourier Analysis[1] is the mathematical tool used to determine the content of a given periodic waveform and a few typical examples are illustrated in Fig. 6.24. It is therefore possible to determine the response of a network to a complex wave by first finding the response to all the

[1] LAWDEN, D. F., *Mathematics of Engineering Systems*, Methuen & Co. Ltd. (1954).

component sinewaves. Then provided we are concerned with a linear network we can apply the Superposition Theorem* to obtain the overall response. This is rather a tedious process, however, and a much simpler and quite satisfactory method of approach is that of studying the waveforms. For example, if we are concerned with the response of a linear network to a rectangular voltage pulse, then we can simplify the problem by considering the pulse to be made up of 2 step functions as illustrated in Fig. 6.25. We define a step function of voltage to be a function that is zero until time $t = t_o$ rises instantaneously to some value V at t_o and is then constant at

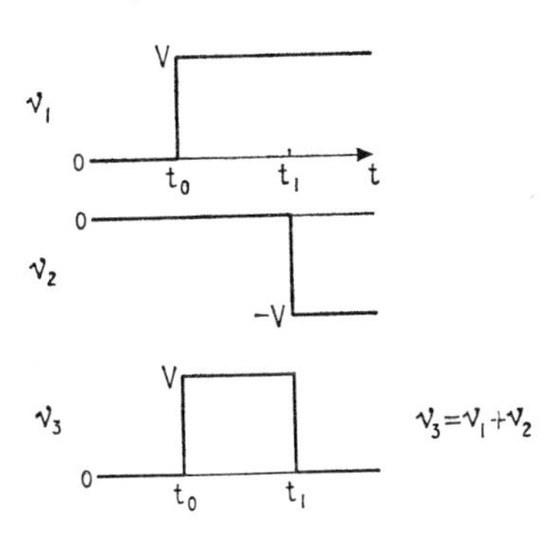

Fig. 6.25. Rectangular pulse as the sum of two step functions

that value. If we can determine the response of the network to the step function, we can again apply the Superposition Thereom to determine the response to the pulse. We thus see that it is only necessary to consider the relatively simple case of the step function. The simplest method of production of a step function is by a zero impedance battery and an ideal switch. The only mathematics therefore required in our study of switching circuits is that required to determine the transient response of a linear network to a suddenly applied d.c. voltage. We can still make use of our usual methods of circuit analysis. Kirchoff's two laws apply at any instant, whether we are dealing with linear or non-linear circuits. Thevenin's Theorem can be made use of in the case of linear circuits only. Fig. 6.26 demonstrates the application of the theorem to a typical circuit, where the capacitor is considered to be the load.

6.5.2. The CR Circuit

Consider the *CR* circuit shown in Fig. 6.27 and assume the capacitor has an initial charge Q_o. If the switch S is closed at time $t = 0$, then

* See Appendix.

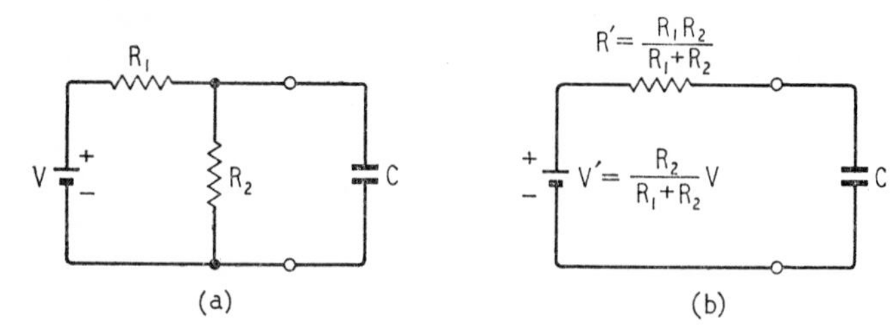

Fig. 6.26. Circuit simplification using Thevenin's Theorem. (a) Original circuit ; (b) Equivalent circuit

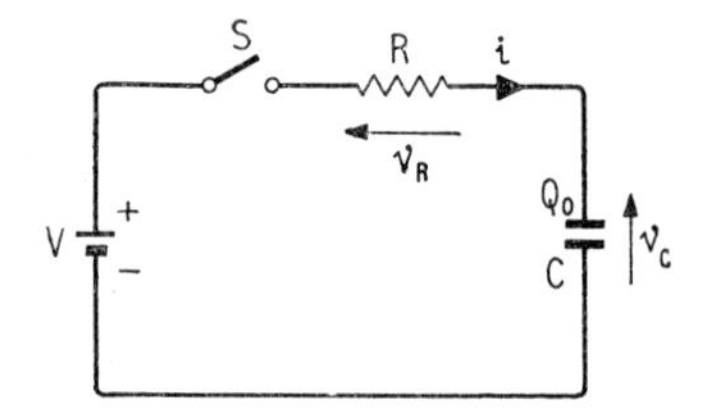

Fig. 6.27. The CR circuit

applying Kirchoff's Voltage Law, we may write at any instant:

$$v_C + iR = V \tag{6.3}$$

But since

$$i = \frac{dq}{dt} \quad \text{and} \quad q = Cv_C$$

we may rewrite Equation 6.3 as

$$v_C + CR\frac{dv_C}{dt} = V \tag{6.4}$$

Rearranging the equation and separating the variables:

$$\frac{dt}{CR} = \frac{dv_C}{V - v_C} \tag{6.5}$$

Integrating both sides of Equation 6.5 gives

$$\frac{t}{CR} = -\log_e(V - v_C) + k_1 \tag{6.6}$$

where k_1 is the integrating constant which may be evaluated from the knowledge of the initial conditions, i.e. at $t = 0, q = Q_o$ hence

$$v_C = \frac{Q_o}{C} = V_o$$

Inserting these values in Equation 6.6 gives

$$k_1 = \log_e(V - V_o)$$

Therefore

$$\frac{t}{CR} = \log_e\left[\frac{V - V_o}{V - v_C}\right]$$

i.e.

$$(V - v_C) = (V - V_o)\,e^{-t/CR}$$

Hence

$$v_C = V(1 - e^{-t/CR}) + V_o\,e^{-t/CR} \tag{6.7}$$

Now

$$i = C\frac{dv_C}{dt}$$

$$= CV\left(0 + \frac{1}{CR}.e^{-t/CR}\right) - \frac{CV_o}{CR}.e^{-t/CR}$$

Therefore

$$i = \frac{V - V_o}{R}.e^{-t/CR} \tag{6.8}$$

Thus the voltage at any instant across the resistor, v_R, is given by

$$v_R = i.R$$

$$= (V - V_o)\,e^{-t/CR} \tag{6.9}$$

The characteristics represented by Equations 6.7 and 6.9 are sketched in Fig. 6.28.

Now consider the case when $V = 0$, i.e. a capacitor C with a charge Q_o discharging through a resistor R. We may obtain directly from the previous results

$$v_C = V_o\,e^{-t/CR} \tag{6.10}$$

$$v_R = -V_o\,e^{-t/CR} \tag{6.11}$$

These characteristics are sketched in Fig. 6.29.

6.5.3. The Circuit Time Constant

Notice that the curves just plotted are exponential in shape and theoretically never reach the final values. Since networks which exhibit this type of response are used extensively in switching circuits we need some criterion of performance, and for this purpose we define a circuit *time constant*. The time taken for the voltage v_C in

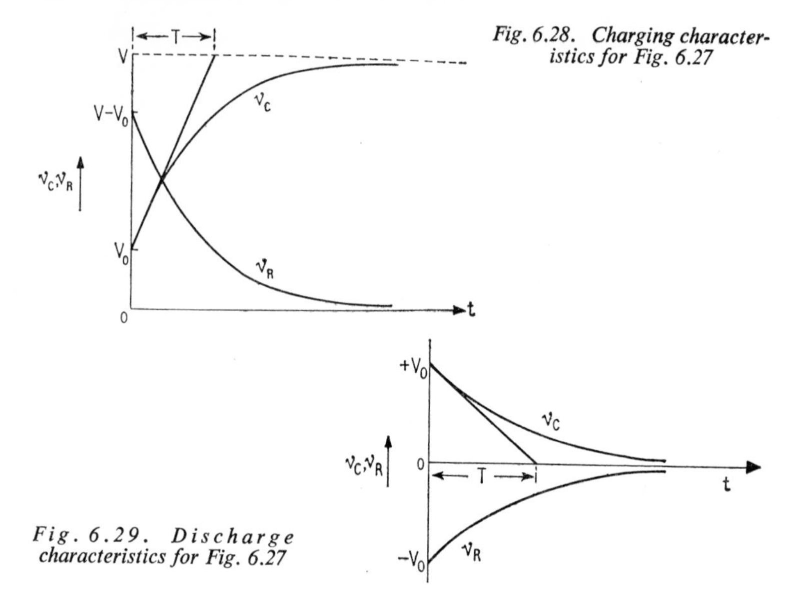

Fig. 6.28. Charging characteristics for Fig. 6.27

Fig. 6.29. Discharge characteristics for Fig. 6.27

Fig. 6.28 to attain its final value, if it continued to increase at the initial rate, is defined as the time constant, T sec., of the circuit.

Considering Fig. 6.28 the initial rate of change of voltage is given by the slope of the tangent to the curve at $t = 0$.

$$\text{Therefore the initial rate of change} = \frac{V-V_o}{T} \tag{6.12}$$

Also from Equation 6.4 at $t = 0$, when $v_C = V_o$ we have

$$\frac{dv_C}{dt} = \frac{V-V_o}{CR} \tag{6.13}$$

Equating Equations 6.12 and 6.13 we obtain

$$T = CR \text{ sec} \tag{6.14}$$

Thus we may rewrite Equations 6.7 and 6.9

$$v_C = V(1-e^{-t/T})+V_o e^{-t/T} \tag{6.15}$$

$$v_R = (V-V_o)e^{-t/T} \tag{6.16}$$

Similar results may be obtained for the discharge circuit. Hence

$$v_C = V_o e^{-t/T} \tag{6.17}$$

$$v_R = -V_o e^{-t/T} \tag{6.18}$$

The time constant, T, is most important in considering the response of circuits such as Fig. 6.27 to rectangular wave excitation. A table of the factors $e^{-t/T}$ and $1-e^{-t/T}$ against time in units of time constant is shown in Table 6.1 and the results are plotted in Fig. 6.30. Notice that after a time equal to 4 time constants, steady state conditions have for all practical purposes been attained.

Table 6.1

t *Units of T*	$e^{-t/T}$	$1-e^{-t/T}$
0	1	0
0·2	0·819	0·181
0·4	0·670	0·330
0·6	0·549	0·451
1·0	0·368	0·632
1·5	0·223	0·777
2·0	0·135	0·865
2·5	0·082	0·918
3·0	0·050	0·950
4·0	0·018	0·982
5·0	0·007	0·993

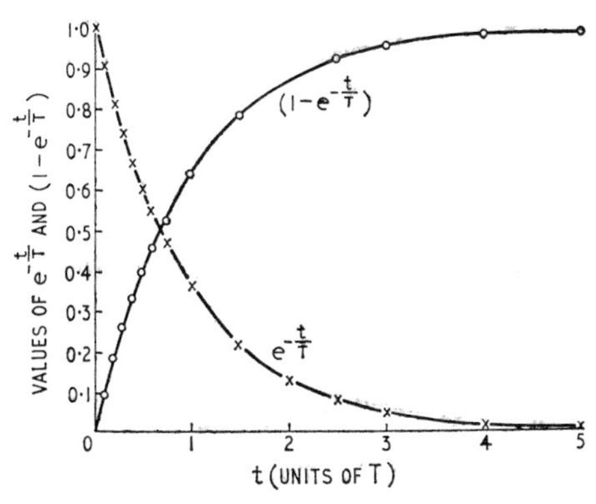

Fig. 6.30. Values of $e^{-t/T}$ *and* $1-e^{-t/T}$

Several very important conclusions may be drawn from the foregoing analysis, which will prove of great use in building up a pictorial representation of the circuit response.

(1) The voltage on a capacitor cannot change instantaneously but varies as an exponential curve with time constant CR sec.

(2) At the instant of change in the conditions in the CR circuit the capacitor acts as a battery and any initial voltage across it acts in series with the circuit voltage.

(3) Any instantaneous change in circuit voltage will appear unchanged on the other side of the capacitor.

(4) In the steady state there is no current flow and hence no d.c. path through the capacitor.

Example 6.3

In the circuit of Fig. 6.31(a) the switch has been on position 1 for a time sufficient for steady state conditions to be established. The switch is now instantaneously changed to position 2. Determine the time which elapses before the output voltage v_o is again zero.

The circuit may be redrawn for the initial condition as in Fig. 6.31(b), which may be simplified by application of Thevenin's Theorem as shown. Thus the initial voltage on the capacitor is 10V, plate a positive, and the output is zero. At time $t = t_1$ the switch is changed to position 2, thus earthing plate a. Since the voltage on the capacitor cannot change instantaneously the output voltage thus drops to -10V. The equivalent circuit representing the new condition is shown in Fig. 6.31(d) and it is seen that the capacitor voltage will change from 10V with plate a positive, to 10V with plate b positive with time constant CR,
where

$$R = R_3 + \frac{R_1 R_2}{R_1 + R_2} = 18\text{k}\Omega$$

We are interested in the instant when the capacitor voltage is zero, at $t = t_2$.

From Equation 6.7

$$v_o = V(1 - e^{-(t-t_1)/CR}) + V_o e^{-(t-t_1)/CR}$$

In this example, when $t = t_2$, $V = 10$V, $V_o = -10$V, and $v_o = 0$

Hence $\quad 0 = 10 - 20\,e^{-(t_2-t_1)/CR}$

i.e. $\quad t_2 - t_1 = CR \log_e 2$

$$= 0{\cdot}08 \times 10^{-6} \times 18 \times 10^3 \times 0{\cdot}693 \text{ sec}$$

$$= 1 \text{ m sec.}$$

The waveform of v_o will be as in Fig. 6.31(f).

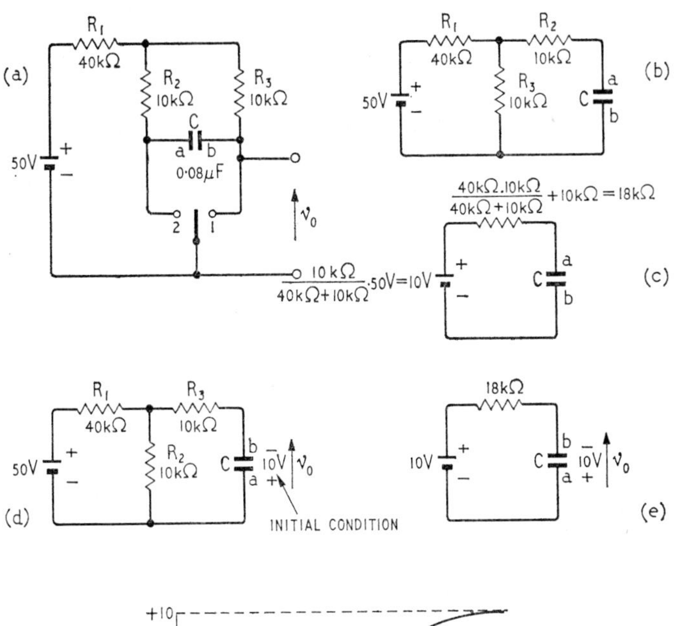

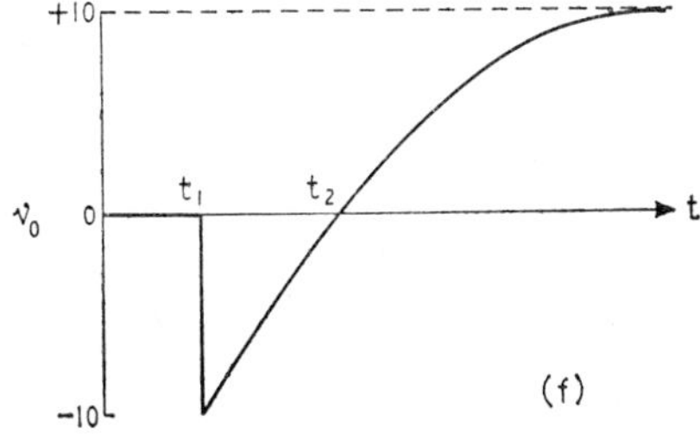

Fig. 6.31. Relevant to Example 6.3. (a) The circuit diagram; (b) Initial condition; (c) Thevenin equivalent of (b); (d) Equivalent circuit after switching; (e) Thevenin equivalent of (d); (f) The resultant waveform

6.5.4. The LR Circuit

Consider next the LR circuit shown in Fig. 6.32 and assume the switch is closed at time $t = 0$.
We can write the equation

$$iR_1 + L\frac{\mathrm{d}i}{\mathrm{d}t} = V \tag{6.19}$$

Rearranging and separating the variables

$$\frac{R_1}{L}.\mathrm{d}t = \frac{\mathrm{d}i}{\dfrac{V}{R_1} - i} \tag{6.20}$$

Integrating

$$\frac{R_1}{L}.t = -\log_e\left(\frac{V}{R_1} - i\right) + k_2 \tag{6.21}$$

where k_2 is the integration constant.

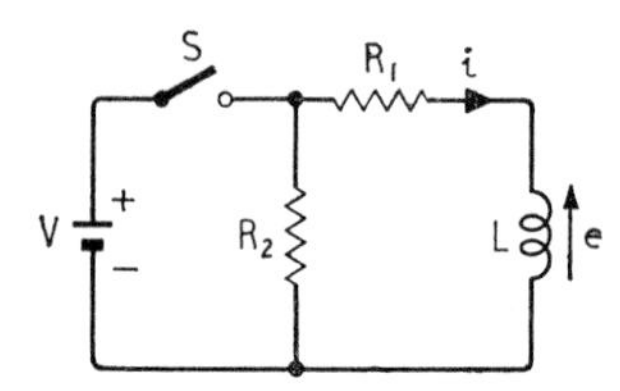

Fig. 6.32. The LR circuit

Since i is zero at $t = 0$ then from Equation 6.21

$$k_2 = \log_e \frac{V}{R_1}$$

Substituting for k_2 in Equation 6.21 and rearranging we have

$$i = \frac{V}{R_1}(1 - \mathrm{e}^{-R_1t/L}) \tag{6.22}$$

The back e.m.f. e induced in the inductor is given by

$$e = -L\frac{\mathrm{d}i}{\mathrm{d}t}$$

$$= -\frac{VL}{R_1}\left(0 + \frac{R_1}{L}\mathrm{e}^{-R_1t/L}\right)$$

Therefore

$$e = -V\mathrm{e}^{-R_1t/L} \tag{6.23}$$

Now assume that the current has reached the steady state value of V/R_1 and that the switch is opened. As the current begins to decay the back e.m.f. induced in the inductive circuit acts to maintain it and drives it round the circuit completed by R_2. Thus the application of Kirchoff's Voltage Law to this loop yields:

$$-L\frac{di}{dt} = (R_1+R_2)\,i \qquad (6.24)$$

i.e.

$$\frac{(R_1+R_2)}{L}dt = -\frac{di}{i}$$

Hence

$$\frac{(R_1+R_2)}{L}.t = -\log_e i + k_3 \qquad (6.25)$$

In this case $t = 0$ gives $i = V/R_1$.

Therefore we find

$$i = \frac{V}{R_1}e^{-(R_1+R_2)t/L} \qquad (6.26)$$

$$e = -L\frac{di}{dt}$$

$$= \frac{R_1+R_2}{R_1}.Ve^{-(R_1+R_2)t/L} \qquad (6.27)$$

By the same reasoning as for the *CR* circuit we may deduce the time constant of the *LR* circuit to be

$$T = \frac{L}{R}\,\text{sec} \qquad (6.28)$$

where R is the total circuit resistance.

The variation of current in an inductive circuit is thus seen to be completely analogous to the variation of voltage in the capacitive circuit and we may draw the following important conclusions:

(1) The current in an inductive circuit cannot change instantaneously but varies exponentially with time constant L/R sec.

(2) The back e.m.f. induced by the changing current always acts to oppose the change.

(3) At the instant of applying a voltage step to the inductive circuit, the induced e.m.f. is exactly equal and opposed to the applied voltage, hence no current flows and the inductor acts as an open circuit.

6.5.5. The LCR Circuit

Conditions sometimes arise in which L, C and R all occur together in a switching circuit. We shall therefore briefly consider the effect of applying a d.c. voltage to a combination of these elements as in Fig. 6.33.

At any instant we have

$$Ri+L\frac{di}{dt}+\frac{q}{C} = V \tag{6.29}$$

But since

$$i = \frac{dq}{dt} = C.\frac{dv_C}{dt}$$

we have

$$CR\frac{dv_C}{dt}+LC\frac{d^2 v_C}{dt^2}+v_C = V$$

i.e.

$$\frac{d^2 v_C}{dt^2}+\frac{R}{L}.\frac{dv_C}{dt}+\frac{1}{LC}.v_C = V \tag{6.30}$$

This is a linear second order differential equation, the solution of which may be obtained by reference to a standard mathematical

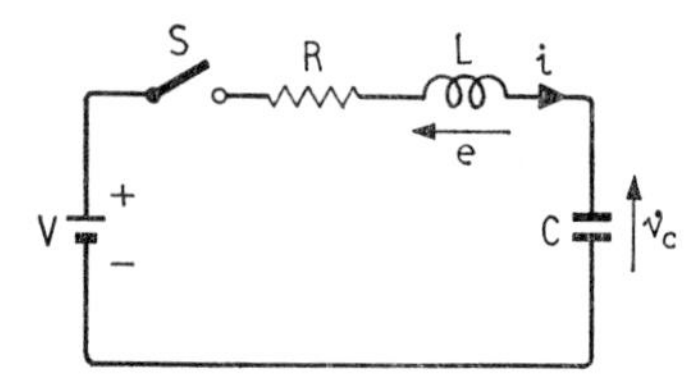

Fig. 6.33. The LCR circuit

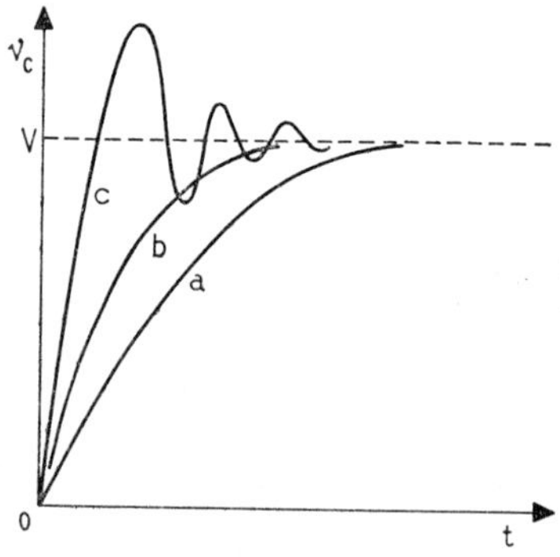

Fig. 6.34. Sketch of the response of Fig. 6.33. (a) Overdamped; (b) Critically damped; (c) Underdamped

textbook. It is the physical appreciation of the response with which we are primarily concerned at this stage.

The response of the network depends upon the relative magnitudes of the coefficients of dv_C/dt and v_C in Equation 6.30. The variation in this damping effect is illustrated in Fig. 6.34 for the following conditions

$\frac{R}{L} > \frac{2}{\sqrt{(LC)}}$ corresponds to an over-damped response

$$\frac{R}{L} = \frac{2}{\sqrt{(LC)}}$$ corresponds to a critically damped response

$$\frac{R}{L} < \frac{2}{\sqrt{(LC)}}$$ corresponds to an under-damped response.

Notice that as the damping is reduced the time constant of response is reduced until finally overshoot occurs, i.e. the capacitor voltage exceeds the steady state value and an oscillatory condition occurs. The rate at which this oscillation dies down is again dependent upon the amount of damping.

6.6. RECTANGULAR WAVE RESPONSE OF THE *CR* CIRCUIT

6.6.1. The Differentiating Circuit

Consider the circuit in Fig. 6.35(a) and assume it to be energised by a voltage step function. The resulting output wave will then be as shown in Fig. 6.35(b). If the step function is inverted, so also will be the output wave. We may therefore deduce the circuit response to a rectangular pulse by applying the Superposition Theorem as in Fig. 6.35(d). The decay or droop of the pulse top depends upon the relationship between the circuit time constant and the pulse duration. The result of transmitting a recurrent square wave through the circuit of Fig. 6.35(a) is shown in Fig. 6.36 for two conditions:

(a) Half period of the wave = circuit time constant

(b) Half period of the wave = $10 \times$ circuit time constant.

Notice that although the input varies between zero and $+V_0$ the output varies symmetrically about zero. This must always be so since we have already seen that the d.c. component is blocked by the capacitor. Thus the average value of the output voltage is zero, and the areas enclosed by the curve above and below the zero axis are equal.

The second version of the circuit of Fig. 6.35(a) (with the time constant very short compared to the wave period) has produced a train of narrow pulses each of amplitude V_0 and alternately positive and negative in sign. Such a circuit has great application for the production of trigger pulses, as we shall see.

In general, in cases similar to the first version of Fig. 6.35(a) it is not so simple to determine the excursions of the output waveform. The procedure is as follows.

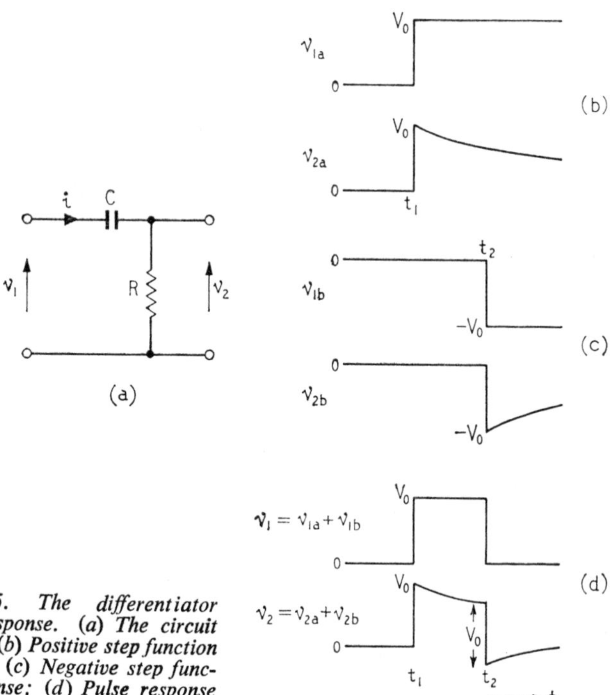

Fig. 6.35. The differentiator circuit response. (a) The circuit diagram; (b) Positive step function response; (c) Negative step function response; (d) Pulse response

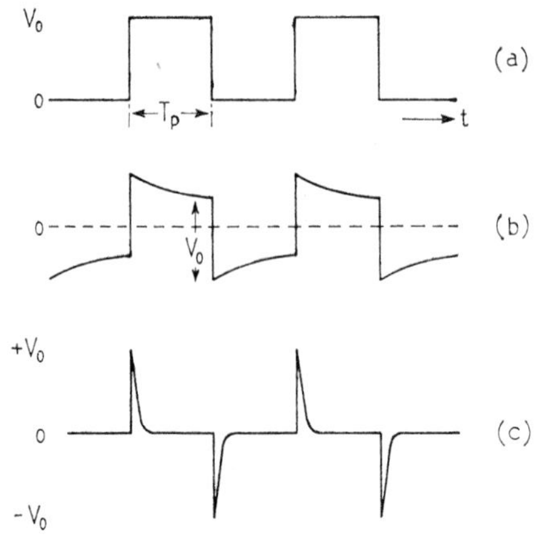

Fig. 6.36. The effect on a square wave of transmission through the differentiator circuit of Fig. 6.35. (a) Input waveform; (b) Output waveform when $T_p = CR$; (c) Output waveform when $T_p = 10\ CR$

Assume that steady state conditions hold and that the peak output voltage at point A is V_1 in Fig. 6.37(b). Then at point B

$$V_2 = V_1 . e^{-T_p/CR}$$

At this point the input drops from V_0 to zero. Therefore since the voltage on a capacitor cannot change instantaneously the output

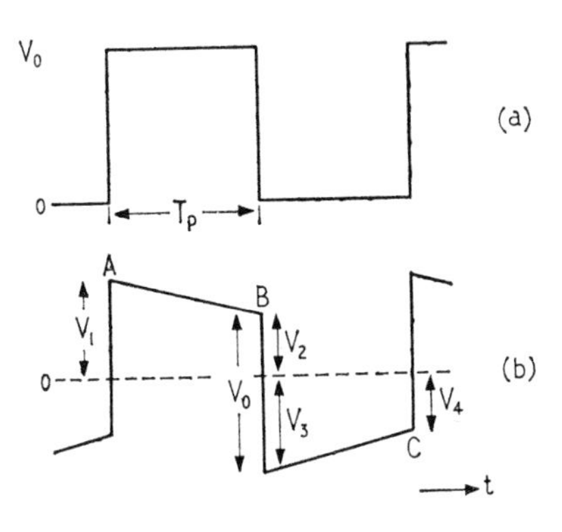

Fig. 6.37. Determination of output waveform when $CR \gg T_p$. (a) Input waveform; (b) Output waveform

voltage must also drop by V_0 to $V_3 = -V_1$, since the wave is symmetrical. Hence at point C, $V_4 = -V_2$. Therefore we have:

$$\left.\begin{aligned} V_1 + V_2 &= V_o \\ V_2 &= V_1 e^{-T_p/CR} \end{aligned}\right\} \quad (6.31)$$

Solving these two equations gives

$$V_1 = \frac{V_o}{1+e^{-T_p/CR}}; \quad V_2 = \frac{V_o}{1+e^{T_p/CR}} \quad (6.32)$$

Now $e^{-x} = 1 - x + \frac{1}{2}x^2 - \ldots$

Therefore if the time constant of the circuit is long compared to the pulse period, i.e. $T_p/CR \ll 1$

$$e^{-T_p/CR} \simeq 1 - \frac{T_p}{CR}$$

$$e^{+T_p/CR} \simeq 1 + \frac{T_p}{CR}$$

Further, by the Binomial Theorem, if x is small

$$\frac{1}{1 \pm x} \simeq 1 \mp x$$

Therefore

$$\left.\begin{aligned} V_1 &= \frac{V_o}{2}\left(1+\frac{T_p}{2CR}\right) \\ V_2 &= \frac{V_o}{2}\left(1-\frac{T_p}{2CR}\right) \end{aligned}\right\} \tag{6.33}$$

The droop in this case is thus linear with time and we may quote it as a percentage of $V_o/2$ as

$$\% \text{ droop} = \frac{V_1 - V_2}{\frac{V_o}{2}} \times 100$$

$$= \frac{100T_p}{CR} \tag{6.34}$$

Example 6.4

The input to the circuit of Fig. 6.38(a) is the waveform of Fig. 6.38(b). Determine the steady state output.

Since $R = 100\text{k}\Omega$, $C = 0{\cdot}01\mu\text{F}$, the circuit time constant is 1msec. The period between pulses is 100μsec and so the shape of the

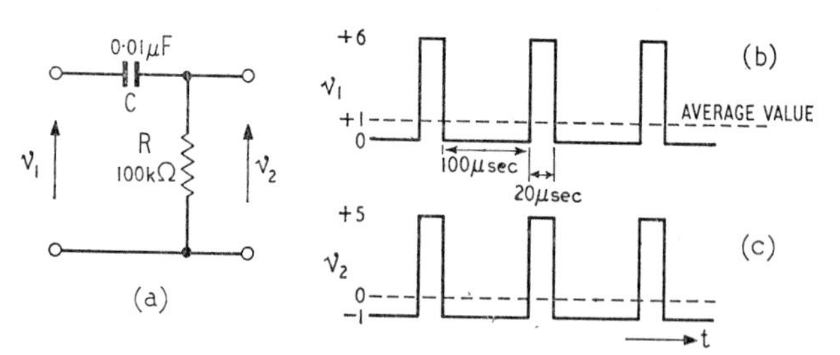

Fig. 6.38. Relevant to Example 6.4. (a) Circuit diagram; (b) Input waveform; (c) Output waveform

waveform will be practically unaltered. The average value of the input wave is

$$\frac{6 \times 20}{120} = 1\text{V}$$

The average value of the output wave must be zero, therefore the wave varies between -1V and $+5$V as shown in Fig. 6.38(c).

Consider again the *CR* circuit as illustrated in Fig. 6.35(a). We may write

$$\left.\begin{aligned} v_1 &= iR + \frac{1}{C}\int i.\mathrm{d}t \\ v_2 &= iR \end{aligned}\right\} \qquad (6.35)$$

Now if the time constant of the circuit is made very short compared to the period of the input waveform the voltage drop across *R* will be very small compared to that across *C*. Hence we have approximately:

$$v_1 = \frac{1}{C}\int i.\mathrm{d}t$$

or

$$i = C.\frac{\mathrm{d}v_1}{\mathrm{d}t}$$

Therefore

$$v_2 = CR.\frac{\mathrm{d}v_1}{\mathrm{d}t} \qquad (6.36)$$

The circuit thus acts as a differentiator, as is demonstrated by Fig. 6.36(c), i.e. the differential is non-zero only at the discontinuities of the input square wave.

6.6.2. The Integrating Circuit

Now consider the circuit of Fig. 6.39(a). Following the same approach as before we may deduce the response of the circuit to a rectangular pulse input as in Fig. 6.39(d). Again the relationship between pulse duration and time constant is most important and in order that the rise time of the pulse is not unduly affected the time constant must be very short compared to the period. Fig. 6.40 illustrates the effect of varying time constant on the transmission of a square wave through the network. For the circuit of Fig. 6.39(a) we have

$$\left.\begin{aligned} v_1 &= iR + \frac{1}{C}\int i.\mathrm{d}_t \\ v_2 &= \frac{1}{C}\int i.\mathrm{d}t \end{aligned}\right\} \qquad (6.37)$$

Now we make the time constant very long compared to the period of the input wave, resulting in a very small voltage across the capacitor.

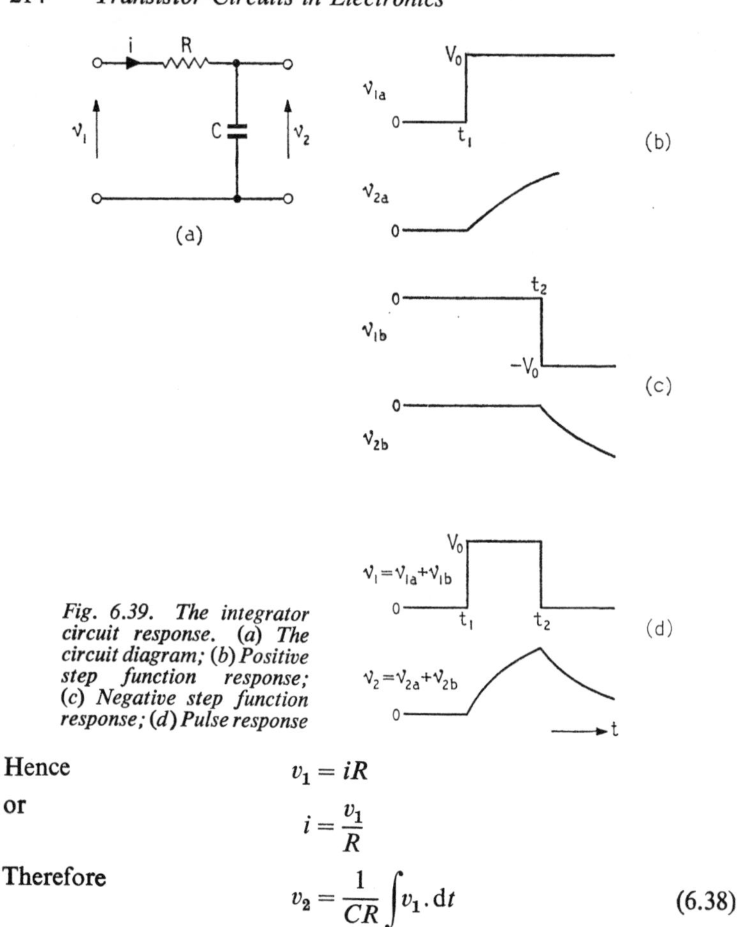

Fig. 6.39. The integrator circuit response. (a) The circuit diagram; (b) Positive step function response; (c) Negative step function response; (d) Pulse response

Hence
$$v_1 = iR$$
or
$$i = \frac{v_1}{R}$$
Therefore
$$v_2 = \frac{1}{CR}\int v_1 . \mathrm{d}t \tag{6.38}$$

Thus this circuit acts as an integrator as is verified by Fig. 6.40(c).

6.7. D.C. RESTORATION

We have seen that transmission of a waveform through the differentiator circuit results in the loss of the d.c. component. This is often a serious drawback in switching circuits since the 2 states of the switch usually correspond to particular voltage levels. For example if a square wave varying between 0V and -10V is passed through such a circuit and the time constant is very long compared to the wave period, then the output varies between $+5$V and -5V and all reference to the input is lost. If it is required to transmit the original reference levels it is necessary therefore to reintroduce the d.c. component. The clue to a possible arrangement is apparent from a study of the distribution of the potentials in the circuit of Fig. 6.41. The capacitor builds up a charge corresponding to the mean d.c. level of the input waveform and thus

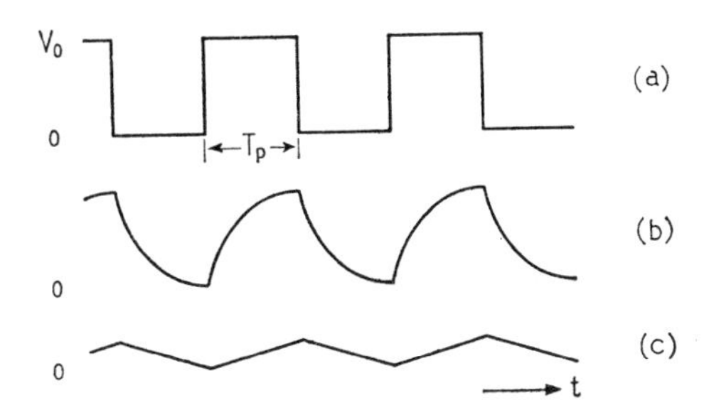

Fig. 6.40. The effect on a square wave of transmission through the integrator circuit of Fig. 6.39(a). (a) Input waveform; (b) Output waveform when $T_p = 5CR$; (c) Output waveform when $T_p = \frac{1}{10}CR$

if we can prevent this happening the output waveform will ideally follow the input.

Consider now the circuit of Fig. 6.42 assumed to be energised by a zero impedance square wave source. We can redraw the circuit as in Fig. 6.43(a) where the source is replaced by a switch alternating between a battery of 10V e.m.f., Position 1, and earth, Position 2. Let the switch be connected to Position 1, when the output will be initially -10V and will decay slightly as the capacitor begins to charge through R with plate a negative. The diode is reverse-biased and therefore has little effect. When the switch is moved to Position 2,

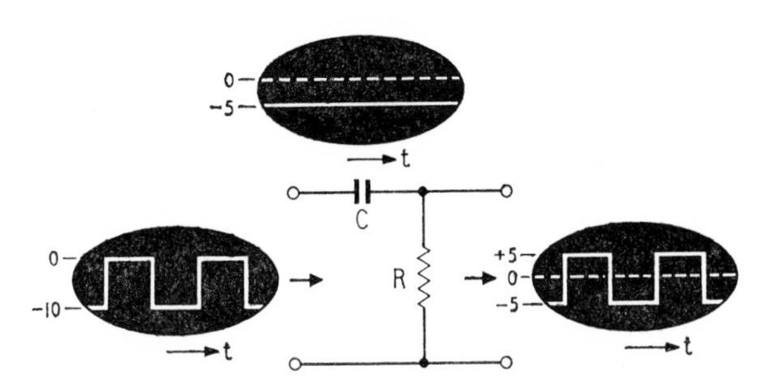

Fig. 6.41. D.c. isolation in the differentiator circuit

plate *a* of the capacitor is earthed and the output voltage will initially be the voltage on the capacitor. However, the polarity of the capacitor voltage is such as to forward-bias the diode and the capacitor quickly discharges through this forward resistance, which should be very low compared to the resistance *R*. For the major part of the time that the input is zero, so also is the output. The

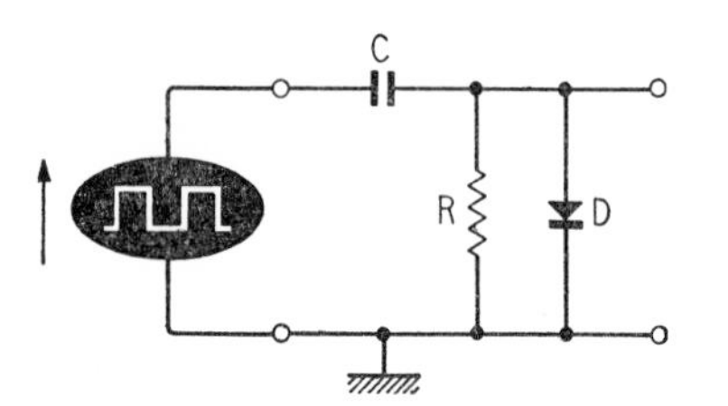

Fig. 6.42. The basic d.c. restorer circuit

cycle repeats as the input varies between 0V and −10V and thus the output varies between the same limits, as demonstrated in Fig. 6.43(b).

The series resistance of the source has been neglected. It has little effect when the diode is not conducting but its resistance must be added to that of the diode when the diode is conducting.

If the diode connections are reversed as in Fig. 6.44(a) then when the switch is connected to the −10V battery, the capacitor rapidly charges to −10V through the forward resistance of the diode, with plate *a* negative. The output is thus zero except for the short time while the capacitor is charging. When the switch is moved to Position 2, plate *a* is earthed and the capacitor voltage appears at the output. The diode is now reverse-biased and will have little effect on the circuit. Thus the output is initially +10V and decays slightly via *R*. The output waveform now varies between 0V and +10V as in Fig. 6.44(b) and the lower edge of the input waveform

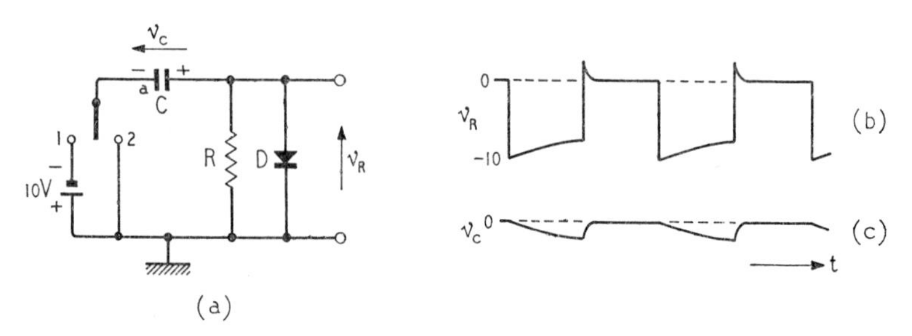

Fig. 6.43. Illustrating d.c. restoration. (a) Equivalent circuit; (b) Output voltage; (c) Capacitor voltage

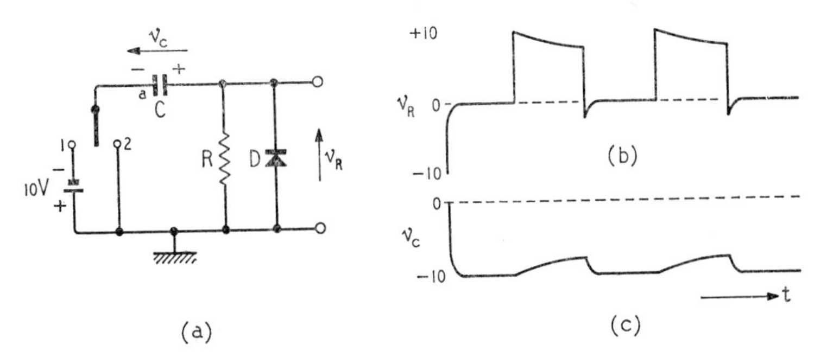

Fig. 6.44. *Effect of reversing the diode.* (*a*) *Equivalent circuit;* (*b*) *Output Voltage;* (*c*) *Capacitor voltage*

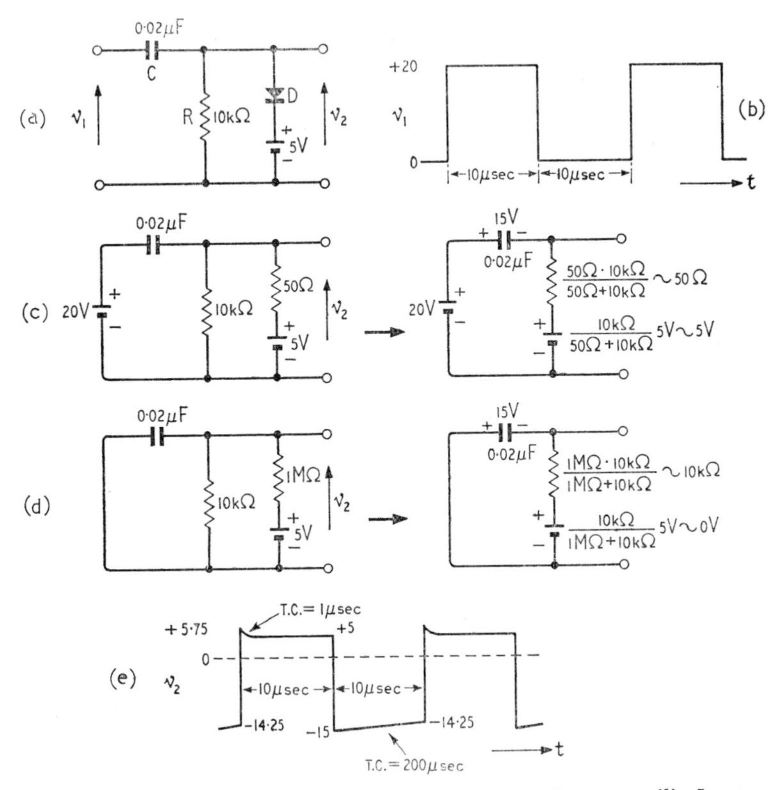

Fig. 6.45. *Relevant to Example 6.5.* (*a*) *Circuit diagram;* (*b*) *Input waveform;* (*c*) *Equivalent circuits for forward-biased diode;* (*d*) *Equivalent circuits for reverse-biased diode;* (*e*) *Output waveform*

has been clamped to 0V, instead of the upper edge as in the previous case. It is possible by inclusion of a battery in series with the diode to obtain clamping levels other than zero.

Example 6.5

The input to the d.c. restorer circuit of Fig. 6.45(a) is the waveform of Fig. 6.45(b). If the diode is assumed to have a reverse resistance of 1MΩ and a forward resistance of 50Ω determine the output waveform, showing all voltage levels and time constants.

At $t = 0$ the equivalent circuit is as in Fig. 6.45(c). Thus the capacitor charges to 15V with polarity as shown and time constant

$$0{\cdot}02 \times 10^{-6} \times 50 = 1\mu\text{sec}$$

For the major part of the first half-cycle the output is therefore clamped to +5V. When v_1 drops to zero, assuming the diode changes state instantaneously, the equivalent circuit is modified as in Fig. 6.45(d). Since the voltage on the capacitor cannot change instantaneously, the output must drop to −15V.

The capacitor voltage will then readjust according to Equation 6.10. Therefore after a further 10μsec, since the time constant is now

$$0{\cdot}02 \times 10^{-6} \times 10 \times 10^{3} = 200\mu\text{sec}$$

and

$$V_o = -15\text{V}$$

Therefore

$$V_c = -15e^{-10/200}$$
$$= -15 \times 0{\cdot}951$$
$$= -14{\cdot}25\text{V}$$

Thus at the instant that the input changes to +20V the output has decayed to −14·25V, and must instantaneously rise to +5·75V. The equivalent circuit is again as in Fig. 6.45(c) and v_2 will readjust to +5V with a time constant of 1μsec. Hence the waveform of v_2 will be as Fig. 6.45(e).

If the decay of the capacitor voltage was linear for Fig. 6.45(d) then after 10μsec

$$v_e = -15 + \frac{10}{200} . 15$$
$$= -14{\cdot}25\text{V}$$

Thus the decay is in fact linear, which is to be expected, since we are only considering that part of the curve between $t = 0$ and $t = 0{\cdot}1$ on the characteristic of Fig. 6.30.

CHAPTER 7

REGENERATIVE SWITCHING CIRCUITS

So far we have considered the use of transistors in circuits analogous to non-locking switches. For example the transistor in Fig. 7.1 is in the *OFF* condition with no input signal. In order to cause a transition to the *ON* condition a negative bias must be applied to the base relative to the emitter. Removal of the negative bias causes a return to the *OFF* condition.

For some purposes a locking type switch is required, that is the switch can be left in either position. Alternatively the requirement

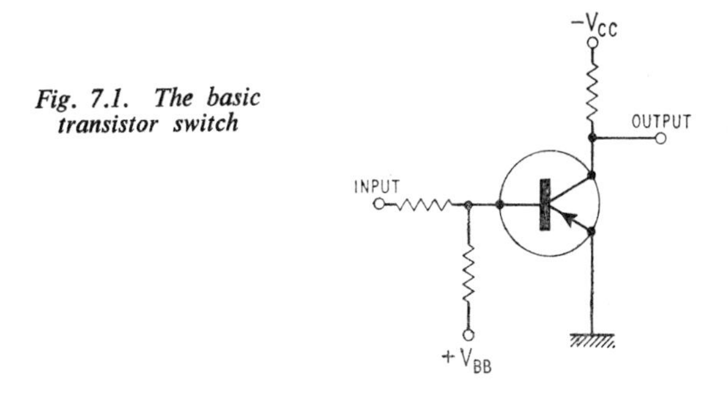

Fig. 7.1. The basic transistor switch

may be that the switch must change from one state to the other for a predetermined time and then revert to the original state.

7.1. MULTIVIBRATORS

A class of switching circuits which will carry out the above functions and others are called multivibrators. The basic multivibrator circuit contains 2 transistors arranged in such a manner that one is always fully conducting and the other is cut-off. As is shown in Fig. 7.2 the 2 transistors are connected in a positive feed-back loop. Assume that an initial disturbance produces a negative going signal at the

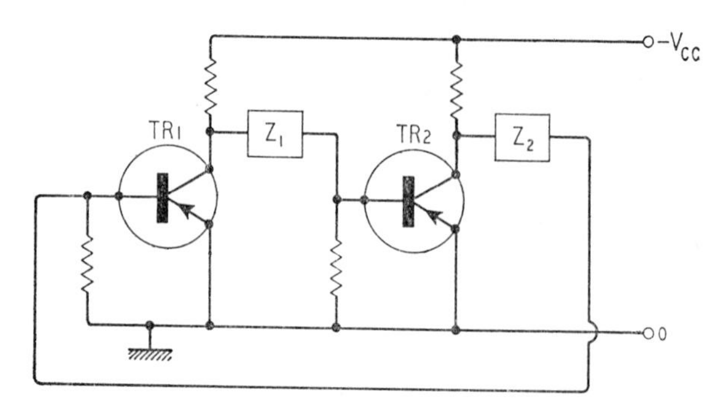

Fig. 7.2. The basic multivibrator circuit

base of transistor TR1. This will produce a positive going voltage at the collector of TR1 which is coupled to the base of TR2, thus tending to cut it off. A reduction of collector current results making the collector of TR2 more negative. This again is coupled back to the base of TR1 with a polarity such as to enhance the original disturbance. The circuit is thus rapidly driven into the state where TR1 is fully conducting and TR2 is cut off. It is the *regenerative switching* nature of the circuit which produces these conditions and drives the circuit into one or other of its *stable states*. Whether or not these states are permanently stable depends upon the nature of the coupling impedances Z_1 and Z_2 in Fig. 7.2, as we shall see.

In order for this regenerative action to take place there is a requirement that the gain round the feedback loop is greater than unity. This requirement is usually easily met and so we shall concern ourselves first with the determination of the stability of the circuit in the two states. There are three important variations of the basic multivibrator circuit of Fig. 7.2 which may be produced by changes in the coupling impedances Z_1 and Z_2. They are as follows.

(1) The *bistable* multivibrator, which stays permanently in either state until caused to change by an external signal.

(2) The *monostable* multivibrator, which has one permanently stable state and one quasi-stable state. If put in the quasi-stable state by an external signal it will return to the permanently stable state after a predetermined time.

(3) The *astable* multivibrator, which is not permanently stable in either state and continually oscillates back and forth.

We shall consider these circuits in turn, in each case illustrating the important points by an example.

7.2. THE BISTABLE CIRCUIT

Consider the basic bistable multivibrator circuit of Fig. 7.3(a) in which the coupling elements are resistors R_3, R_4. The circuit appears to be completely symmetrical, but the slightest unbalance in collector current is sufficient to initiate the regenerative switching action. We must now determine whether the circuit has got permanently stable states and will stay in either one indefinitely when put there. In other words, under normal operating conditions is one transistor fully conducting and the other cut off? With ideal transistors as in Fig. 7.3(b) let us assume that transistor TR1 of Fig. 7.3(a) is fully

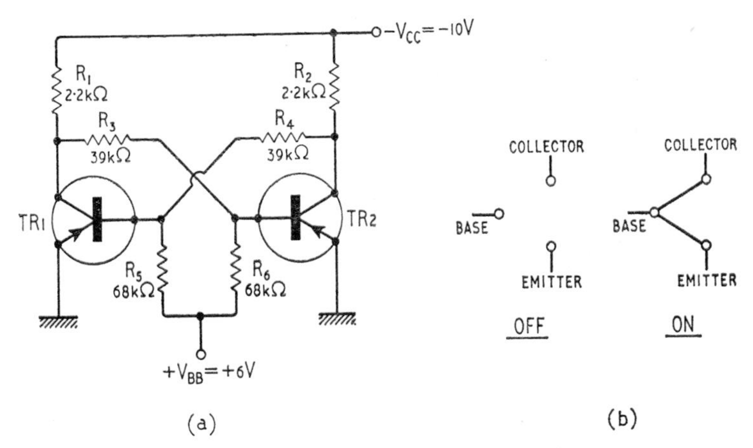

Fig. 7.3. (*a*) *The symmetrical bistable multivibrator;* (*b*) *The ideal transistor switch*

conducting and TR2 is cut off and, by analysis of the circuit conditions, determine if the original assumptions were correct. Consider first the conditions applying to the base of TR2 as shown by the equivalent circuit of Fig. 7.4(a).

Hence
$$V_{BE_2} = \frac{R_3}{R_6 + R_3} . V_{BB}$$
$$= \frac{39}{68+39} . 6$$
$$= 2{\cdot}2\text{V}$$

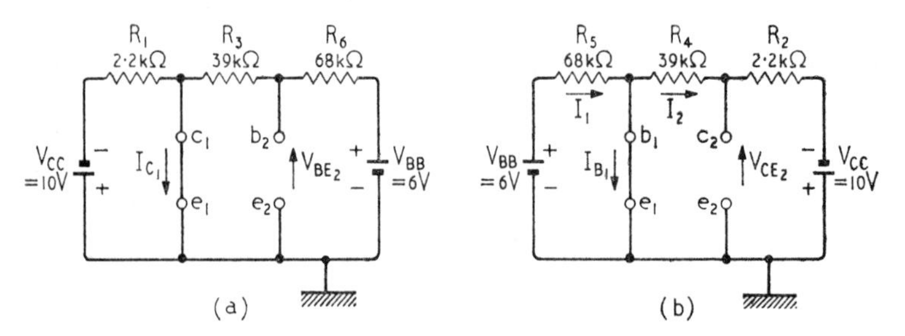

Fig. 7.4. Equivalent circuits of the bistable multivibrator with TR1 ON *and TR2* OFF. *(a) Circuit for determining V_{BE2}; (b) Circuit for determining V_{CE2} and I_{B1}*

Since a base–emitter voltage of $+0{\cdot}2$V is sufficient to cut off the transistor, TR2 is definitely cut off.

Next consider the conditions applying to the base of TR1, illustrated in Fig. 7.4(b)

$$I_1 = \frac{V_{BB}}{R_5}$$

$$= \frac{6}{68}\text{mA}$$

$$= 88\mu\text{A}$$

$$I_2 = \frac{V_{CC}}{R_2 + R_4}$$

$$= \frac{10}{2{\cdot}2+39}\text{mA}$$

$$= 243\mu\text{A}$$

Therefore

$$I_{B1} = I_1 - I_2$$

$$= -155\mu\text{A}$$

To determine if TR1 is in saturation it is necessary to superimpose the load-line for the stage on the common emitter output characteristics. This is shown in Fig. 7.5(b) where the necessary information is given in the equivalent circuit of Fig. 7.5(a) obtained by use of Thevenin's Theorem. Notice that little error is introduced by assuming the effective load is simply 2·2kΩ.

From Fig. 7.5(b) we deduce that TR1 will be saturated if $|I_{B1}| > 100\mu$A. Since we have calculated I_{B1} for this circuit to be -155μA, then TR1 is definitely in saturation.

Finally we have:

$$V_{CE2} = -\frac{R_4}{R_2+R_4}.V_{CC}$$

$$= -\frac{39}{2{\cdot}2+39}.10$$

$$= -9{\cdot}5\text{V}$$

The circuit has thus taken up one of its stable states with voltage and current values as summarised below:

$$V_{BE1} \simeq V_{CE1} \simeq 0\text{V} \qquad V_{BE2} = +2{\cdot}2\text{V}$$

$$I_{B1} = -155\mu\text{A} \qquad V_{CE2} = -9{\cdot}5\text{V}$$

$$I_{C1} \simeq -4{\cdot}5\text{mA} \qquad I_{B2} \simeq I_{C2} \simeq 0$$

The other stable state will be with transistor TR1 in its *OFF* condition

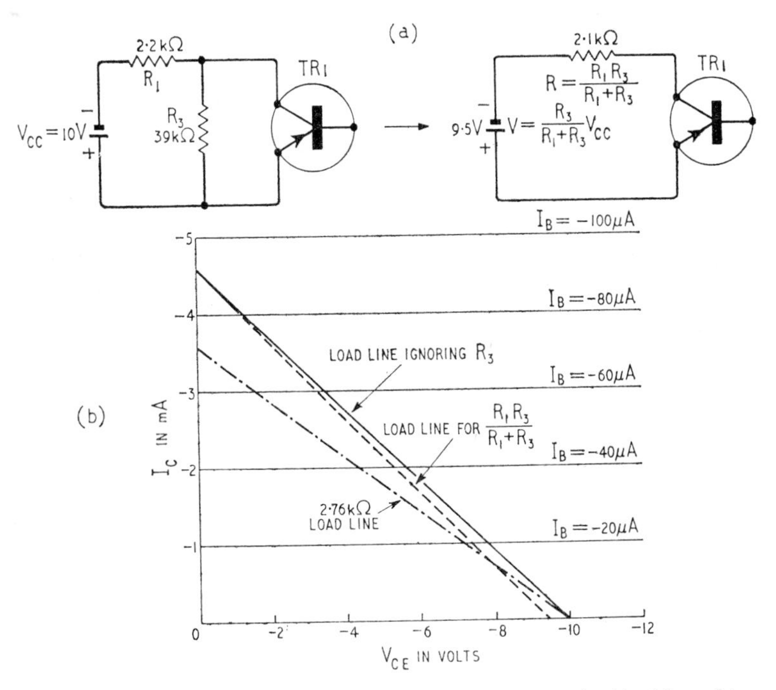

Fig. 7.5. Determining the states of the transistors in the bistable multivibrator. (a) Thevenin equivalent circuit for transistor TR1; (b) Idealised common emitter output characteristics

and TR2 in its *ON* condition, resulting in a set of values as above but with subscripts 1 and 2 interchanged.

7.2.1. The Self-biased Circuit

It is not necessary to provide a separate positive base bias supply V_{BB} as indicated in Fig. 7.3(a). If a common emitter resistor is used as

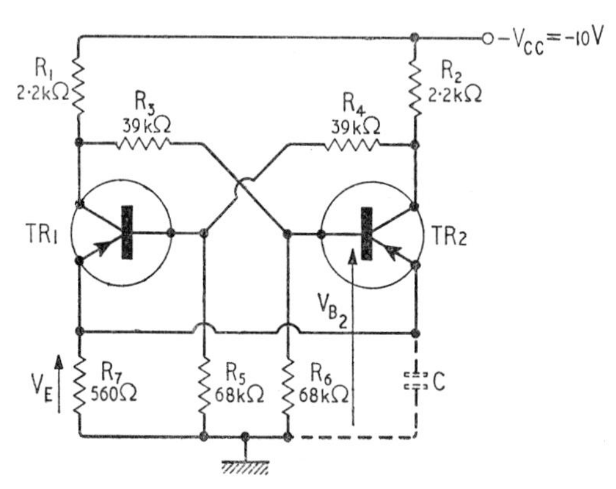

Fig. 7.6. The self-biased bistable multivibrator

in Fig. 7.6 self-bias is obtained. This is illustrated in the following example.

As before assume that transistor TR1 is on and TR2 is off. Ignoring the cross-coupling resistors, the effective load on TR1 is now

$$2{\cdot}2\text{k}\Omega + 560\Omega = 2{\cdot}76\text{k}\Omega$$

Drawing this load-line on the characteristic shows that a base current of -80μA will drive TR1 into saturation, with a collector current of $-3{\cdot}6$mA. The emitter voltage V_E is then approximately -2V, as are the collector and base voltages.
Therefore

$$\begin{aligned} V_{BE2} &= V_{B2} - V_E \\ &= -\frac{R_3}{R_3+R_6} V_E \\ &= \frac{39}{39+68} \times 2 \\ &= 0{\cdot}73\text{V} \end{aligned}$$

Hence TR2 is cut off.

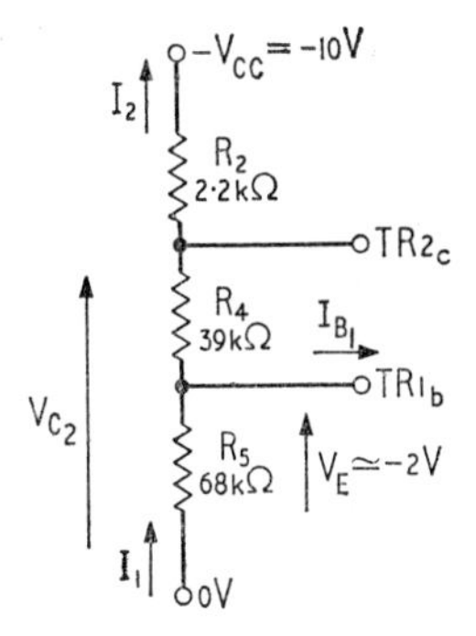

Fig. 7.7. Approximate equivalent circuit at the base of ON *transistor TR1*

The base current of TR1 can be found from a consideration of the approximate equivalent circuit of Fig. 7.7.

$$I_1 = -\frac{V_E}{R_5}$$

$$= \frac{2}{68}\,\text{mA}$$

$$= 29\mu\text{A}$$

$$I_2 = \frac{V_{CC}+V_E}{R_2+R_4}$$

$$= \frac{8}{2{\cdot}2+39}\,\text{mA}$$

$$= 194\mu\text{A}$$

Therefore

$$I_{B1} = I_1 - I_2$$

$$= -165\mu\text{A}$$

Since a minimum of -80μA is required, TR1 is definitely in saturation.

Also

$$V_{C2} = -V_{CC} + R_2 I_2$$

$$= -10 + 2{\cdot}2 \times 0{\cdot}194$$

$$= -9{\cdot}6\text{V}$$

Notice that the contribution of the base current of the *ON* transistor to the total emitter current has been ignored. Since we are only concerned with determining if the circuit is stable this small inaccuracy is of no account.

Although ideally the emitter current will be the same in either stable state, during a transition it will be changing. In order to maintain the emitter voltage constant during the transition it is necessary to shunt the emitter resistor with a capacitor of such a value as to make the time constant of the combination very long

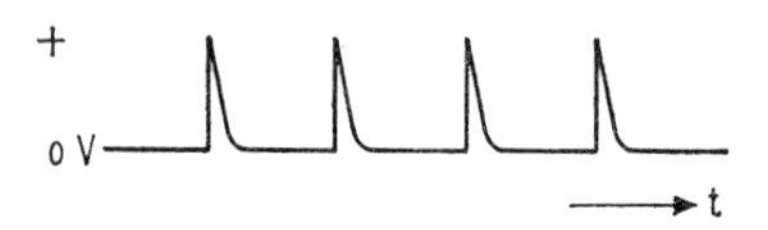

Fig. 7.8. A train of positive trigger pulses

compared to the time taken to make a transition. For instance, if the value of the capacitor were 0·5μF, giving a time constant of 280μsec and a transition was completed in 2μsec the emitter voltage would be maintained practically constant. The presence of the capacitor has no effect, of course, on the previous steady state calculations.

7.2.2. Triggering the Bistable Circuit

Having determined that the circuit of Fig. 7.3(a) has two stable states we must now consider means of causing a transition from one state to the other, or *triggering* the circuit. The most usual requirement is that a transition is to be made on reception of a sharp voltage pulse or *trigger pulse* as in Fig. 7.8. On reception of a second trigger pulse, on the same or a different input, the circuit is to revert to its original condition. Thus a succession of pulses causes the circuit to switch back and forth between its two stable states.

Before considering the actual means of introducing the trigger pulse to the circuit, we must make sure the circuit will respond to such a pulse. Suppose for instance that we produce a transition by cutting off TR2. A negative-going voltage step will thus be produced at the collector of TR2 which must be transmitted via R_4 and R_5 to the base of TR1 in order to bring it into saturation. In going from cut-off to saturation the transistor passes through the active region where amplification occurs, thus producing the regenerative switching action. It also produces a large effective input capacitance due to the Miller effect as discussed in Section 3.9, with the result that the voltage changes exponentially at the base of TR1 as in Fig. 7.9(a). The effect of this capacitance may be

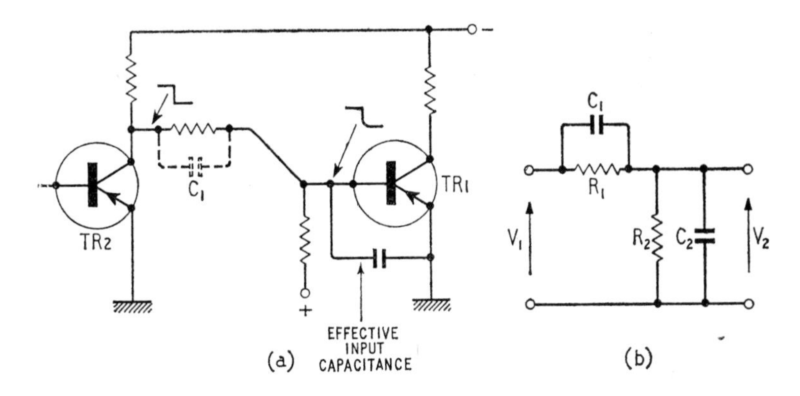

Fig. 7.9. (*a*) *Effect of input capcitance ;* (*b*) *The compensated circuit overcoming the effect of input capacitance*

such that the trigger pulse has ceased before the circuit is able to respond. This difficulty may be overcome by the addition of a compensating capacitor as shown dotted in Fig. 7.9(a). Its function may be explained as follows, with reference to Fig. 7.9(b).

In order that a rectangular wave may be transmitted unaltered in shape through the coupling network all its frequency components must be attenuated equally, that is the circuit must be frequency independent.

From Fig. 7.9(b)

$$V_2 = \frac{Z_2}{Z_1+Z_2}.V_1 \tag{7.1}$$

where

$$Z_1 = \frac{R_1}{1+j\omega C_1 R_1}$$

$$Z_2 = \frac{R_2}{1+j\omega C_2 R_2}$$

Therefore

$$V_2 = \frac{\dfrac{R_2}{1+j\omega C_2 R_2}}{\dfrac{R_1}{1+j\omega C_1 R_1}+\dfrac{R_2}{1+j\omega C_2 R_2}}.V_1$$

$$= \frac{R_2}{R_2+R_1\dfrac{1+j\omega C_2 R_2}{1+j\omega C_1 R_1}}.V_1 \tag{7.2}$$

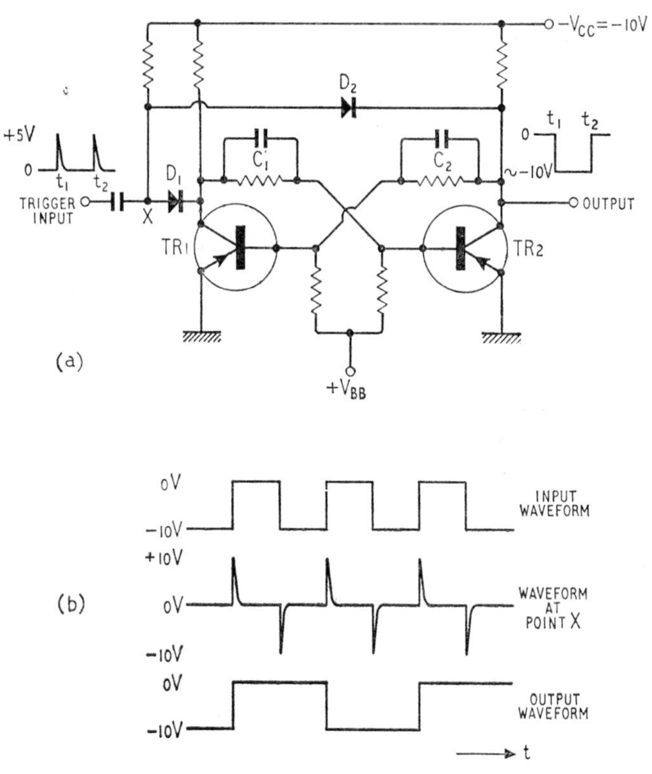

Fig. 7.10. (a) The bistable multivibrator with symmetrical trigger circuit; (b) Waveforms in the triggered bistable multivibrator

If now we make $C_1 R_1 = C_2 R_2$ then

$$V_2 = \frac{R_2}{R_1 + R_2} \cdot V_1 \tag{7.3}$$

i.e. the relationship between V_1 and V_2 is independent of frequency. Notice that this simple analysis applies only in the active region*. Thus the circuit may be made to respond to a trigger pulse more quickly by the use of so-called *speed-up* or *commutating capacitors.*

The actual value is not at all critical, but if it is made too large then a longer time must be allowed for the circuit to reach its steady state, before making another transition.

* For a more detailed analysis see HAYKIN, S. S., *Junction Transistor Circuit Analysis*, Iliffe Books Ltd. (1962), Chapter 11.

A possible arrangement for the introduction of triggering pulses is the symmetrical method shown in Fig. 7.10 in which the circuit only responds to positive pulses.

Consider that initially TR1 is *OFF* and TR2 is *ON*. This means that ideally $V_{CE1} = -9{\cdot}5$V and $V_{CE2} = 0$V as previously calculated. Diodes D_1 and D_2 are thus reverse-biased and point X is at -10V. Now assume that at time t_1 a trigger pulse of amplitude $+5$V is applied at the trigger terminal. The potential at point X rises instantaneously to -5V, forward-biasing D_1, but not D_2 since V_{CE2} is at 0V. D_1 thus provides a path for the trigger pulse to the collector of TR1 and thence through the speed-up capacitor C_1 to the base of TR2. This positive pulse on the base will cut off TR2 and V_{CE2} will change from 0V to $-9{\cdot}5$V. A negative-going transition is thus coupled to the base of TR1 causing it to come out of cut-off into saturation. The function of the diodes D_1 and D_2 is to guide the pulse to the base of the conducting transistor, hence they are often referred to as *steering diodes*. On reception of the next trigger

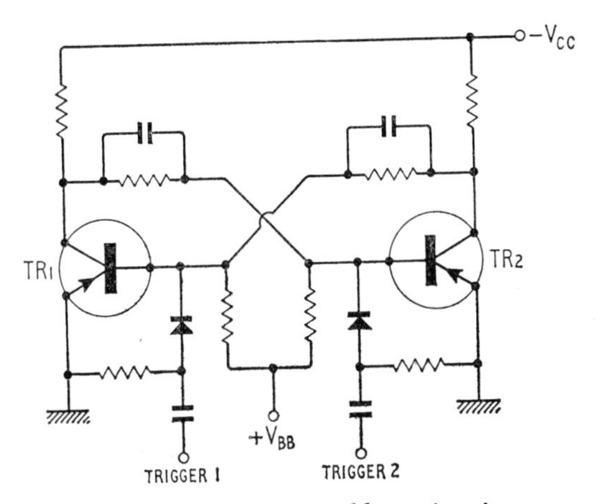

Fig. 7.11. Asymmetrical base triggering

pulse, at time t_2, D_2 will conduct but not D_1, since V_{CE1} is now at 0V. The pulse thus appears at the base of TR1 causing it to cut off and the circuit reverts to its original state. Two input pulses are therefore required to cause a complete switching cycle of the circuit, a factor which makes the bistable circuit most useful in scaling and counting applications. The trigger pulses could have been introduced directly onto the bases of the transistors in Fig. 7.10(a). Also instead of the symmetrical arrangement, in which all trigger pulses are introduced at the same point, the asymmetrical arrangement of Fig. 7.11 can be used. In this arrangement positive trigger pulses must be introduced at each base in turn in order to cause successive transitions.

7.2.3. Loop Gain Requirements

Now let us make an approximate investigation of the loop gain requirements. To simplify the analysis we shall assume that the

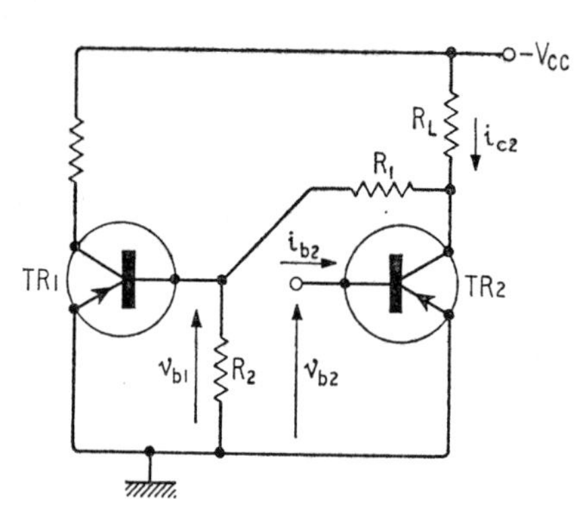

Fig. 7.12. Approximate circuit for investigation of loop gain

circuit is operating in its linear region and that the frequency of operation is sufficiently low to justify neglecting high frequency effects. Thus if the load is small the input resistance of each transistor stage is nearly equal to h_{ie} as was shown in Section 3.6. Normally the resistor R_2 is large compared to h_{ie} and we shall neglect its shunting effect on h_{ie}. Therefore if a voltage v_{b2} is applied to the base of transistor TR2 in Fig. 7.12 we have

$$i_{b2} \simeq \frac{v_{b2}}{h_{ie}}$$

$$i_{c2} \simeq \frac{h_{fe} . v_{b2}}{h_{ie}}$$

and

$$v_{c2} \simeq \frac{h_{fe} v_{b2} R_L}{h_{ie}}$$

where R_1 is assumed large compared to R_L so that R_L is the effective load of each stage.

Hence the voltage appearing at the base of transistor TR1 will be

$$v_{b1} \simeq \frac{h_{ie}}{h_{ie}+R_1} \cdot v_{c2} \tag{7.4}$$

Thus

$$\frac{v_{b1}}{v_{b2}} \simeq \frac{h_{fe} \cdot R_L}{h_{ie}+R_1} \tag{7.5}$$

The loop gain will be the square of this term and must be greater than unity for regenerative switching to occur. Therefore to ensure a transition we must have

$$R_1 \leqslant h_{fe} R_L - h_{ie} \tag{7.6}$$

In the circuit of Fig. 7.3(a), if the transistors have typical values of $h_{fe} = 50$, $h_{ie} = 1\text{k}\Omega$, this condition is well satisfied.

7.2.4. The Bistable Multivibrator as a Divider

If a rectangular voltage wave is applied at the trigger input terminal of Fig. 7.10(a) and the time constant of the *RC* network is very short compared to the period of the wave, then a train of positive and negative pulses will appear at point *X*, Fig. 7.10(b). The circuit will not respond to the negative pulses since they produce reverse bias on both D_1 and D_2. The circuit will respond to the positive pulses, however, as previously described, and a transition at the output will occur at each positive pulse. The resulting output waveform is as shown in Fig. 7.10(b), and it is obvious that the frequency of the input waveform has been halved, i.e. the circuit has divided by *2*.

Now consider a chain of 4 bistable circuits in cascade, Fig. 7.13(a), the output of each stage coupled symmetrically via a differentiating circuit to the succeeding stage.

Consider that initially the righthand transistor of each stage is fully conducting. If a train of positive pulses is fed into B_o the waveforms at various points in the circuit will be as shown in Fig. 7.13(b). Notice that B_o makes a transition for each input pulse and that succeeding stages only make a transition when the output transistor of the preceding stage switches *ON*, thus producing a positive pulse. The effective output of B_3 is thus a single positive pulse occurring after the sixteenth input pulse. All stages are then set in their original conditions and a complete cycle of switching has occurred. The circuit has divided the number of input pulses by 16, since each stage has divided by 2, total division being $2^4 = 16$. Thus if there were n stages the dividing factor would be 2^n.

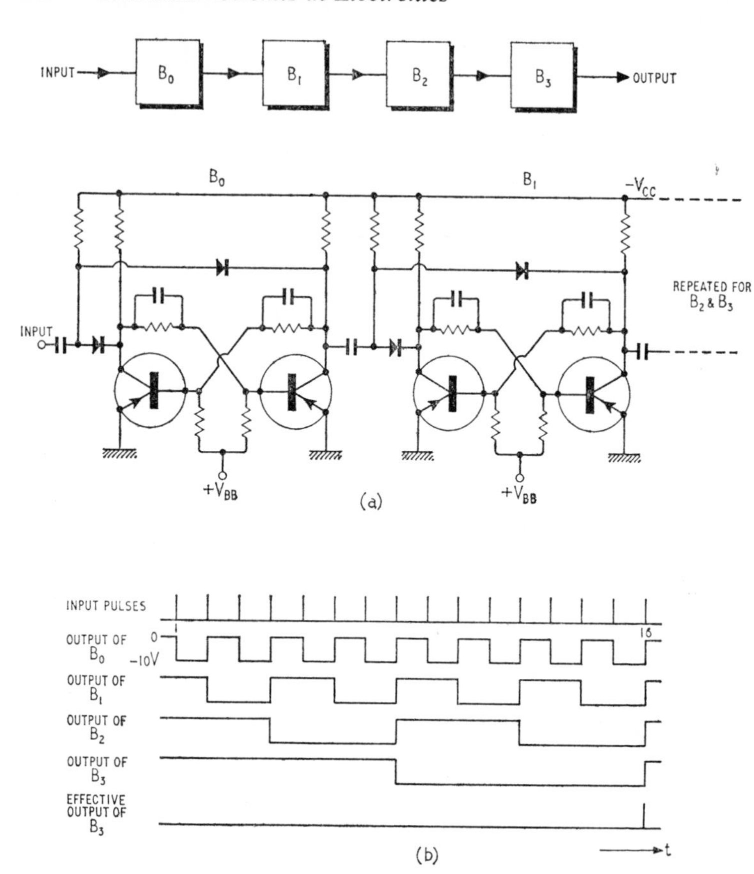

Fig. 7 13. The bistable multivibrator as a divider. (a) A chain of 4 bistable multivibrators; (b) Waveforms in (a)

7.2.5. The Bistable Multivibrator as a Counter

Notice that during the above process the 4 stages occupied all possible combinations of the two states. Table 7.1 shows the states of the stages after each input pulse, where a 0 corresponds to the output transistor of each stage being *ON*, i.e. its collector is at approximately 0V and a 1 corresponds to it being *OFF*, i.e. its collector at approximately −10V.

We shall now show that the states of the stages at any time indicate the number of pulses which have been received, i.e. the circuit is counting the input pulses.

Table 7.1

Pulses	States of the BMV				Pulses	States of the BMV			
	B_3	B_2	B_1	B_0		B_3	B_2	B_1	B_0
1	0	0	0	1	9	1	0	0	1
2	0	0	1	0	10	1	0	1	0
3	0	0	1	1	11	1	0	1	1
4	0	1	0	0	12	1	1	0	0
5	0	1	0	1	13	1	1	0	1
6	0	1	1	0	14	1	1	1	0
7	0	1	1	1	15	1	1	1	1
8	1	0	0	0	16	0	0	0	0

In the decimal number system, the number is represented by an expansion in powers of 10 with coefficient $0, 1, 2, \ldots, 9$,

e.g. $\quad 109{\cdot}32 = 1\times 10^2 + 0\times 10^1 + 9\times 10^0 + 3\times 10^{-1} + 2\times 10^{-2}$

The base of the decimal system is thus 10. It is not essential to use 10 as the base for the system, however, since although 10 is useful for counting on the fingers it does not lend itself to simple representation by electronic circuits. If we choose the base to be 2, which means we have only two coefficients, 0 and 1, then we have a number system which is ideally suited to switching circuits. For example, the 2 states of the bistable multivibrator can be designated 0 or 1.

Thus in the number system with base 2, or the *binary system* a number is represented as an expansion in powers of 2 with coefficients 0 or 1. For example:

$$
\begin{aligned}
&14{\cdot}625 \text{ (decimal)}\\
&= 1\times 2^3 + 1\times 2^2 + 1\times 2^1 + 0\times 2^0 + 1\times 2^{-1} + 0\times 2^{-2} + 1\times 2^{-3}\\
&= 1110{\cdot}101
\end{aligned}
$$

Table 7.2

Decimal Number	*Binary Equivalents*	*Decimal Number*	*Binary Equivalents*
0	0 0 0 0 0	9	0 1 0 0 1
1	0 0 0 0 1	10	0 1 0 1 0
2	0 0 0 1 0	11	0 1 0 1 1
3	0 0 0 1 1	12	0 1 1 0 0
4	0 0 1 0 0	13	0 1 1 0 1
5	0 0 1 0 1	14	0 1 1 1 0
6	0 0 1 1 0	15	0 1 1 1 1
7	0 0 1 1 1	16	1 0 0 0 0
8	0 1 0 0 0		

Table 7.2 gives the binary equivalents of decimal numbers up to 16 and, comparing this with Table 7.1, we see that after each input pulse the circuit gives a binary representation of the number of input pulses received to that point. Thus the circuit is counting in the binary code. After the sixteenth pulse the stages are reset to 0 and the counting cycle starts again. We have thus produced a *scale of* 16 counter and obviously n stages will give a *scale of* 2^n counter.

7.2.6. Scale-of-ten Counter

It may be of course that, although using the cascade of bistable circuits we require a *scale of* 10 counter. Since this is not directly possible, it is necessary to modify the basic arrangement of the

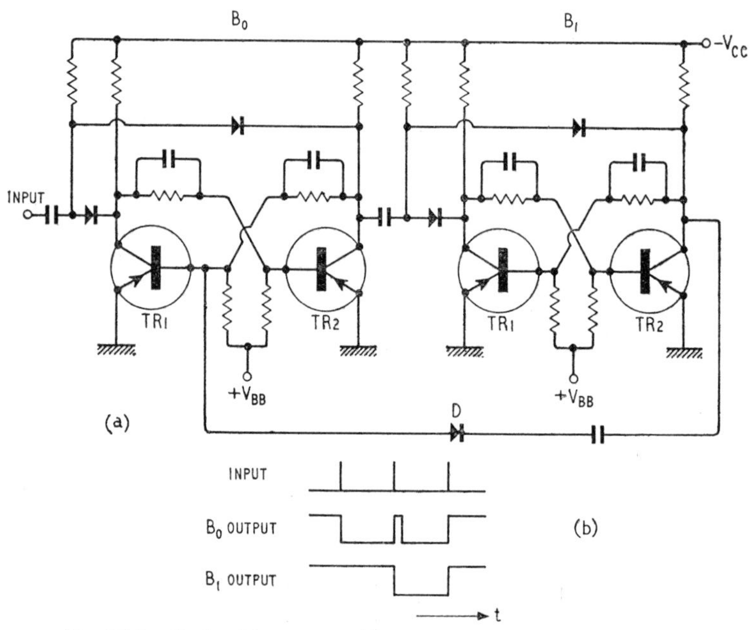

Fig. 7.14. Scale-of-3 counter. (a) Circuit diagram; (b) Waveforms

scale of 16 counter so that the circuit resets after counting 10 pulses, that is, we must advance the count by 6. This may be done by the use of feedback connections as follows.

Consider first 2 stages of the bistable multivibrator in cascade, but with a feedback connection between the collector of transistor TR2 in bistable circuit B_1 and the base of transistor TR1 in B_0 as

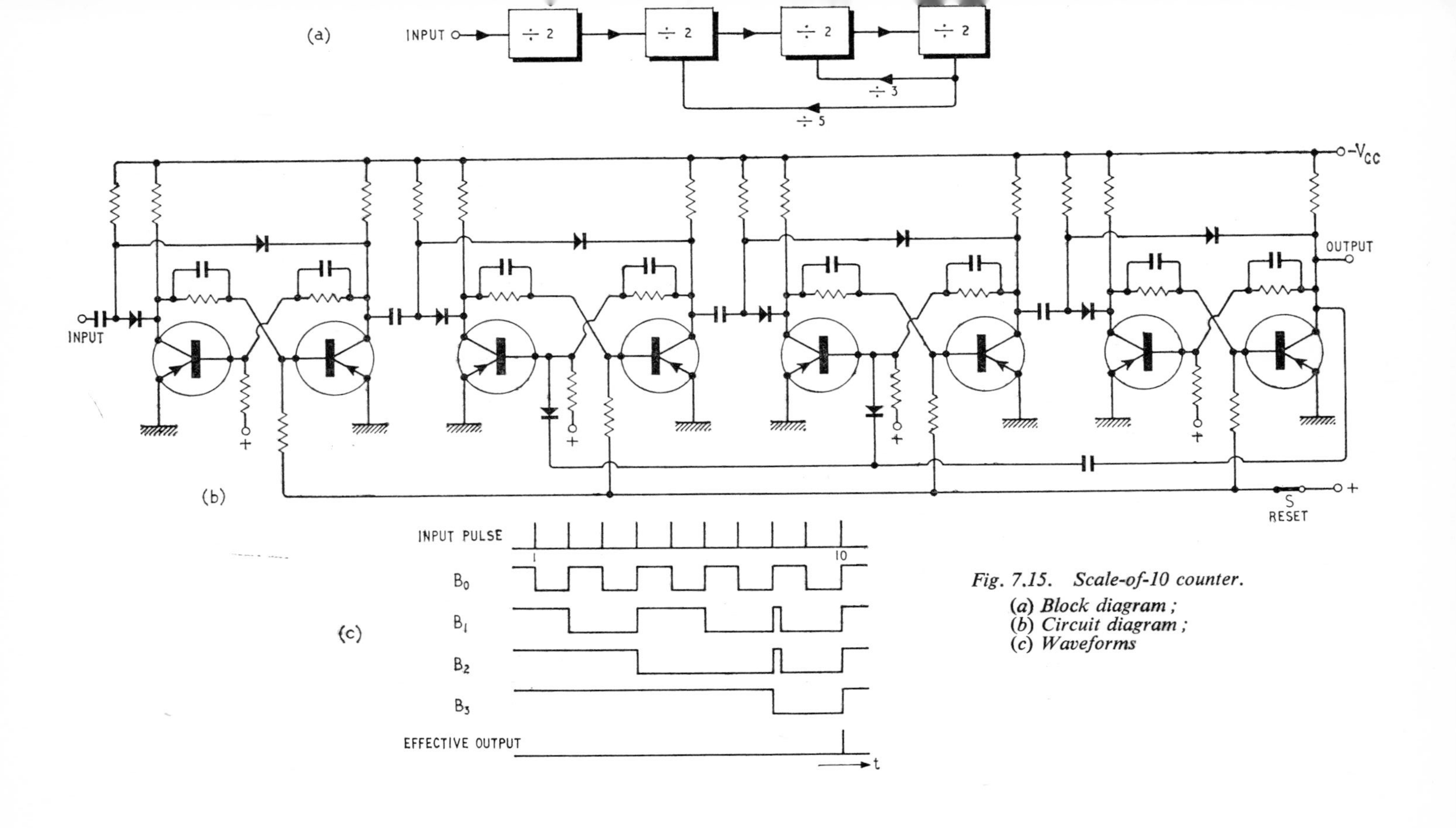

Fig. 7.15. *Scale-of-10 counter.*
(*a*) *Block diagram ;*
(*b*) *Circuit diagram ;*
(*c*) *Waveforms*

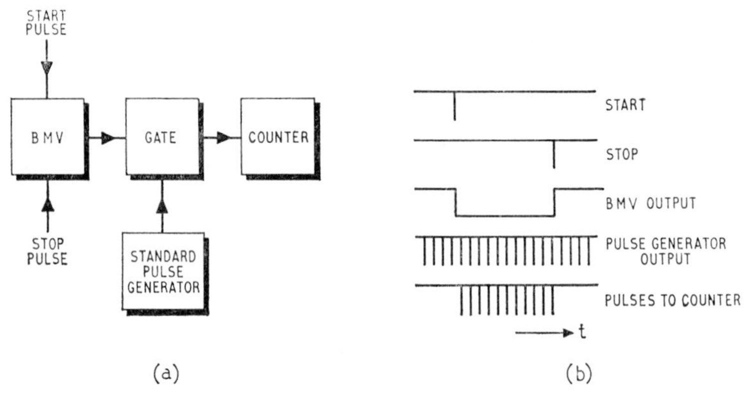

Fig. 7.16. Measurement of time interval.
(a) Block diagram; (b) Waveforms

in Fig. 7.14(a). After the first pulse B_o goes to 1 as before. After the second pulse B_o goes to 0 thus causing B_1 to go to 1. This produces a negative voltage step at the collector of TR2 of B_1 which is coupled back via the diode D to the base of TR1 of B_o, immediately returning B_o to state 1. Then after the third pulse both B_o and B_1 revert to state 0. The circuit has thus reset after 3 input pulses instead of 4, the resulting waveforms being as in Fig. 7.14(b). Now connect another stage in cascade, making the scale 6, and add an overall feedback loop, reducing the scale to 5. A further stage in cascade makes a scale of 10 counter and the circuit resets after the tenth input pulse. The basic circuit and relevant waveforms are shown in Fig. 7.15.

We have referred to the necessity for resetting the counter and a facility for doing this is included in Fig. 7.15. Opening the switch S removes the positive bias from all right-hand transistors which will switch on, thus giving 0 at the output.

7.2.7. Measurement of Time Interval

If the counter is fed with a train of pulses of known repetition rate, the number of pulses counted may be converted to a time interval. Thus the counter may be adapted for the measurement of time intervals between the occurrence of particular events, e.g. the closing and opening of a switch. A typical circuit arrangement is shown in block schematic form in Fig. 7.16. The gate is a circuit which will be described in detail in Chapter 8, but which gives a negative voltage output only if both inputs are negative. Initially the bistable multivibrator BMV gives 0V out. On reception of the input *START* pulse the BMV changes state to give −10V out. While the BMV is in this state negative pulses from the standard pulse generator pass to the counter. The *STOP* pulse resets the BMV to 0V and no further pulses are passed to the counter. The state of the counter is then a measure of the time interval. Obviously time can only be measured in integral units of the standard pulse interval so the pulse period must be small compared to the period measured to ensure reasonable accuracy.

7.2.8. Measurement of Frequency

By a small rearrangement of the previous circuit the counter may be adapted for frequency measurement. Now we open the gate for a known length of time and count the number of pulses passed to the counter in that time. A typical arrangement is as shown in Fig. 7.17. The standard pulse generator frequency is divided down to some relatively long period, say 1 sec. If the unknown source of frequency were a sine wave it would be necessary to square it and differentiate

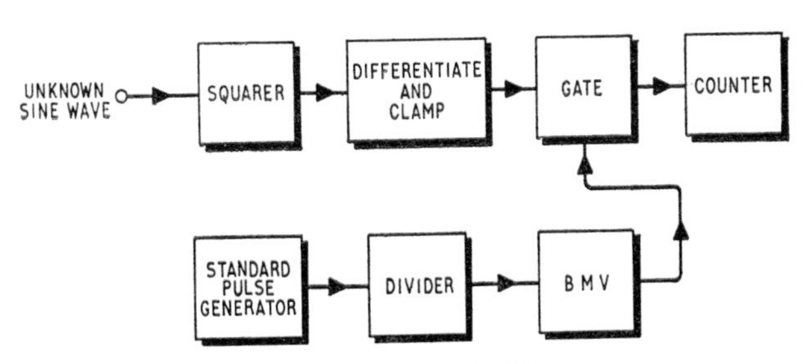

Fig. 7.17. Measurement of frequency

before passing to the gate. Clamping of the differentiator output eliminates the positive pulses, resulting in one negative pulse per cycle of the input sine wave.

7.3. THE MONOSTABLE CIRCUIT

This second variation on the basic multivibrator circuit of Fig. 7.2 is obtained by making one of the coupling impedances a resistor (plus a speed-up capacitor) and the other a capacitor. The result is that this circuit has only one permanently stable state as is illustrated in the following example.

Consider the monostable multivibrator circuit of Fig. 7.18(a). Since there is no direct coupling between the collector of transistor

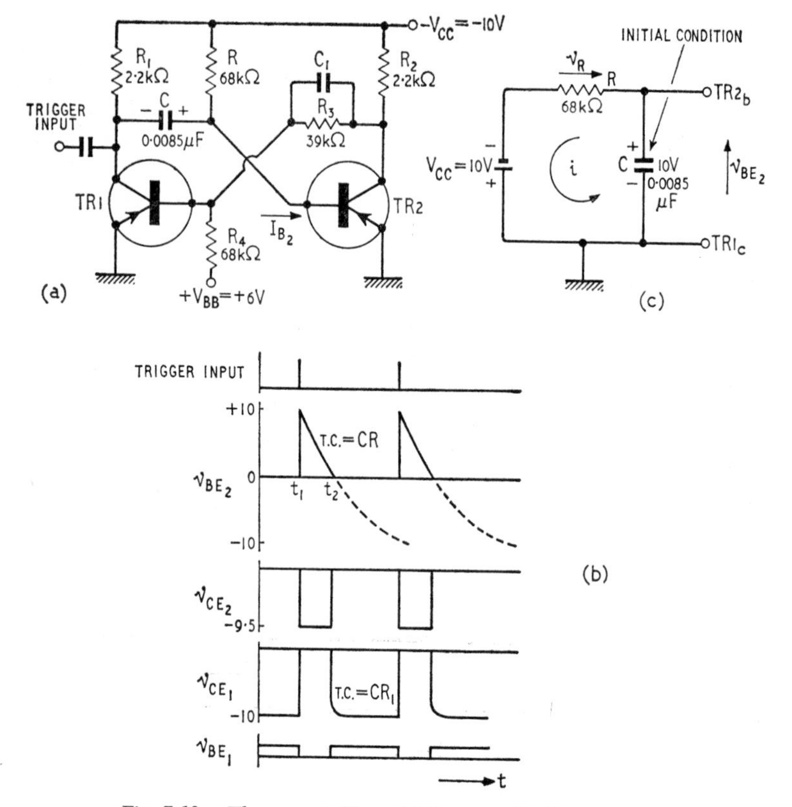

Fig. 7.18. The monostable multivibrator. (a) Basic circuit; (b) Waveforms; (c) Equivalent circuit for determining v_{BE2}

TR1 and the base of TR2, which is returned via R to the negative supply, then in the permanently stable state TR2 will be *ON* and TR1 will be *OFF*. Therefore, assuming again that the transistors are ideal,

$$I_{B2} = -\frac{V_{CC}}{R}$$

$$= -\frac{10}{68}\,\text{mA}$$

$$= -147\mu\text{A}$$

With the 2·2kΩ load we know that TR2 will saturate with a base current of -100μA, therefore it is hard *ON*.

Then
$$V_{BE_1} = \frac{R_3}{R_3+R_4} . V_{BB}$$
$$= \frac{39}{39+68} . 6$$
$$= 2{\cdot}2\text{V}$$

Therefore TR1 is cut off and $V_{CE1} = -10$V.

A positive trigger pulse applied at the input terminal at time t_1, will cause TR2 to switch off and a transition to the quasi-stable state will occur with TR1 *ON* and TR2 *OFF*.

The conditions at the base of TR1 are exactly as for the bistable multivibrator of Fig. 7.3 and therefore the base current will be -155μA. Thus TR1 is saturated.

As TR1 switches on its collector goes from -10V to 0V. Since the voltage on the capacitor C, which was 10V with the polarity shown in Fig. 7.18(a) in the permanently stable state, cannot change instantaneously, the voltage at the base of TR2 will change from 0V to $+10$V. Thus TR2 is cut-off, and $V_{CE2} = -9{\cdot}5$V.

The capacitor C will now discharge via R and would eventually charge to the opposite polarity. However, as v_{BE2} goes through zero towards a negative value at time t_2, TR2 again comes into the active region and a regenerative switch occurs to the original conditions. The action of the circuit is illustrated by the waveforms of Fig. 7.18(b). The small delay occurring in v_{CE1} returning to -10V is due to the capacitor C recharging via R_1.

In response to a trigger input this circuit thus produces a voltage pulse at the output whose duration is completely determined by the circuit elements. It therefore forms the basis of a simple pulse generator.

In order to determine the pulse duration, consider the approximate equivalent circuit of Fig. 7.18(c), describing the conditions at the base of TR2 just after the arrival of the trigger pulse. Taking account of the initial potential of 10V on C with polarity shown, the charging current i at time t will be

$$i = \frac{20}{R} . \mathrm{e}^{-(t-t_1)/CR} \tag{7.7}$$

The circuit switches when $v_{BE2} = 0$ at $t = t_2$. This will be so when the charging current produces a voltage drop v_R across R of 10V, as shown in Fig. 7.18(c).

Then
$$10 = R.\frac{20}{R}e^{-(t_2-t_1)/CR}$$

from which we obtain

$$t_p = t_2 - t_1$$
$$= CR.\log_e 2\,\text{sec} \tag{7.8}$$

The pulse duration is thus determined by the circuit elements C and R. For the circuit of Fig. 7.18(a)

$$t_p = 0{\cdot}0085 \times 10^{-6} \times 68 \times 10^3 \times 0{\cdot}693$$
$$= 400\mu\text{sec}$$

The circuit will thus produce a train of pulses of duration t_p (= 400μsec in this example) in response to a train of trigger

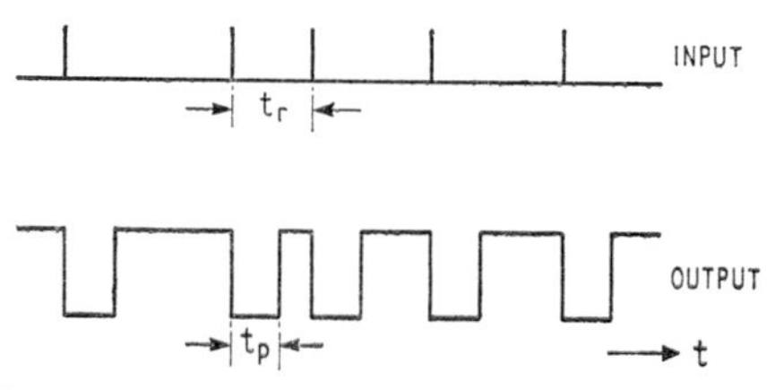

Fig. 7.19. Pulse forming in the monostable multivibrator

pulses, the repetition rate being determined by the trigger pulse interval t_r as in Fig. 7.19. Notice that $t_{r\,\min}$ must be greater than t_p.

7.3.1. The Monostable Circuit as a Delay Unit

A very useful application of the monostable multivibrator is as an adjustable time delay unit. The principle is illustrated in Fig. 7.20. The positive-going edge of the input pulse triggers the first monostable circuit MMV1 which returns to its stable state after time t_p sec. The output of MMV1 thus produces a positive trigger pulse after time t_p which is applied to the second monostable circuit MMV2. The output of this stage is a pulse of duration adjusted to be equal to the input pulse duration, but delayed in time by t_p. The time delay may be varied by varying the time constant CR of the first MMV or by returning the resistor R to an adjustable negative supply, rather than to $-V_{CC}$. This principle will be explained in more detail in connection with the astable multivibrator.

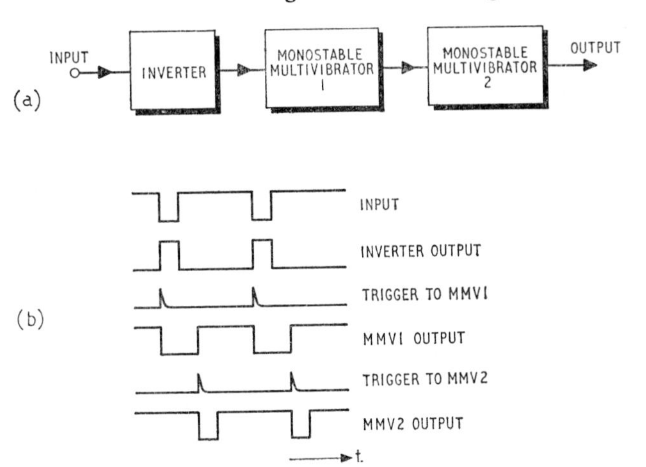

Fig. 7.20. The MMV as a delay unit; (a) Block diagram; (b) Waveforms

7.4. THE ASTABLE CIRCUIT

If both the coupling impedances in the basic circuit of Fig. 7.2 are capacitors then the circuit will not have a permanently stable state, since there is no d.c. coupling between stages. Instead the circuit

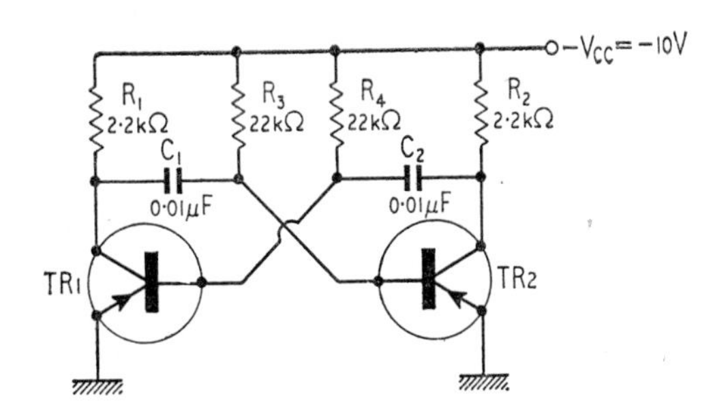

Fig. 7.21. The astable multivibrator

will switch back and forth between the 2 quasi-stable states in which either transistor TR1 is *ON* and TR2 is *OFF* or vice versa. The period for which the circuit remains in either state is determined by the particular circuit arrangement. We thus have an oscillator producing a good approximation to a square wave and for this reason the astable circuit is an important member of this class of switching circuits. The action of the circuit is best illustrated by means of an example.

Consider the circuit of Fig. 7.21 and again assume perfect transistors. Commence the analysis by assuming that at time t_1, transistor TR1 has just switched *ON*. Thus the collector voltage v_{CE1} rises instantaneously from -10V to 0V. This step is transmitted through the capacitor C_1 to the base of TR2 and the base

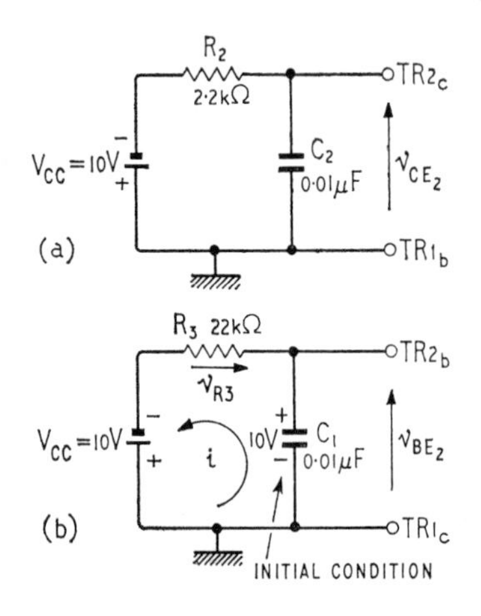

Fig. 7.22. Equivalent circuits for the OFF *transistor. (a) Circuit for determining* v_{CE2}*; (b) Circuit for determining* v_{BE2}

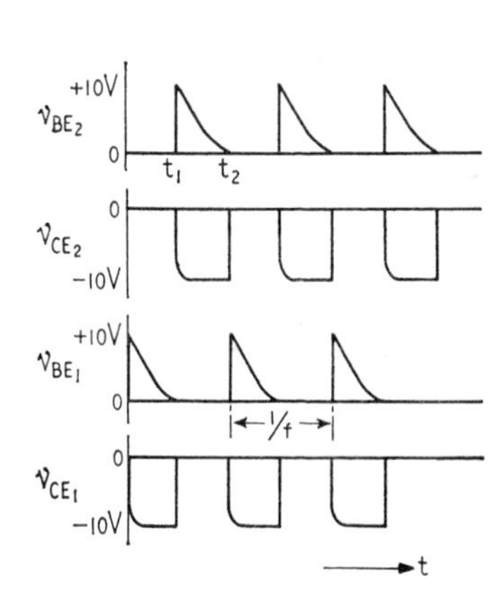

Fig. 7.23. Waveforms in the astable multivibrator

voltage v_{BE2} must therefore change from its *ON* value of 0V to $+10$V. Thus TR2 cuts-off. The equivalent circuits describing the conditions at the base and collector of TR2 at this time are as in Fig. 7.22. Capacitor C_2 has an initial voltage of 0V, and will charge exponentially to -10V via R_2. v_{CE2} will thus drop to -10V with a time constant

$$C_2 R_2 = 2{\cdot}2 \times 10^3 \times 10^{-8} \text{ sec}$$
$$= 22\mu\text{sec}$$

Capacitor C_1 has an initial voltage of 10V with polarity as shown and will begin to charge via R_3 to -10V. Thus v_{BE2} will begin to decrease from $+10$V toward -10V with time constant

$$C_1 R_3 = 22 \times 10^3 \times 10^{-8} \text{ sec}$$
$$= 220\mu\text{sec}$$

However, when v_{BE2} passes through 0V to a negative value, at time t_2, TR2 will switch *ON*. TR1 will thus receive a +10V step at its base and switch *OFF*. The cycle of events will then repeat as above.

Charging current i in Fig. 7.22(b) is given by

$$i = \frac{20}{R_3} e^{-(t-t_1)/220} \tag{7.9}$$

where t and t_1 are in μsec. v_{BE2} is zero, at time t_2, when the voltage drop v_{R3} across R_3 is 10V directed as shown,

$$10 = R_3 . \frac{20}{R_3} . e^{(t_2-t_1)/220}$$

Hence

$$(t_2 - t_1) = 220 \log_e 2 \mu\text{sec}$$

$$= 152 \mu\text{sec}$$

A complete cycle of events will take twice this period, since the circuit is symmetrical. Hence the frequency of the rectangular output wave is given by

$$f = \frac{10^6}{2 \times 152} \text{c/s}$$

$$\simeq 3{\cdot}3\text{kc/s}$$

Waveforms obtainable at various points in the circuit are shown in Fig. 7.23.

7.4.1. Frequency Adjustment

The initial frequency of the astable multivibrator is obviously determined by suitable choice of the elements R_3, R_4 and C_1, C_2. However, it is possible to vary the frequency of a given circuit by varying the voltage to which the resistors R_3, R_4 are returned, as shown in Fig. 7.24(a). This arrangement alters the slope of the base voltage curve and thus changes the time taken for the base voltage to return from $+V_{CC}$ to 0V. The equivalent circuit describing the condition at the base of an *OFF* transistor is approximately as in Fig. 7.24(b), where it is assumed that the voltage source V_{BB} has zero impedance.

$$\text{Charging current} = \frac{V_{BB}+V_{CC}}{R} . e^{-(t-t_1)/CR} \tag{7.10}$$

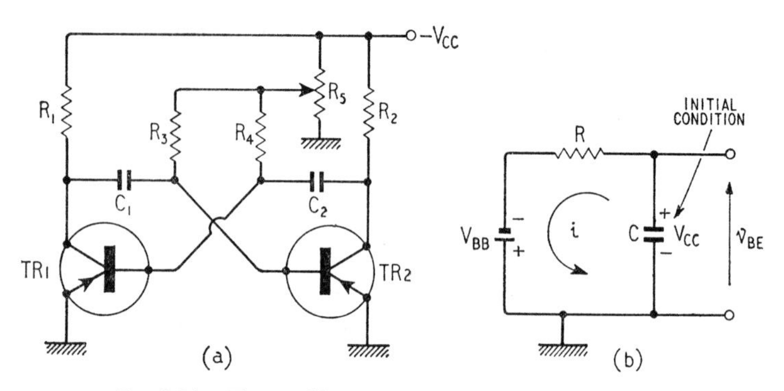

Fig. 7.24. The astable multivibrator with frequency control

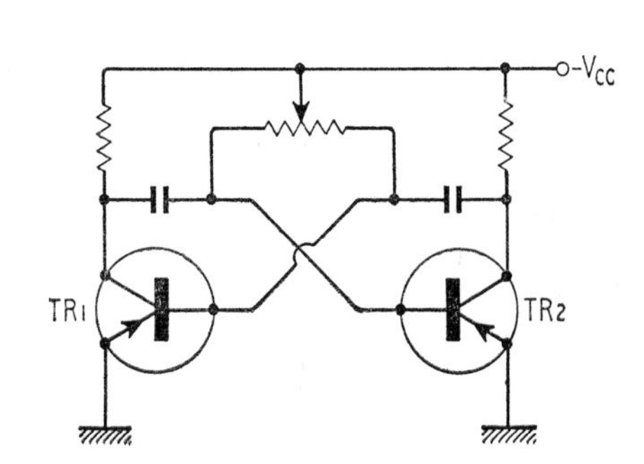

Fig. 7.25. Symmetry adjustment in the astable multivibrator

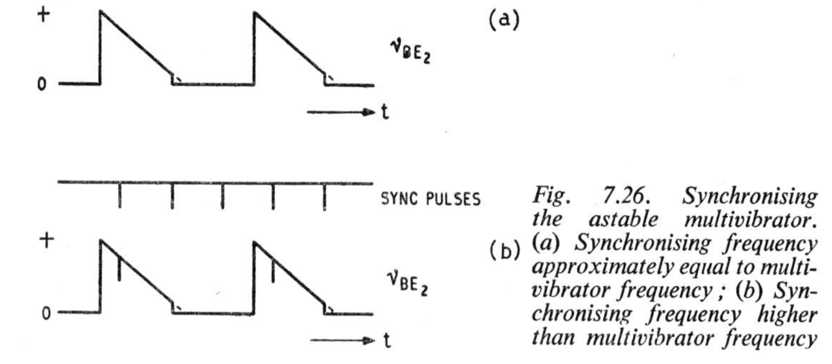

Fig. 7.26. Synchronising the astable multivibrator. (a) Synchronising frequency approximately equal to multivibrator frequency; (b) Synchronising frequency higher than multivibrator frequency

Therefore v_{BE} is 0 when

$$V_{BB} = R.\frac{V_{BB}+V_{CC}}{R}.e^{-(t_2-t_1)/CR}$$

i.e.

$$t_2-t_1 = CR\log_e(1+V_{CC}/V_{BB}).$$

Therefore

$$f = \frac{1}{2CR\log_e\left(1+\frac{V_{CC}}{V_{BB}}\right)} \tag{7.11}$$

7.4.2. Symmetry Adjustment

The symmetrical nature of the circuit of Fig. 7.21 results in a square wave output. A rectangular wave, in which the duration of the *OFF* times of the 2 transistors is unequal, may be obtained either by making the time constants $C_1 R_3$ and $C_2 R_4$ different or by returning the base resistors R_3 and R_4 to different voltages. A possible circuit arrangement for the first method is shown in Fig. 7.25.

7.4.3. Synchronising the Astable Circuit

A complex switching system may incorporate several different astable circuits as square wave generators and if the accuracy of frequency required is very great it may be necessary to synchronise the individual astable multivibrators to a master pulse train. This may be done as follows.

Consider again the circuit of Fig. 7.21 and introduce a negative pulse at the base of TR2 just prior to the time when v_{BE2} goes through zero. If the pulse is of sufficient amplitude it will drive the base negative and thus determine the switching instant. Therefore if a train of pulses with a repetition frequency slightly higher than the natural frequency of oscillation is introduced at this point the astable multivibrator will be locked to the incoming signal as shown in Fig. 7.26(a). By suitable choice of the input pulse amplitude the astable circuit can be synchronised to a higher frequency as indicated in Fig. 7.26(b). Thus the astable multivibrator will act as a frequency divider.

7.5. THE EMITTER-COUPLED TRIGGER CIRCUIT (SCHMITT TRIGGER)

A variation on the basic bistable multivibrator may be obtained by replacing one of the cross-coupling resistance networks by a common

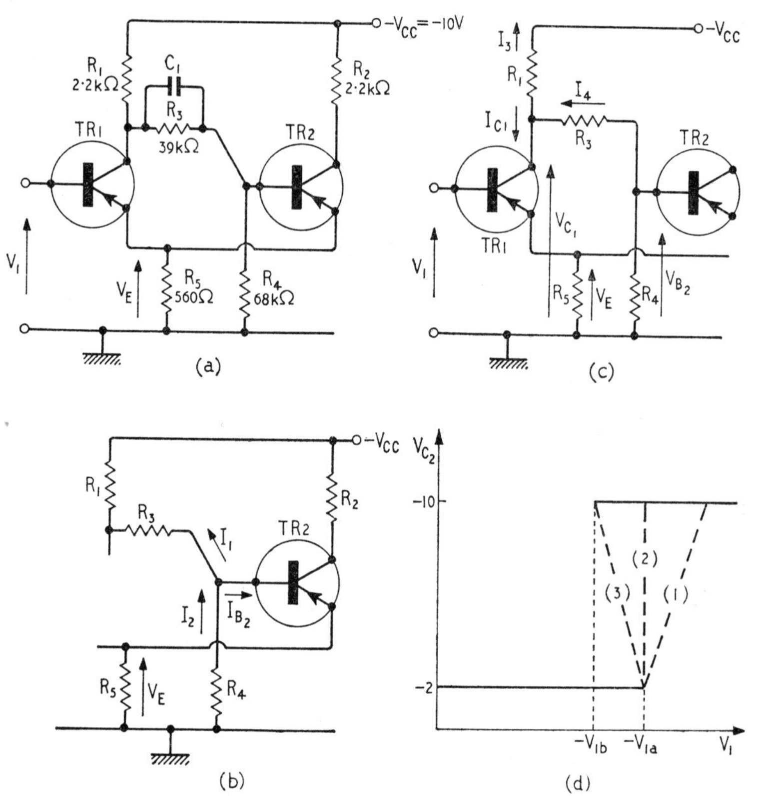

Fig. 7.27. The Schmitt Trigger circuit. (a) Circuit diagram; (b) TR1 cut off; (c) TR2 cut off; (d) Hysteresis effect—(1) loop gain < unity, (2) loop gain = unity, (3) loop gain > unity

emitter resistor as illustrated in Fig. 7.27(a). This will produce a positive feedback arrangement and the state of the circuit will now depend on the voltage V_1 at the base of transistor TR1.

Consider that initially V_1 is zero. Then as for the self-biased bistable multivibrator (Section 7.2) we find that $V_E = -2V$ and $I_{B2} = -165\mu A$. Thus TR2 is saturated and TR1, with its emitter 0·73V negative relative to its base is cut off. The circuit will remain in this condition as V_1 is increased from 0V in a negative direction until it exceeds $-2V$. TR1 will then come out of cut-off and begin to conduct. This will produce a positive voltage step at its collector which is coupled to the base of TR2 tending to cut it off. V_E will reduce thus increasing the negative base–emitter voltage of

TR1 and driving it further into conduction. Positive feedback is thus obtained via the common emitter coupling and, if the loop gain is greater than unity a regenerative switch results, TR1 conducting and TR2 cutting off. Reducing V_1 causes a return to the original condition.

This circuit exhibits a form of hysteresis in its switching action as may be verified by the following simplified analysis.

With reference to Fig. 7.27, ignoring the effect of capacitor C_1, consider the conditions when transistor TR2 is in the *ON* condition and TR1 is *OFF*. The circuit may be redrawn as in Fig. 7.27(b) from which we obtain

$$I_1 = I_2 - I_{B2} \tag{7.12}$$

But, ignoring the small voltage drop V_{BE} of transistor TR2

$$I_1 = \frac{V_{CC}+V_E}{R_1+R_3} \tag{7.13}$$

$$I_2 = -\frac{V_E}{R_4} \tag{7.14}$$

$$I_{B2} = \frac{V_E}{R_5}(1-\alpha_B) \tag{7.15}$$

From Equations 7.12 to 7.15, V_E is obtained as

$$V_E = \frac{-V_{CC}}{1+\frac{1}{R_4}(R_1+R_3)+\frac{1}{R_5}(R_1+R_3)(1-\alpha_B)} \tag{7.16}$$

Hence provided that V_1 is less than the value of V_E given in Equation 7.16 the circuit will remain in the condition with transistor TR2 *ON* and TR1 *OFF*.

Next consider the conditions when V_1 is of such a value that the circuit has changed state, i.e. transistor TR1 is conducting but not necessarily in saturation and TR2 is cut off. Fig. 7.27(c) shows the conditions which now apply. Hence

$$I_3 = -I_{C1} + I_4 \tag{7.17}$$

But, ignoring the voltage drop V_{BE} of TR1 and assuming α_B is close to unity,

$$I_3 = \frac{V_{CC}+V_{C1}}{R_1} \tag{7.18}$$

$$I_{C1} \simeq \frac{V_E}{R_5} \simeq \frac{V_1}{R_5} \tag{7.19}$$

$$I_4 = -\frac{V_{C1}}{R_3+R_4} \tag{7.20}$$

From Equations 7.17 to 7.20 we obtain

$$\frac{V_{CC}+V_{C1}}{R_1} = -\frac{V_1}{R_5} - \frac{V_{C1}}{R_3+R_4} \tag{7.21}$$

i.e.

$$V_{C1} = \frac{-\dfrac{V_{CC}}{R_1} - \dfrac{V_1}{R_5}}{\left(\dfrac{1}{R_1} + \dfrac{1}{R_3+R_4}\right)} \tag{7.22}$$

But

$$V_{B2} = \frac{R_4}{R_3+R_4} \cdot V_{C1} \tag{7.23}$$

Hence from Equations 7.22 and 7.23

$$V_{B2} = \frac{R_4}{R_3+R_4} \left[\frac{-\dfrac{V_{CC}}{R_1} - \dfrac{V_1}{R_5}}{\dfrac{1}{R_1} + \dfrac{1}{R_3+R_4}} \right] \tag{7.24}$$

Transistor TR2 will be cut off as long as V_{B2} is less than V_1. Thus the limiting condition for V_1 from Equation 7.24 is when $V_{B2} = V_1$, i.e.

$$V_1 = \frac{-V_{CC}}{1 + \dfrac{1}{R_4}(R_1+R_3) + \dfrac{R_1}{R_5}} \tag{7.25}$$

Comparison of Equation 7.16 and 7.25 shows that in general the circuit switches at different values of V_1 depending on whether V_1 is increasing or decreasing. In order that the two voltage levels are equal we must have,

$$\frac{R_1}{R_5} = \frac{1}{R_5}(R_1+R_3)(1-\alpha_B)$$

i.e.

$$R_1 \alpha_B = R_3(1-\alpha_B)$$

Hence

$$R_1 = \frac{1-\alpha_B}{\alpha_B} . R_3$$

$$= \frac{R_3}{\alpha_E} \tag{7.26}$$

It can be shown that this condition corresponds to unity loop gain in the circuit and in this case the circuit switches at the same value of V_1 whether increasing or decreasing. If R_1 is less than the value given in Equation 7.26 the loop gain is less than unity and a regenerative switching action is not obtained. If, however, R_1 is greater than the value given in Equation 7.26 the loop gain exceeds unity and the hysteresis effect illustrated in Fig. 7.27(d) is obtained, i.e. the circuit switches at $-V_{1a}$ when V_1 is increased from zero and at $-V_{1b}$ when V_1 is decreased from some value greater than $-V_{1a}$ towards zero.

7.6. BLOCKING OSCILLATORS

7.6.1. Pulse Transformers

Transformers find application in switching circuits for various reasons, some similar to the uses in analogue circuits, but others particular to switching circuits. For instance, they may be used to change the impedance level at a particular point, for voltage transformation or d.c. isolation. On the other hand, they may be used to perform a switching function, e.g. differentiation or pulse forming.

As with the *RC* coupling circuit we are mainly interested in waveform response and we can deduce the important effects from consideration of an equivalent circuit. The full equivalent circuit is shown in Fig. 7.28(a). In a typical pulse transformer, however, the winding resistance is very small and the core loss resistance is very large. Very little accuracy will be lost if we ignore them for the purpose of this description. Further, the capacitances are relatively small, and so, assuming a unity turns ratio transformer, we may study the approximate equivalent circuit of Fig. 7.28(b).

Assume that the circuit is energised by a rectangular pulse as shown. What will be the output waveshape? The clue to the approach lies in the relative values of L_1 and L_2. In a well-designed pulse transformer the shunt inductance L_2 is very much greater than the leakage inductance L_1. Thus during the time of response of the circuit to the leading edge of the pulse, L_2 may be considered as an open circuit. The equivalent circuit then simplifies to that of

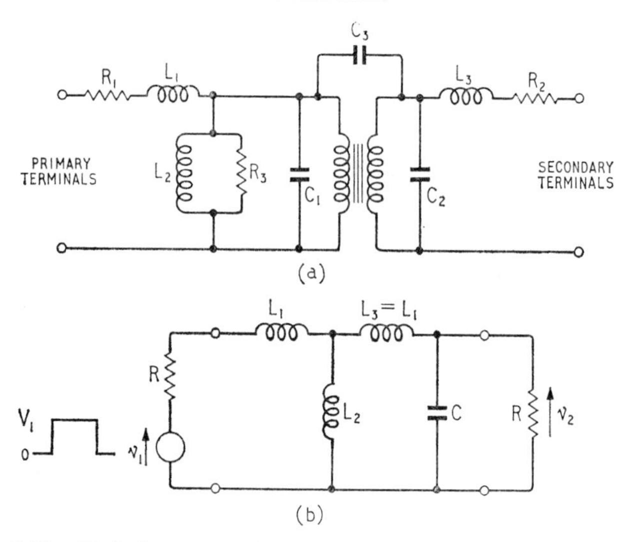

Fig. 7.28. Equivalent circuits for the pulse transformer. (a) Full equivalent circuit, R_1, R_2—winding resistances, L_1, L_3—leakage inductances, C_1, C_2, C_3—winding capacitances, R_3—core loss resistance, L_2—magnetising inductance; (b) Simplified equivalent circuit

Fig. 7.29(a). This circuit is a variation on the basic *LCR* circuit introduced in Section 6.5, and the response depends upon the damping as shown in Fig. 7.29(b). A certain amount of overshoot is acceptable, even desirable, since it shortens the rise time.

During the time of the flat top of the pulse the only element of the equivalent circuit which will significantly affect the response is the shunt inductance. Current will begin to flow in the shunt inductance and thus produce an exponential fall in output voltage. The amount of droop will depend on the time constant formed by L_2 and the parallel combination of the source and load resistance (Fig. 7.30).

Finally we must consider the trailing edge. In general the pulse transformer will be in the collector circuit of a transistor so the source impedance will become practically infinite when the transistor is cut off. The resulting equivalent circuit is in Fig. 7.31. Again we have a resonant circuit, but with conditions different from those at the leading edge, and here also the response depends on damping. The overall pulse response may thus be similar to Fig. 7.32. Notice the undershoot at the trailing edge which, as in the case of the high pass *RC* circuit, must always be present to some extent, due to the

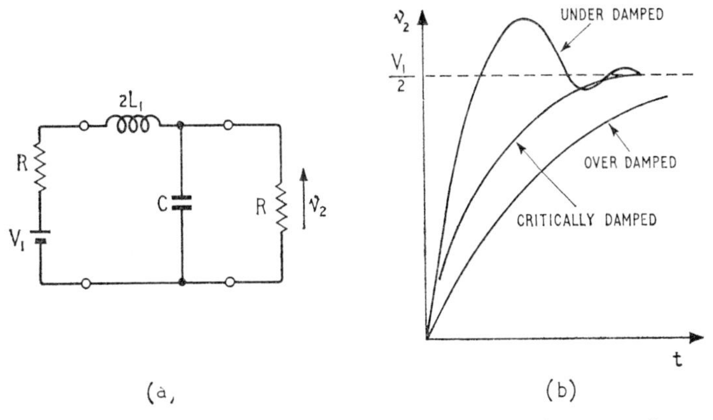

Fig. 7.29. *Determining the leading edge.* (*a*) *Equivalent circuit for leading edge ;* (*b*) *Response to the leading edge*

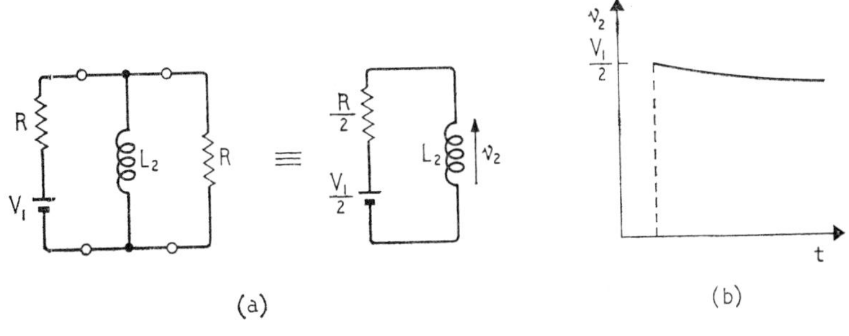

Fig. 7.30. (*a*) *Thevenin's equivalent circuit for the pulse top ;* (*b*) *Response to the pulse top*

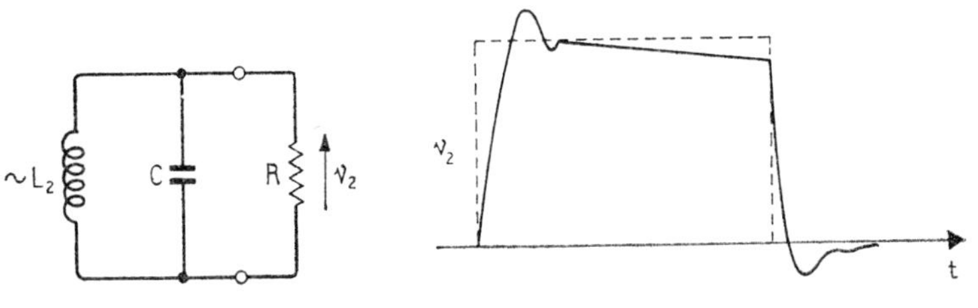

Fig. 7.31. Equivalent circuit for the trailing edge

Fig. 7.32. Typical overall response of pulse transformer

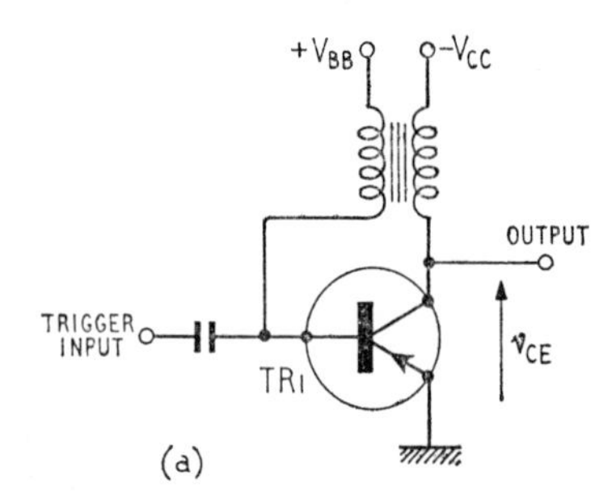

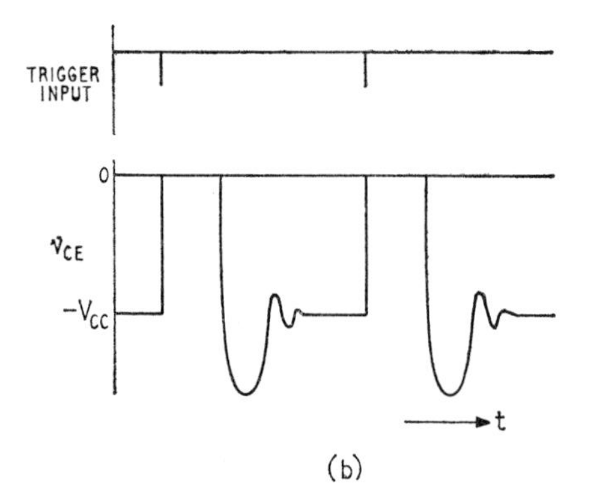

Fig. 7.33. The monostable blocking oscillator. (a) Circuit diagram; (b) Output waveform

fact that the d.c. component is removed and the areas of the pulse above and below the zero voltage axis must be equal. D.c. restoration may again be introduced.

7.6.2. The Monostable Blocking Oscillator

The monostable blocking oscillator circuit of Fig. 7.33(a) illustrates a typical application of the pulse transformer to switching circuits.

Consider the sequence of events in the circuit of Fig. 7.33(a). In the stable condition TR1 is cut off by the positive base bias. If now a negative pulse is applied to the base with sufficient amplitude to cause TR1 to begin conducting, collector current will commence to flow. This changing collector current will induce a voltage in the secondary winding of the transformer and the winding connections are made so that this voltage is returned to the base with a polarity such as to increase the collector current. Thus positive feedback occurs, producing a regenerative switching action and provided the loop gain exceeds unity, drives TR1 into saturation. As soon as

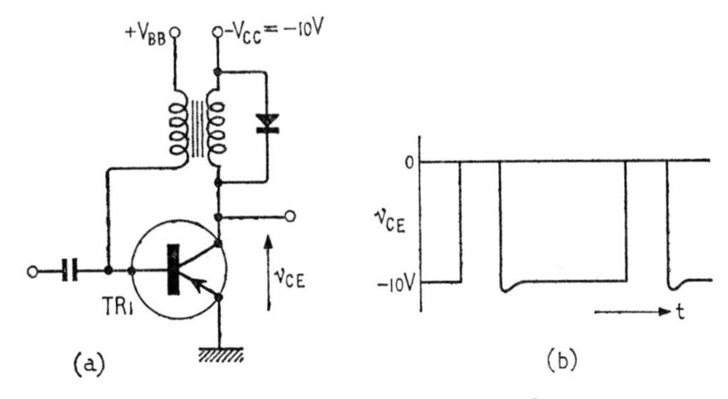

Fig. 7.34. Clamping the collector voltage.
(a) Circuit diagram; (b) Waveform

TR1 saturates, collector current ceases to change, the loop gain is less than unity, and we have the quasi-stable state. The secondary voltage cannot now be maintained across the transformer, but due to the presence of the magnetising inductance will decay exponentially in a manner similar to the decay of the pulse top previously described. When the voltage has decayed sufficiently for the transistor to come out of saturation the collector current begins to decrease. A changing current again induces a secondary voltage but, since now the change is in the opposite direction, the voltage applied to the base will be of opposite polarity, tending to cut the transistor off. Thus a regenerative switch takes place in the opposite

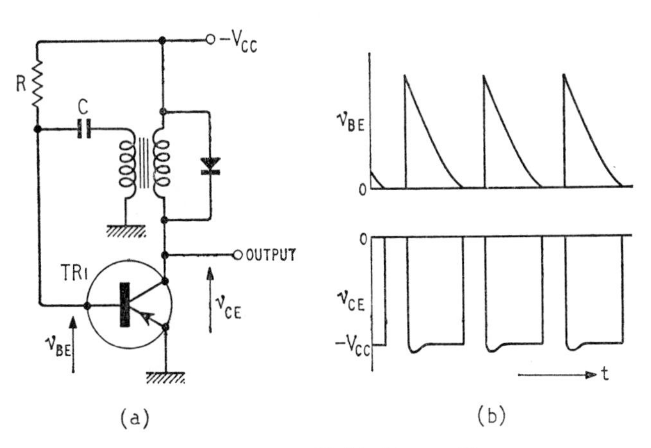

Fig. 7.35. The astable blocking oscillator
(a) Circuit diagram; (b) Waveforms

direction and the transistor is driven into cut-off by a large positive base voltage. Depending upon the transformer parameters a large overshoot in collector voltage may be obtained, as indicated in Fig. 7.33(b). It is possible that this overshoot is of such a magnitude as to exceed the maximum collector voltage $V_{C\,max}$ for the transistor. In such circumstances it is necessary to include a diode across the transformer winding, with polarity as shown in Fig. 7.34. As long as the collector voltage is less than the supply voltage this diode is reverse-biased, but as soon as -10V is exceeded the diode conducts and clamps the collector to approximately -10V.

7.6.3. The Astable Blocking Oscillator

If the previous circuit is rearranged as shown in Fig. 7.35 the result is an astable circuit in which relatively narrow pulses are produced at regular intervals.

Consider this circuit at a time when we have a potential on capacitor C making the base of TR1 positive, thus cutting it off. This potential will discharge via R and the capacitor would eventually attain a voltage of 10V with opposite polarity. However, as soon as the base of TR1 goes negative the transistor will begin to conduct, and the transformer connections are made as before so that a regenerative switch to saturation takes place. The action determining the pulse period is also exactly as for the monostable circuit and at the end of the sequence, when TR1 is cut off, the capacitor C is left with a charge giving positive voltage at the base of TR1. C then begins to discharge via R and the sequence of events commences again. Thus the pulse duration is determined mainly by the parameters of the pulse transformer and the transistor, and the pulse repetition rate by the time constant CR.

CHAPTER 8

LOGIC CIRCUITS

8.1. INTRODUCTION

We have seen that two-state switching circuits are well suited to counting in the binary number system and in this chapter we shall further develop their capabilities of carrying out operations in binary arithmetic. The synthesis of switching circuits for such purposes can become quite complex, but the adaption of a special form of algebra to represent the mathematical relationships between switching functions has led to much more economic and efficient systems, compared to those developed by cut-and-try methods.

This algebra is adapted from Boolean Algebra, a class of mathematics originally introduced by George Boole[1] in connection with his study of logical reasoning, where simple statements which may be true or false, were related to form certain propositions.

Of Boole's propositional logic, 3 basic propositions form the basis of the application of logic to switching systems.

8.1.1. Conjunction

This is the statement formed by connecting 2 or more statements by the connective *AND*. It is usually represented in the algebra by multiplication. Thus if 2 symbols A and B represent logical propositions, their product AB represents the proposition A *AND* B.

8.1.2. Alternation or Disjunction

This is the statement formed by connecting 2 or more statements by the connective *OR*. It is usually represented by addition. Thus the sum $A+B$ represents the proposition A *OR* B.

[1] BOOLE, G., *An Investigation of the Laws of Thought*, Reprint by Dover Publications Inc. (1951).

8.1.3. Negation or Complementation

This is the statement which is true when a given statement is false. It is usually represented by a primed symbol or group of symbols. Thus the proposition A' represents the negation of the proposition A, i.e. *NOT A*.

In the algebra the constants 1 and 0 are used to represent truth and falsity respectively. In this way algebraic equations may be written down which represent simple propositions.

Example 8.1

Let P represent the proposition:
'The sun rises in the east'.
Let Q represent the proposition:
'It always rains on Sunday'.
Then:

$$P+Q = 1 \tag{8.1}$$

i.e. P or Q is true, since P is true

$$PQ = 0 \tag{8.2}$$

i.e. P and Q together are false, since Q is false.

It is apparent immediately that in Boolean Algebra the variables can have 1 of only 2 values and it was this particular characteristic which Shannon[1] appreciated in his interpretation of the algebra in

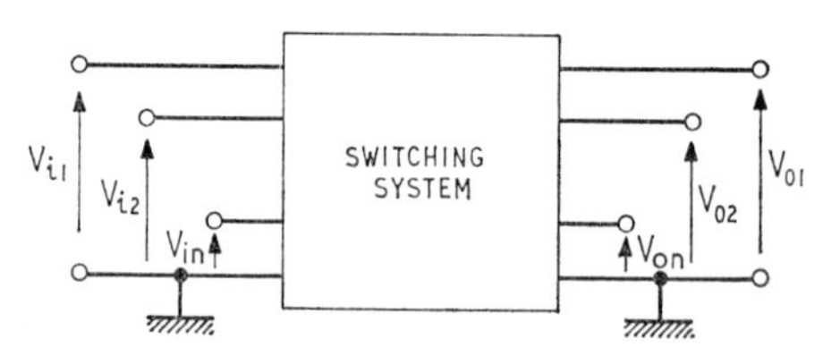

Fig. 8.1. The general switching system connecting input potentials V_i and output potentials V_o

relay switching circuits. A variable A, say, was made to represent a make contact on a relay switch, and the negation of A, that is A', represented a break contact on the same switch. The constant 0 represented a closed switch and 1 an open switch. (Notice that this allocation of 0 and 1 is purely a question of definition and in

[1] SHANNON, C. E., 'A symbolic analysis of relay and switching circuits', *Trans. A.I.E.E.*, 57, 713 (1938).

actual fact we shall use the opposite definitions for reasons which will be clear later.) It is obvious, however, that the application of Boolean Algebra is not restricted to relay circuits but may be applied to any 2-state system. In an electronic switching system for instance the values of the variables may represent a high or low voltage at a particular point, the presence or absence of a pulse in a certain interval of time, etc.

8.2. INTERPRETATION OF BOOLEAN ALGEBRA IN SWITCHING SYSTEMS

In general a switching system, whether it be mechanical or electronic, may be represented as in Fig. 8.1 with a variety of input and output

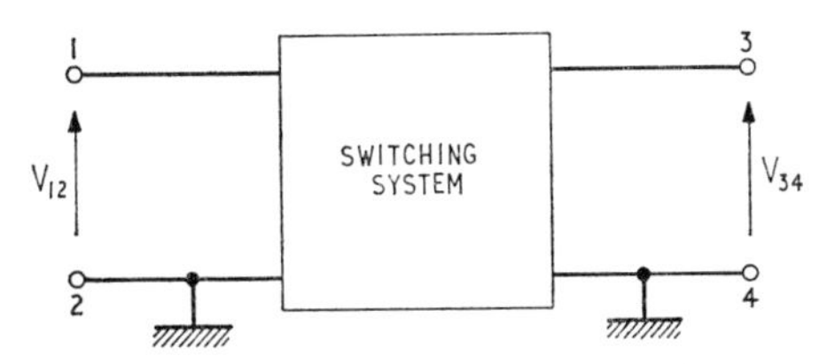

Fig. 8.2. A system with single input and output

connections, and usually, but not always, we are concerned with the relationships between potentials at input and output.

Consider for a moment a system with a single input and output connection as shown in Fig. 8.2, where the input voltage is V_{12} and the output is V_{34}.

If the switching system is perfect then, regardless of its complexity, when there is no connection between terminals 1 and 3, V_{34} is zero. When there is a connection $V_{34} = V_{12}$. Normalising with respect to V_{12} we have the 2 conditions:

No connection

$$\frac{V_{34}}{V_{12}} = 0 \tag{8.3}$$

Connection

$$\frac{V_{34}}{V_{12}} = 1 \tag{8.4}$$

These 2 states correspond to the 2 values of the variables in Boolean Algebra.

In general it is much easier to see connecting paths in relay contact networks than in an electronic switching system, and so, although the electronic system is our main interest, we shall develop the algebra using simple 2-terminal contact networks as illustrations.

We have already seen that a variable is chosen to represent a make contact, the negation of that variable then representing a break contact on the same switch. We shall allocate the constant 0 to represent an open switch and 1 to represent a closed switch.

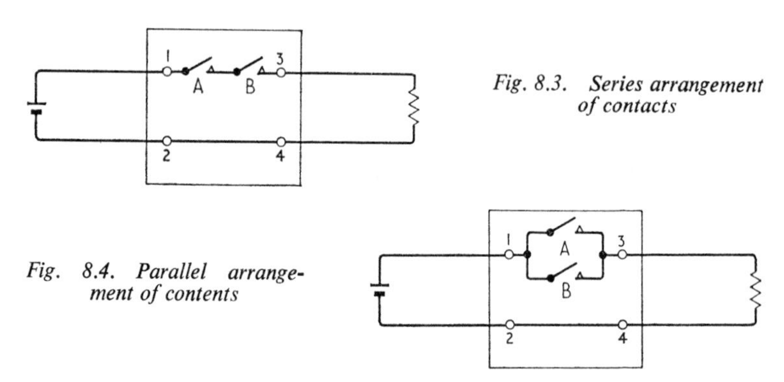

Fig. 8.3. Series arrangement of contacts

Fig. 8.4. Parallel arrangement of contents

Thus we see that the simplest switching system we could fit in Fig. 8.2 would be a make contact between terminals 1 and 3.

If 2 separate make contacts, *A* and *B*, say, are connected in series between terminals 1 and 3 of the switching system as shown in Fig. 8.3, then there will be a connection between terminals 1 and 3 only if contacts *A* and *B* are closed simultaneously. Thus multiplication of the variables corresponds to series connection. Conversely, if the make contacts are connected in parallel as in Fig. 8.4, then a connection will be made if either *A* or *B* operates. Thus addition corresponds to parallel connection.

In this way it is possible to write an algebraic expression corresponding to any series-parallel arrangement of a 2-terminal network.

8.3. FUNDAMENTALS OF BOOLEAN ALGEBRA

In this section the basic rules and theorems of Boolean Algebra applicable to switching circuits will be given. Rigorous proofs are not included but the validity of various statements will be demonstrated, where relevant, by reference to contact circuit examples.

Since the variables can assume only 1 of 2 values it is also possible to verify the relationships between functions involving relatively few variables by evaluating the functions for all possible combinations

Table 8.1. TRUTH TABLE FOR THE BASIC OPERATIONS

A	B	*AND* AB	*OR* $A+B$	*NOT* A'
0	0	0	0	1
0	1	0	1	1
1	0	0	1	0
1	1	1	1	0

of values of the variables. When this is carried out in tabular form, as illustrated in Table 8.1, the result is termed a *truth table.* For n variables there will be 2^n possible combinations and, as is demonstrated in Table 8.1, these combinations will correspond to the binary numbers 0 to 2^n-1. It is thus a simple matter to construct the truth table. We shall see later that the truth table has further applications as a means of writing a particular switching requirement, from which the algebraic relationships may be deduced.

Bearing in mind that the function of subtraction and division do not exist in Boolean Algebra, the various expressions may be dealt with in a manner similar to ordinary numerical algebra, i.e. terms may be rearranged, multiplied out, factored and combined according to the usual rules. Due to the restrictions on the values of the variables in Boolean Algebra there are a number of rules which permit simplification of expressions and which do not apply in normal algebra.

Boolean Expression *Circuit Interpretation*

$F=0$

$F=1$

$F=A$ (switch A)

$F=A'$ (switch A')

$F=AB$ (switches A, B)

$F=A+B$ (switches A, B) (8.5)

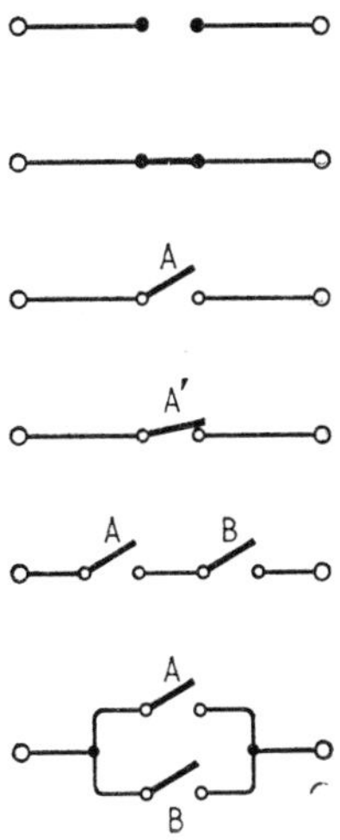

Boolean Expression	*Circuit Interpretation*	
$F = A[B+C(D+E')]$	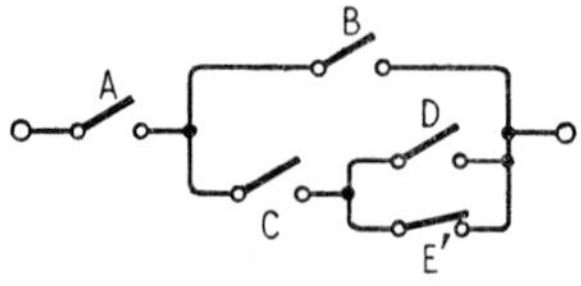 A, B, C, D, E′	(8.5)
$0.0 = 0$	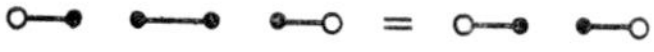	(8.6)
$0+0 = 0$	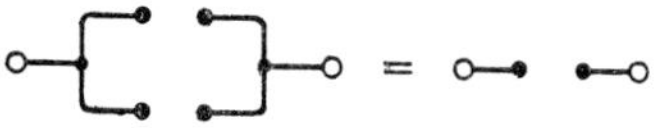	
$1.1 = 1$		
$1+1 = 1$	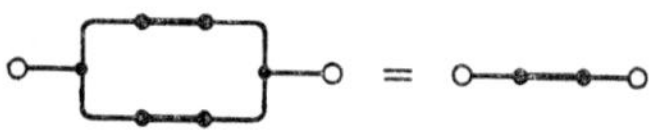	
$1.0 = 0$		
$1+0 = 1$	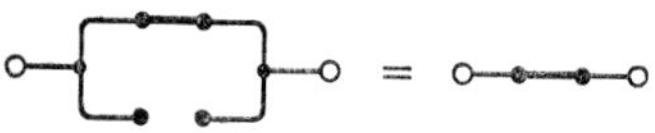	
$0' = 1$		
$1' = 0$		
$A+0 = A$	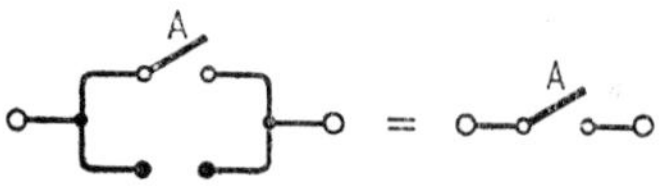 A = A	(8.7)
$A+1 = 1$	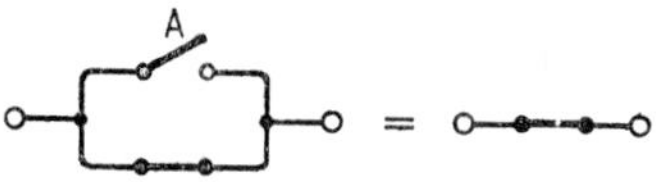 A	
$A.0 = 0$	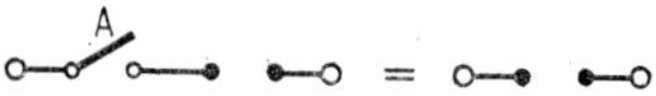 A	

Boolean Expression *Circuit Interpretation*

$$A.1 = A$$

$$A.A = A$$

$$A + A = A$$

$$A + A' = 1$$

$$A.A' = 0$$

$$(A')' = A$$

(8.7)

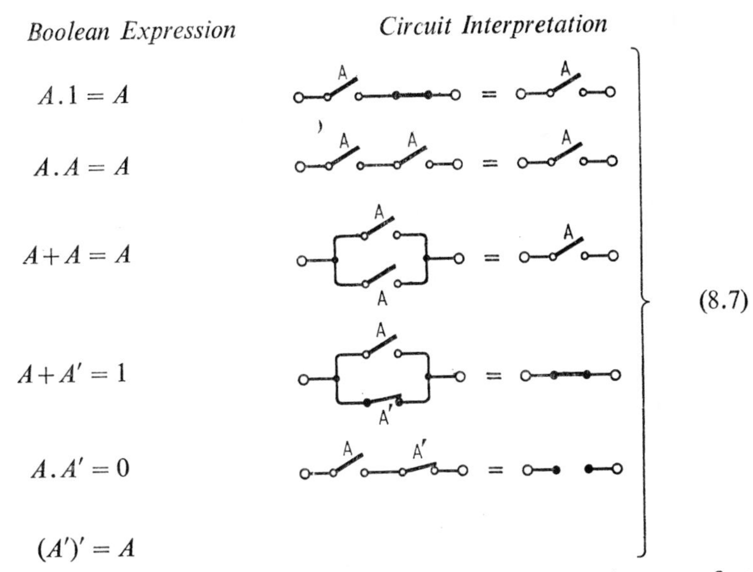

The important points to notice from these expressions are first that there is no requirement for numerical coefficients or indices, since in general:

$$A + A + A + \ldots = nA = A$$

$$A.A.A.A.\ldots = A^n = A$$

Further, we see that the sum of a variable and its complement is always 1 and the product of a variable and its complement is always 0. Next consider relationships involving more than one variable.

As in normal algebra the laws of commutation, association and distribution apply

$$A + B = B + A$$

$$A + B + C = A + (B + C) = (A + B) + C$$

$$ABC = A(BC) = (AB)C$$

$$A(B + C) = AB + AC$$

(8.8)

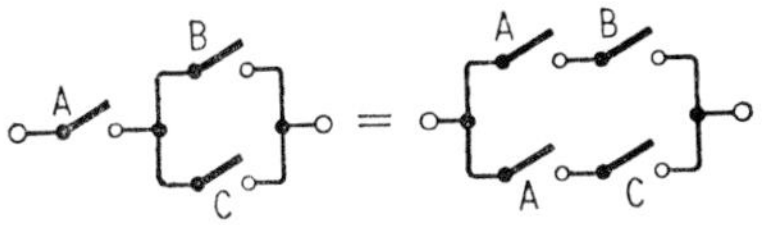

Of the relationships which are peculiar to Boolean Algebra and to which no parallel exists in normal algebra the following are the most important in switching systems:

Boolean Expression *Circuit Interpretation*

$A + AB = A$
since $1 + B = 1$ (8.9a)

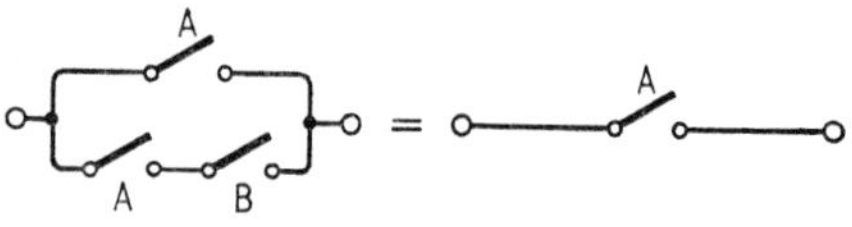

$A(A + B) = A$
since $AA = A$
and $1 + B = 1$ (8.9b)

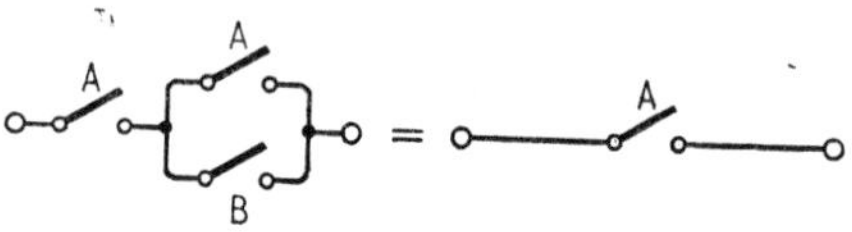

$A(A' + B) = AB$
since $AA' = 0$ (8.9c)

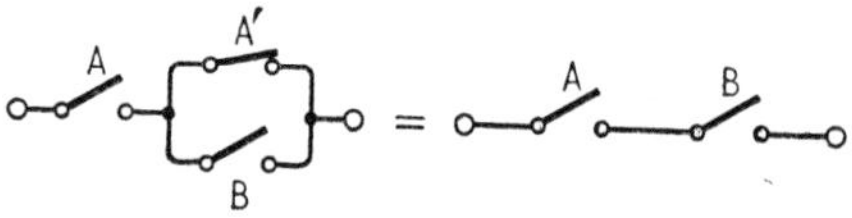

$A + A'B$
$= (A + A')(A + B)$
$= A + B$
since
$AA = A, AA' = 0,$
$1 + B = 1, A + A' = 1$ (8.9d)

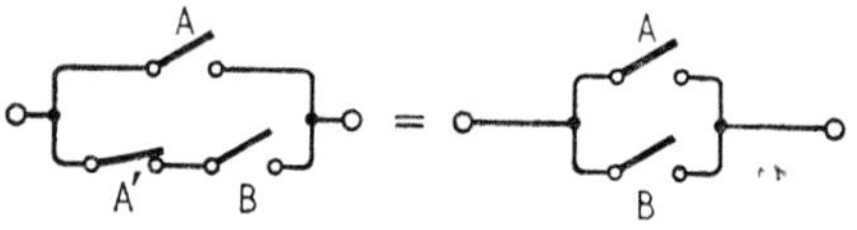

$A + BC$
$= (A + B)(A + C)$
since
$AA = A,$
$1 + B = 1$ (8.9e)

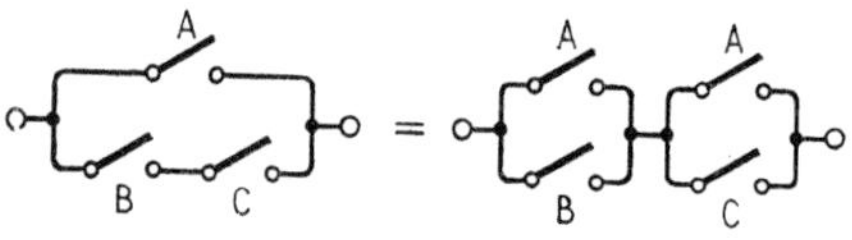

Finally:

$$(AB)' = (A' + B') \tag{8.10a}$$

since

$$(AB)(A' + B') = AA'B + ABB' = 0.B + 0.A = 0 + 0 = 0$$

and

$$AB + A' + B' = A' + B + B' = A' + 1 = 1$$

$$A'B' = (A + B)' \tag{8.10b}$$

since $A'B'\ (A + B) = AA'B' + A'B'B = 0 + 0 = 0$

and $A'B' + A + B = A + B' + B = A + 1 = 1$

Table 8.2. TRUTH TABLE VERIFYING DE MORGAN'S THEOREM FOR TWO VARIABLES E.G. $(AB)' = A'+B'$; $(A+B)' = A'B'$

A	B	A'	B'	AB	$A'B'$	$(A+B)$	$(A'+B')$	$(AB)'$	$(A+B)'$
0	0	1	1	0	1	0	1	1	1
0	1	1	0	0	0	1	1	1	0
1	0	0	1	0	0	1	1	1	0
1	1	0	0	1	0	1	0	0	0

Equations 8.10(a) and (b) are also verified by the truth table in Table 8.2.

These last two most important relationships are known as the Laws of De Morgan[1] from whom we obtain the following Theorem :

De Morgan's Theorem

'To negate a Boolean expression, prime every unprimed symbol, unprime every primed symbol and interchange sum and product signs throughout.'

Example 8.2

Three girls, Mary, Jane and Alice are blonde, brunette and auburn, but not necessarily in that order. Of the following statements only one is true :

Mary is blonde.

Jane is not blonde.

Alice is not brunette.

What colour hair do the girls have?

Let X, Y and Z represent Mary, Jane and Alice, respectively, and let the subscripts a, b and c indicate, respectively blonde, brunette and auburn. For example :

$$\text{Jane is auburn} = Y_c$$

Then, since Mary, Jane or Alice is blonde

$$X_a + Y_a + Z_a = 1 \tag{8.11}$$

[1] DE MORGAN, A., *Syllabus of a Proposed System of Logic*, Walton & Maberly (1860).

Also Mary is either blonde, brunette or auburn.

Hence
$$X_a + X_b + X_c = 1 \tag{8.12}$$

Both Mary and Jane cannot be blonde, so

$$X_a Y_a = 0 \tag{8.13}$$

Since Mary cannot be both blonde and brunette

$$X_a X_b = 0 \tag{8.14}$$

Now from the information given

$$X_a Y_a Z_b + X_a' Y_a' Z_b + X_a' Y_a Z_b' = 1 \tag{8.15}$$

But since $X_a Y_a = 0$ the first term is zero.
Hence

$$X_a' Y_a' Z_b + X_a' Y_a Z_b' = 1$$

i.e.

$$X_a'(Y_a' Z_b + Y_a Z_b') = 1 \tag{8.16}$$

Therefore $X_a' = 1$, i.e. $X_a = 0$ and Mary is not blonde.
Now from Equations 8.16 and 8.11

$$Y_a' Z_b + Y_a Z_b' = 1 \tag{8.17}$$

$$Y_a + Z_a = 1 \tag{8.18}$$

Multiply Equation 8.17 by Equation 8.18 and cross out zero terms.
Therefore

$$Y_a Z_b' = 1 \tag{8.19}$$

Thus the given statement ' Alice is not brunette ' is true. Also Jane is blonde, so Alice must be auburn, Mary being brunette.

8.4. ALGEBRA OF SETS

A further section of Boolean Algebra that is useful in a study of switching systems is that known as the Algebra of Sets, or Classes. A set is defined as a collection of elements with a particular characteristic, for instance the characteristic of roses. Notice that the actual number of roses is of no consequence.

Included in this set there may be a subset; for instance the set of red roses is included in the set of roses.

In defining the set of roses we must automatically define a second set, the set of everything except roses, i.e. the complement of the set of roses. There is obviously no element which is not in one of these sets and so we can define a third set which includes both of these sets, in other words it contains everything and is called the universal set. The complement of the universal set is obviously an empty set, the null set.

In order to illustrate the laws of the algebra of sets a diagrammatic representation of the sets can be built up. This concept was introduced by John Venn[1], hence the diagrams are referred to as Venn diagrams.

A rectangle of arbitrary size is used to represent the universal set and circles within the rectangle represent sets included in the

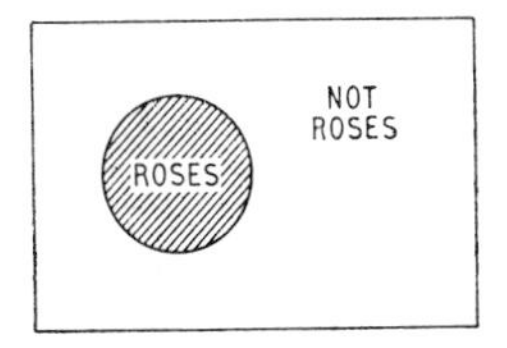

Fig. 8.5. Venn diagram illustrating the set of roses

universal set. Thus the circle shaded in Fig. 8.5 represents the set of roses and the unshaded area represents the set of everything but roses. Now consider a set of all red elements. Some roses are red and so we obtain an *intersection* of the two sets to give red roses. We have thus obtained a new subset which contains elements which are both roses *AND* red, as is illustrated in Fig. 8.6.

We have in this example thus divided the universal set into 4 subsets:

(1) The set of elements which are not red and not roses.
(2) The set of elements which are red and not roses.
(3) The set of elements which are roses and not red.
(4) The set of elements which are red and roses.

If we represent these sets by algebraic symbols and use the function of multiplication as previously defined we see immediately the advantage of these diagrams as illustrations of Boolean expressions. We shall go on further to demonstrate their applications in the simplification of Boolean functions.

Thus if A represents the set of roses
B represents the set of red elements
1 represents the universal set
0 represents the null set
then we have for the previous example

$$(1)\ A'B',\quad (2)\ A'B,\quad (3)\ AB',\quad (4)\ AB$$

These sets are illustrated in Fig. 8.7.

[1] VENN, J., *Symbolic Logic* (2nd Ed.) Macmillan & Co. Ltd. (1894).

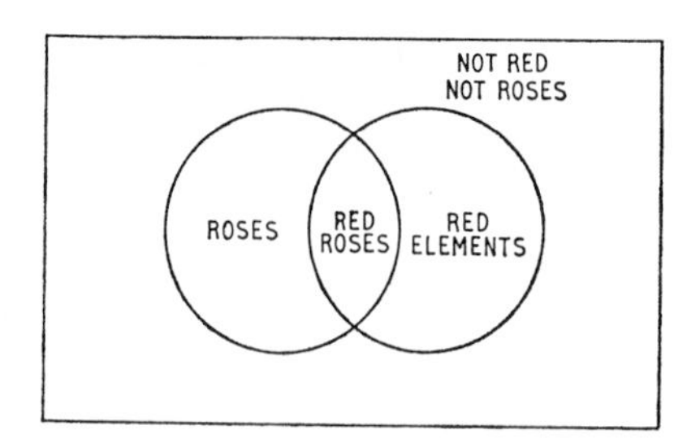

Fig. 8.6. Venn diagram illustrating intersection

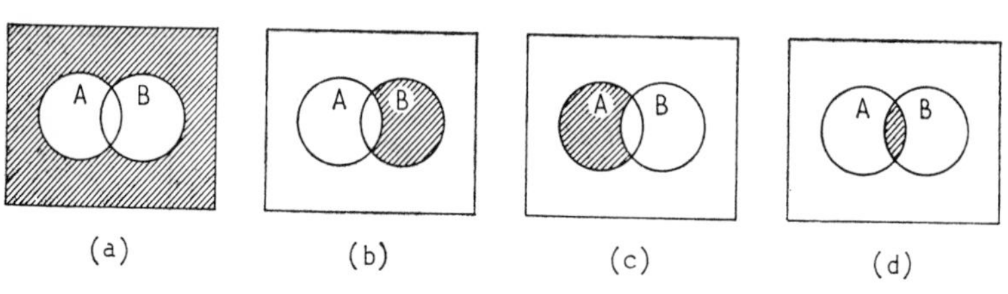

Fig. 8.7. The 4 subsets included in Fig. 8.6.
(a) $A'B'$; (b) $A'B$; (c) AB'; (d) AB

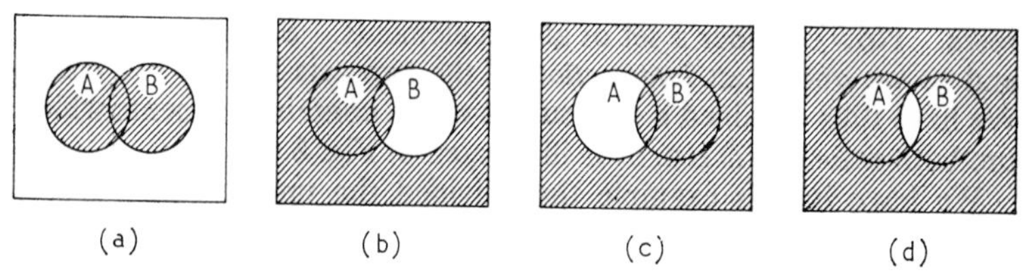

Fig. 8.8. The 4 subsets involved in the union of A and **B**.
(a) $A+B$; (b) $A+B'$; (c) $A'+B$; (d) $A'+B'$

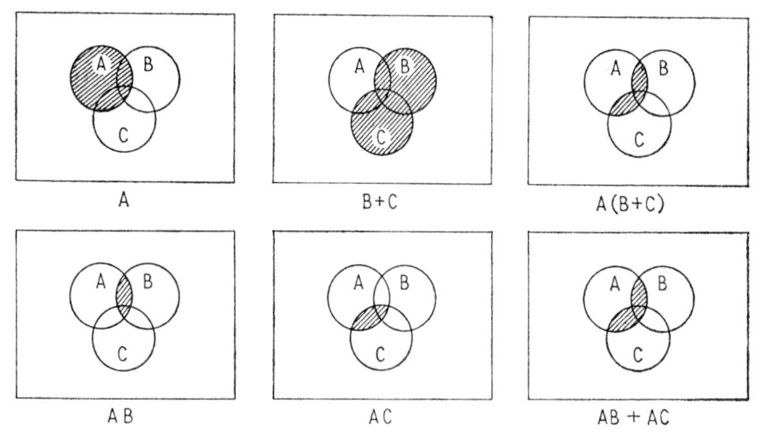

Fig. 8.9. Venn diagrams illustrating the distributive law

Consider next the set of elements with the characteristic of being either roses or red. This set obviously contains the sets of all roses, whatever the colour, and all red elements whether roses or not. We thus obtain the *union* of the set of roses and the set of red elements, which we represent in the algebraic notation as $A+B$. The only elements not included in this set are those which are neither red nor roses. Thus we can again divide the universal set into four subsets, as illustrated in Fig. 8.8.

(1) Elements which are red or roses, $A+B$.
(2) Elements which are roses or not red, $A+B'$.
(3) Elements which are red or not roses, $A'+B$.
(4) Elements which are not roses or not red, $A'+B'$.

It is obvious from the foregoing that if we have a set A, represented by a circle within the rectangle which is the universal set, then that part of the rectangle outside the circle represents the complement or negation of A, i.e. A'. Thus the set $A+A'$ contains all elements of A and of A', i.e. all elements.
Thus

$$A+A' = 1$$

We have thus demonstrated one of the important theorems of Boolean Algebra.

Similarly there are no elements common to A and A', thus

$$A.A' = 0$$

In general the complement of any given set is the set containing all elements not in the original set. Thus the complement of AB is represented on the Venn diagram by the area not covered by AB, i.e. it is the shaded area of Fig. 8.8(d). Thus

$$(AB)' = A'+B'$$

which demonstrates the validity of De Morgan's Theorem.

Further, from Fig. 8.7 and Fig. 8.8.

$$A'B' = (A+B)'$$
$$A'B = (A+B')'$$
$$AB' = (A'+B)'$$

So far we have illustrated the use of Venn diagrams in considering the relationships between two sets within the universal set. It can be extended quite easily to deal with 3 variables by considering the general case of 3 circles drawn to produce all possible intersections between them.

Thus in Fig. 8.9 we demonstrate the distributive law:

$$A(B+C) = AB+AC$$

8.5. ELECTRONIC GATES

In electronic switching circuits it is possible to build up the system from basic logical circuits called *gates* which give the required relationship between input and output signals. There are 3 basic gates corresponding to the Boolean operations of multiplication, addition and negation. They are referred to as *AND*, *OR* and *NOT* gates respectively. The inputs to these gates will consist of signals which are two-valued in nature. For the purpose of description we shall consider that the two values will be either nominally zero voltage, corresponding to 0 in the algebra, or −10V, corresponding to 1. These voltage levels may be steady state levels or represent the extremes of amplitude of a train of pulses of a certain time duration.

In the following section we shall describe simple and, in general, idealised circuits, using diodes and/or transistors which will perform the basic operations.

8.5.1. The NOT Gate

There will be one input to this gate and the output will be 1 (i.e. −10V) only if the input is 0 (0V). Thus, in algebraic notation,

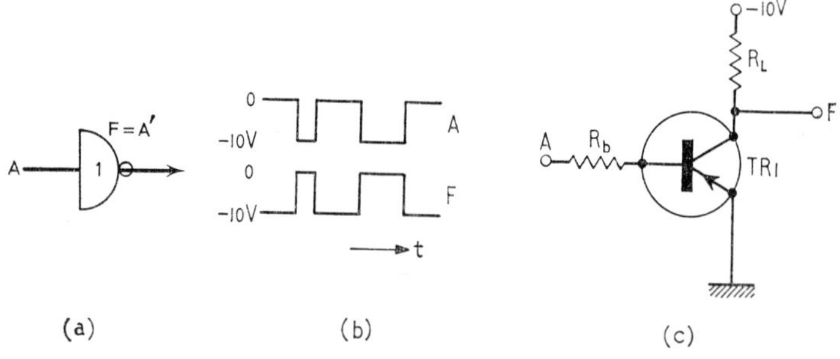

Fig. 8.10. The NOT *gate. (a) Logical symbol; (b) Typical waveforms; (c) Circuit diagram*

if a variable A represents the input and the function F represents the output, then F is 1 if A is 0, i.e. *NOT* 1. Therefore

$$F = A' \tag{8.20}$$

The logical representation of the *NOT* gate we shall use is as shown in Fig. 8.10 with typical input and output waveforms.

It is not possible to realise a circuit to carry out this function using diodes since we are in fact simply inverting the input signal. The transistor in common emitter mode is an inverting amplifier and the circuit of Fig. 8.10(c) will thus give the *NOT* operation. Assuming the transistor is ideal, if the input is 0V, collector current is zero and the output is −10V. If the input is −10V and R_b is such that sufficient base current flows to saturate the transistor, then the output voltage is 0V.

8.5.2. The OR Gate

There may be any number of inputs to this gate but only 1 input is required to be energised in order to obtain an output signal. Thus the output will be −10V if any 1 input is −10V. If variables

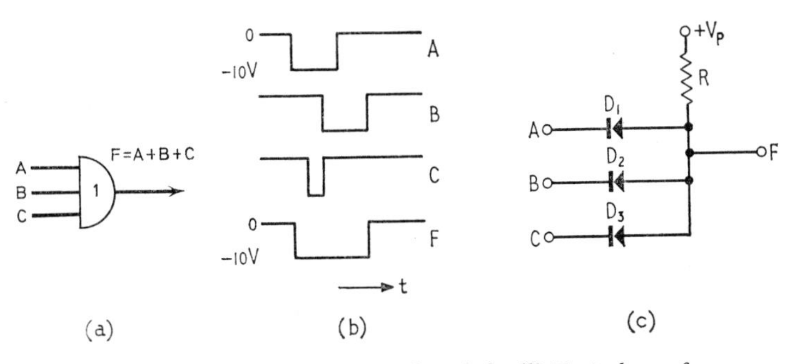

Fig. 8.11. The OR *gate. (a) Logical symbol; (b) Typical waveforms, (c) Diode* OR *circuit*

A, *B*, *C* represent 3 inputs to the *OR* gate then the output *F* is 1 if *A* is 1 or *B* is 1 or *C* is 1,

i.e.
$$F = A + B + C \tag{8.21}$$

The logical representation of the *OR* gate is as shown in Fig. 8.11(a). The number in the semi-circle indicates the number of inputs which must be energised in order to obtain an output. Typical waveforms for the 3 input *OR* gate are shown in Fig. 8.11(b).

The simplest method of realising an electronic *OR* gate is by means of diodes and resistors. Fig. 8.11(c) shows a possible circuit for a 3-input gate. Assume for the purposes of description that the diodes are ideal, that V_P is greater than 10V and that the input signals on *A*, *B* and *C* are either 0V or −10V. If all inputs are at 0V all diodes are forward-biased and the output is clamped to 0V. Now if any 1 input, *A* say, falls to −10V, the others remaining at 0V, a study of the circuit shows that the output will be clamped at −10V, with D_1 conducting and all other diodes reverse-biased.

Though this is a simple circuit and carries out the *OR* function satisfactorily, in practice, due to the forward diode drop and the loading effect of succeeding stages, attenuation of the signal occurs. Thus it is not possible to cascade many gates without including amplification.

A more efficient arrangement uses directly coupled transistors. Consider the circuit of Fig. 8.12 in which 2 transistors are paralleled

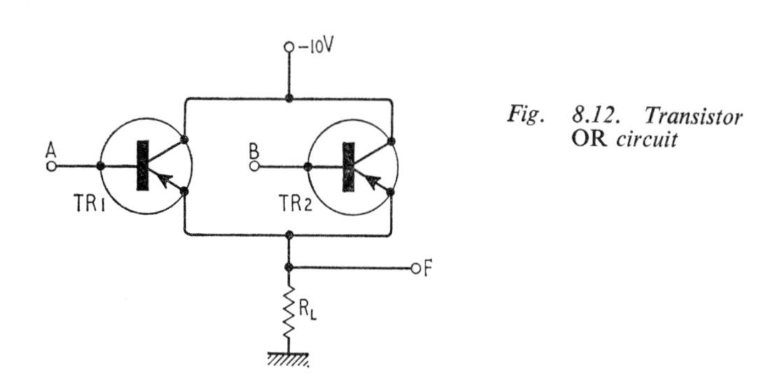

Fig. 8.12. *Transistor* OR *circuit*

in common collector mode. If A or B is -10V then, assuming ideal transistors, the output is also -10V because the emitter potential follows that of the base.

Thus $F = A+B$, giving an *OR* circuit.
There is still some attenuation but the loading is greatly reduced compared to the diode circuit, due to the current gain available.

Consider next 2 transistors paralleled in the common emitter mode, Fig. 8.13(b). The output voltage will now be -10V only if both inputs are 0V, i.e. both transistors cut off.

Therefore F is 1 if A is 0 *AND* B is 0

$$F = A'B'$$

By De Morgan's Theorem $A'B' = (A+B)'$.
Therefore

$$F = (A+B)' \qquad (8.22)$$

This circuit thus gives an *INVERTED-OR* function and is usually referred to as the *NOR* circuit. To complete the *OR* function it is thus necessary to follow up with a *NOT* gate. It is possible to incorporate the *NOR* function into the algebraic expressions as we shall see, and for this reason we give it the logic diagram as in Fig. 8.13(a).

A more economical *NOR* circuit may be obtained by combining the inputs by equal resistors, R_b at the base of a single transistor as in Fig. 8.13(c). Again when all inputs are 0V the output is −10V and when any one input is −10V, the value of R_b is such that sufficient base current flows to saturate the transistor, giving 0V output. In practice, this circuit is not as simple to design as it appears, the tolerance on resistance values being very close for more than 3 or 4 inputs. Also when all of n inputs are 1, the transistor must accept a base current n times the saturation value.

8.5.3. The AND Gate

There may be any number of inputs to this gate also, but a signal will be obtained at the output only if there is a signal simultaneously

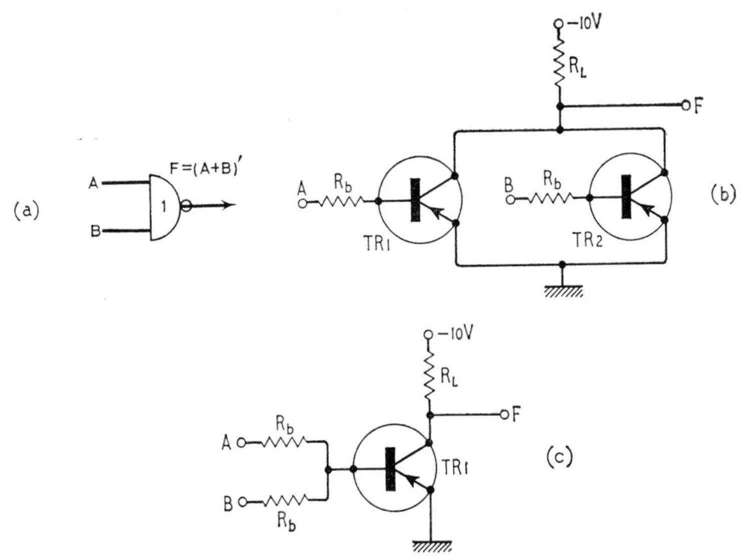

Fig. 8.13. The NOR *gate. (a) Logical symbol; (b) Multi-transistor circuit; (c) Single transistor circuit*

on all inputs. For this reason it is also sometimes referred to as a *COINCIDENCE* gate. Thus if variables A, B and C represent 3 inputs to the *AND* gate, the output F will be −10V only if A *AND* B *AND* C are simultaneously −10V.

Hence

$$F = ABC \tag{8.23}$$

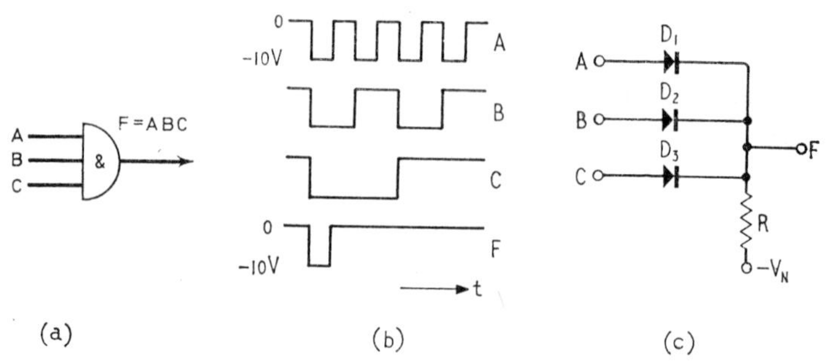

Fig. 8.14. The AND *gate.* (*a*) *Logical symbol;* (*b*) *Typical waveforms;* (*c*) *Diode* AND *circuit*

The logical representation for the *AND* gate we shall use, with typical input and output waveforms, is shown in Fig. 8.14. As before, the number in the semi-circle indicates the number of inputs which must be energised in order to obtain an output.

Again the simplest method of realising the *AND* gate is with diodes and resistors as in Fig. 8.14(c), where the diodes are assumed

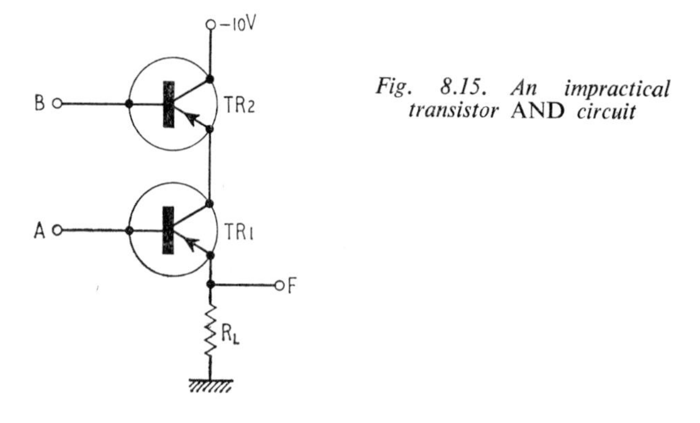

Fig. 8.15. An impractical transistor AND *circuit*

ideal and $V_N \gg 10$V. If all inputs are 0V, all diodes are forward-biased and the output is clamped to 0V. Now let input A fall to -10V, the other inputs remaining at 0V. The only effect of this is to reverse-bias D_1 by 10V, the output remaining clamped to 0V. The same condition holds if B also falls to -10V. If now C falls to -10V the output must follow it and the *AND* operation is complete.

Consider again the directly coupled transistor circuit. It is not possible to obtain the *AND* function directly, since if we series connect in the common collector mode as in Fig. 8.15, the output will follow the A input regardless of the B input condition. The arrangement of Fig. 8.16(b) is practical but must also give inversion.

The output voltage will be -10V if A is 0V (TR1 cut off) or B is 0V (TR2 cut off), i.e. F is 1 if A is 0 *OR* B is 0, or

$$F = A' + B'$$

Again by De Morgan's Theorem

$$F = (AB)' \tag{8.24}$$

This circuit thus gives an *INVERTED AND* function and is usually referred to as the *NAND* circuit, with a logic diagram as in Fig. 8.16(a).

As for the *NOR* circuit it is possible to *NAND* several inputs with one transistor, Fig. 8.16(c), but the circuit arrangement is much more critical. This is so since for n inputs, with all n at -10V, the transistor must be saturated and for $n-1$ inputs at -10V the transistor must be cut off. Thus even assuming ideal transistors a positive base bias is necessary.

The circuits just discussed are rather simplified, but they will serve to enable us to approach the synthesis of switching circuits logically, making use of Boolean Algebra to formulate the initial switching functions and to simplify functions where possible. We shall draw logic diagrams interconnecting the symbols previously defined and finally construct the electronic circuit, sometimes using diodes and transistors, sometimes transistors only. The switching circuits will be confined to those useful in digital computers.

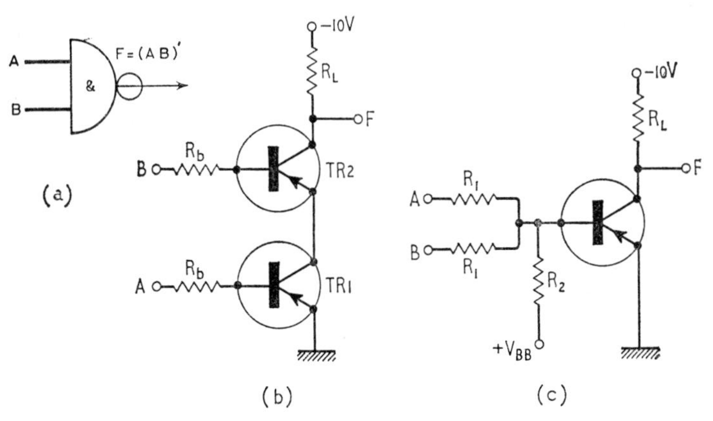

Fig. 8.16. The NAND *gate. (a) Logical symbol; (b) Multi-transistor circuit; (c) Single transistor circuit*

Example 8.3

Develop a switching circuit to carry out the function expressed in the truth table of Table 8.3. Realise the circuit (a) with the minimum possible transistors, (b) with transistors only.

Table 8.3. TRUTH TABLE FOR EXAMPLE 8.3

A	B	F
0	0	1
0	1	0
1	0	0
1	1	1

(a) Inspection of the truth table shows that F is 1 if A and B are both 0 or both 1.

i.e.
$$F = AB + A'B' \tag{8.25}$$

The logic diagram for this function is in Fig. 8.17(a) and requires 2 transistors for the inverters. Application of De Morgan's Theorem to the second term of Equation 8.25 gives

$$F = AB + (A+B)' \tag{8.26}$$

This now requires 1 inverter and so the required circuit is as in Fig. 8.17(c).

(b) The most economic transistor scheme will use the *NOR* circuit with resistive inputs, therefore the equations derived above must be rearranged to accommodate the *NOR* function.

Equation 8.26 may be rewritten by factorising as

$$F = [A + (A+B)'][B + (A+B)']$$

This equation, by De Morgan's Theorem is equivalent to

$$F = \{[A + (A+B)']' + [B + (A+B)']'\}' \tag{8.27}$$

Each of the terms in this equation is arranged in a form suitable for the *NOR* function and the logic diagram and resulting circuit are shown in Fig. 8.18.

8.6. ELECTRONIC DIGITAL COMPUTERS

The electronic digital computer has come into great use in recent years as a device for doing arithmetic. It cannot do anything which

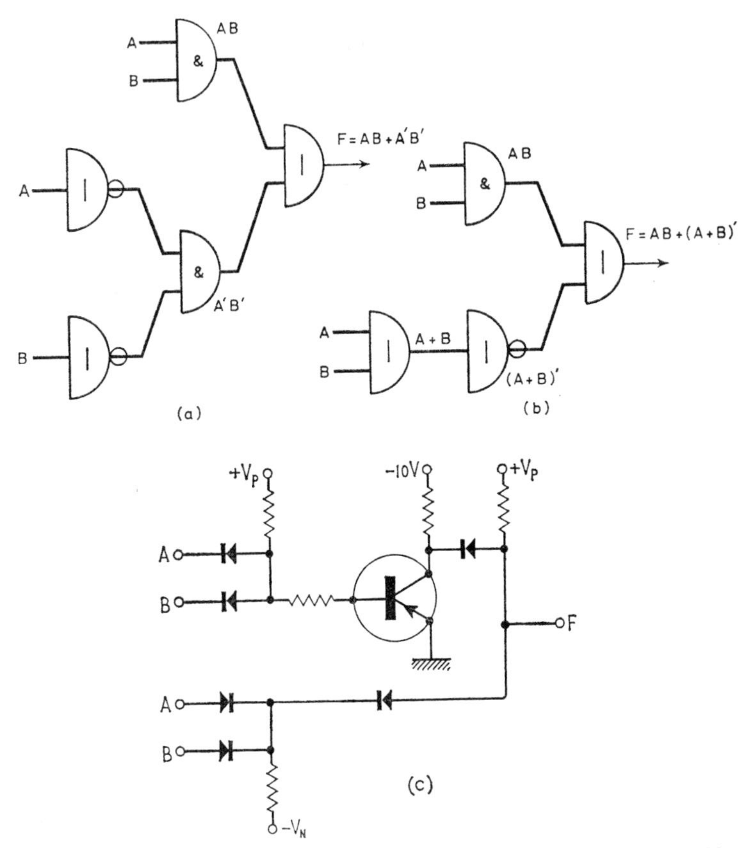

Fig. 8.17. Relevant to Example 8.3(a). (a) Logic diagram for Equation 8.25; (b) Logic diagram for Equation 8.26; (c) Circuit for logic diagram (b)

a trained mathematician is not capable of doing, but its great advantage is the speed at which it can work. We have seen that the simplest and most reliable method of operation of an electronic switch is as a two-state device, and it is for this reason that most modern machines operate with binary arithmetic rather than decimal.

We saw in Chapter 7 that in decimal code a number is represented as an expansion in powers of 10 with coefficients having any value from 0 to 9. For example:

$$4305{\cdot}87 = 4\times10^3+3\times10^2+0\times10^1+5\times10^0+8\times10^{-1}+7\times10^{-2}$$

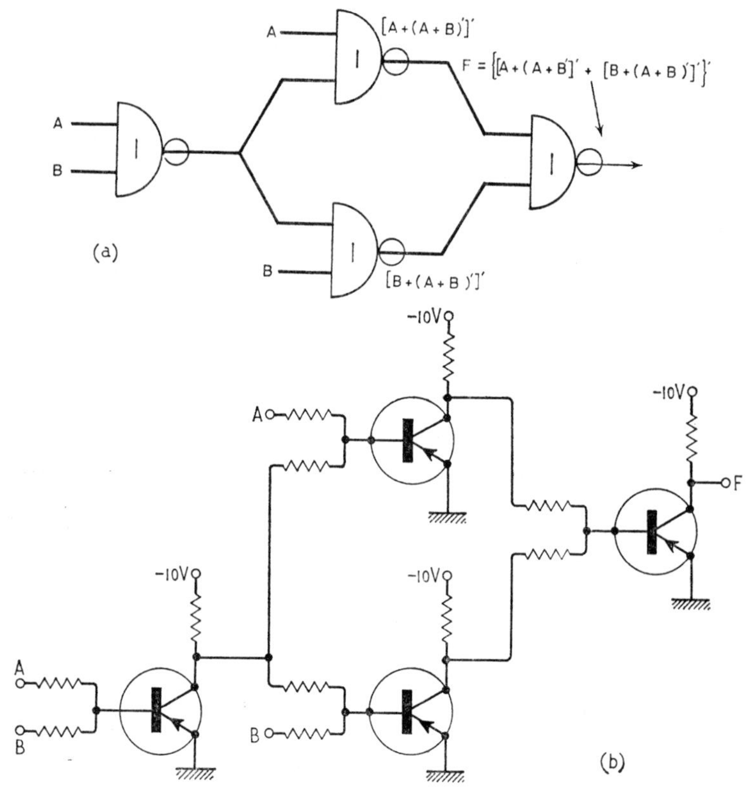

Fig. 8.18. Relevant to Example 8.3(b). (a) Logic diagram for Equation 8.27; (b) Circuit for (a)

The significance of any digit is thus indicated by its position in the number.

In binary code a number is represented as an expansion in powers of 2, with coefficients having one of only two values, 0 or 1, for example:

$$
\begin{aligned}
1011{\cdot}01 \text{ (binary)} &= 1\times 2^3+0\times 2^2+1\times 2^1+1\times 2^0+0\times 2^{-1}+1\times 2^{-2} \\
&= 8+0+2+1+0+0{\cdot}25 \\
&= 11{\cdot}25 \text{ (decimal)}
\end{aligned}
$$

As before the significance of a digit is indicated by its position.

The electronic digital computer carries out the function of arithmetic very much in the way in which we do it with a pencil and paper. The first thing we do when adding two numbers, for instance, is to write the two numbers on the piece of paper. In the computer there must be equipment to effect input and to hold, or store the numbers. Next we carry out the actual function of addition including carrying over of digits from one column to the next. The computer must thus have an arithmetic unit. Having written down the answer we can read it off the paper or use it in a subsequent calculation. Thus the computer must return the number to the store or effect an output. Whereas we keep mental control

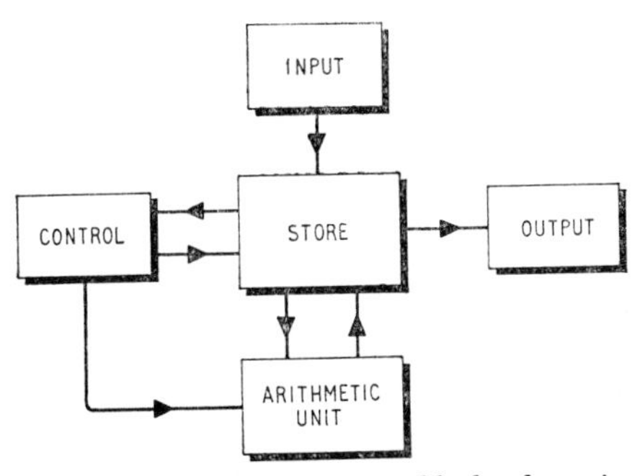

Fig. 8.19. Typical computer block schematic

of the operation, the computer must have control orders fed in to direct the sequence of operations. These orders are also in binary coded form. Therefore the number 1001 can either represent the actual number 9 or the order—'add the two numbers together'. A basic block schematic of a typical computer might therefore be as shown in Fig. 8.19.

As examples of electronic switching circuits and applications of Boolean Algebra we shall only discuss the Arithmetic Unit and Storage devices.

8.7. BINARY ARITHMETIC

Before going on to consider electronic circuits which can do binary arithmetic we must first familiarise ourselves with the rules of the arithmetic. The actual processes of addition, subtraction, multiplication and division are completely analogous to those in decimal arithmetic. A few examples should therefore serve to make the operations clear.

8.7.1. Addition

Each time the digits in a given column add up to 2 a carry to the next higher order is necessary.

I.e.

$$0+0=0$$

$$0+1=1$$

$$1+1=10, \text{ i.e. } 0 \text{ and carry } 1$$

Example 8.4

Add 11001 to 10101

$$\begin{array}{rrl} & 11001 & (= \text{dec. } 25) \\ & 10101 & (= \text{dec. } 21) \\ \hline \text{Sum digit} & 01100 & \\ \text{Carry digit} & 1\ \ \ \ 1\ \ & \\ \hline \text{Final sum} & 101110 & (= \text{dec. } 46) \\ \hline \end{array}$$

8.7.2. Subtraction

In each column, when the digit to be subtracted is greater than that being subtracted from it is necessary to borrow from the next higher order,

i.e.

$$0-0=0$$

$$1-0=1$$

$$1-1=0$$

$$0-1=1 \text{ and borrow } 1$$

Example 8.5

Subtract 10100 from 11011

$$\begin{array}{lrl} & 11011 & (= \text{dec. } 27) \\ & 10100 & (= \text{dec. } 20) \\ \hline \text{Difference digit} & 01111 & \\ \text{Borrow digit} & 1\ \ \ \ \ & \\ \hline \text{Final difference} & 00111 & (= \text{dec. } 7) \end{array}$$

8.7.3. Multiplication

Multiplication in binary arithmetic is a very simple process, being a process of repeated shifting of the multiplicand (the number to be multiplied) to the left (or right for numbers less than unity) and adding when called for by a 1 in the multiplier.

$$0 \times 0 = 0$$

$$1 \times 0 = 0$$

$$1 \times 1 = 1$$

Example 8.6

Multiply 1011 by 1010

```
                        1011   (= dec. 11)
                        1010   (= dec. 10)
                     -------
                   (       0
                   |    1011
Partial products  <       0
                   |1011
                     -------
Product             1101110   (= dec. 110)
```

8.7.4. Division

Division is carried out by right shift and subtraction.

Example 8.7

Divide 101101 by 101 $(45 \div 5)$

```
101|101101|1001
    101
    ---
     001
     000
     ---
      010
      000
      ---
       101
       101
       ---
       000
```

Answer 1001 (= dec. 9)

8.7.5. Binary Complements

Consider a calculation involving binary arithmetic carried out in a machine limited to 6 digits. Thus zero will be

$$000000$$

Now subtract 1

$$\begin{array}{r} 000000 \\ -000001 \\ \hline 111111 \end{array}$$

Thus the negative number, -1 is represented by the positive number 111111 which is the *binary complement* of the negative number. Thus it is possible to carry out the function of subtraction without use of negative numbers, but by *complement addition.*

Example 8.8

Subtract 10100 (= dec. 20) from 11011 (= dec. 27)
To find the binary complement of 10100, subtract from 0,

i.e.

$$\begin{array}{rr} & 00000 \\ & 10100 \\ \hline \text{Difference} & 01100 \end{array}$$

The binary complement is thus 01100 which must be added to 11011

i.e.

$$\begin{array}{r|l} & 11011 \\ & 01100 \\ \hline & 00111 \quad (=7) \end{array}$$

This is a most important function in computer circuits as we shall see later.

8.8. ELECTRONIC CIRCUITS FOR PERFORMING BINARY ARITHMETIC

In the electronic digital computer the digits of the binary numbers are usually represented by zero voltage (0) or a pulse of voltage of some fixed amplitude (1), which we shall assume is -10V. These pulses must be placed in order, to distinguish the significance of the

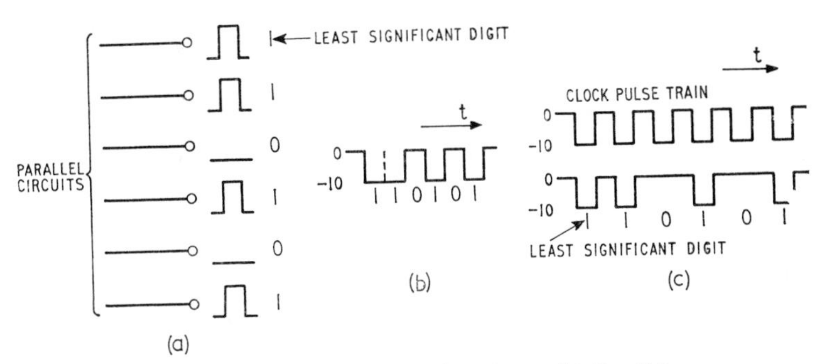

Fig. 8.20. Representation of numbers. (*a*) *Parallel;* (*b*) *Serial d.c. logic;* (*c*) *Serial a.c. logic*

digits they represent. There are 2 fundamentally different methods of representation of the significance of the digit. We can transmit the individual pulses representing digits of a number through different circuits in the machine all at the same time, the significance of the digit being determined by the circuit in which the pulse appears. This is termed *parallel* operation. Alternatively we can transmit a train of pulses in sequence through a common circuit, the significance of the digit now depending on the instant in time that it appears at a particular point. This is termed *serial* operation.

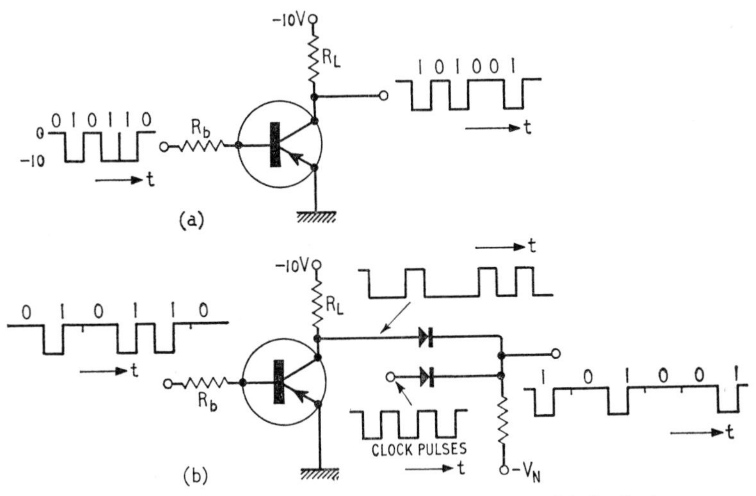

Fig. 8.21. Simple inverter circuits. (*a*) *Inversion with d.c. logic;* (*b*) *Inversion with a.c. logic*

Both methods of operation use the same basic circuits but, for instance when carrying out addition, a parallel machine would need an adding circuit for each digit, the addition time being fundamentally 1 pulse interval. The serial machine is much more economical, since only 1 adding circuit is needed, but it is also much slower, since the addition time will be the product of the number of digits in the number and the pulse interval.

Binary numbers may thus be represented as shown in Fig. 8.20. Notice that 2 distinctly different representations are shown for serial signals. In Fig. 8.20(b) we have a direct relationship between the waveform and the binary digits, which leads to simple circuitry. With this type of signal the *NOT* gate will give inversion direct, as shown in Fig. 8.21(a). Notice that inversion gives the *digital complement*, i.e. 1 replaced by 0 and vice versa, as opposed to the binary complement discussed previously. The disadvantage is that there is no synchronism between different pulse trains in a system. The waveform of Fig. 8.20(c) is obtained from a basic a.c. waveform called the *clock pulse* train. In this system the whole of a computer can be synchronised to a standard time base, giving greater accuracy and allowing for regeneration of signals from the clock pulse train, as shown in Fig. 8.22. During transmission through the various switching circuits, the pulses in general become somewhat distorted in shape. By gating these distorted pulses with the clock pulses in an *AND* gate it is a simple matter to regenerate the original pulse train. Inversion is not directly obtained from the *NOT* gate, which must be followed by an *AND* gate with clock pulses injected on a second input as shown in Fig. 8.21(b).

Notice in Fig. 8.20(b) and (c) that, since time is increasing from the left, the least significant digit occurs first in the pulse train.

8.8.1. Circuits for Addition

Consider the following example of the addition of two numbers X and Y

	X	0101
	Y	0011
Total sum		1000

In this sum we have automatically carried a 1 into the next more significant column where required. In the computer it is necessary to break the process down into two stages, first producing a sum S

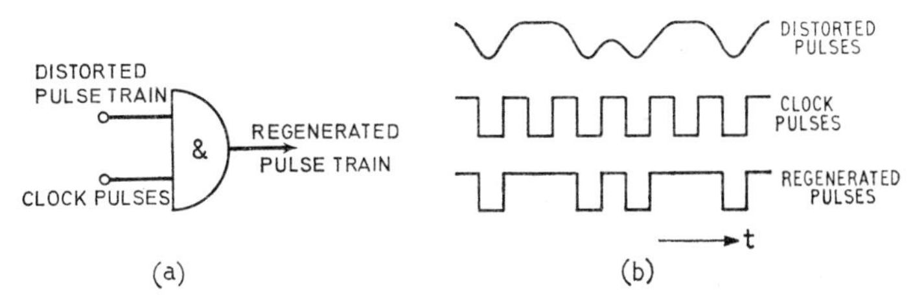

Fig. 8.22. Regeneration of pulses. (a) A simple logical circuit; (b) Relevant waveforms

and carry C. The carry digit is then shifted to the more significant position and added to the sum

$$\begin{array}{ll} X & 0101 \\ Y & 0011 \\ \hline S & 0110 \\ C & 0001 \end{array}$$

Inspection of the above table, which is a form of truth table, shows that S is 1 if X is 1 and Y is 0 or if X is 0 and Y is 1. C is 1 if X and Y are both 1. Writing these conditions in algebraic form we have:

$$S = XY' + X'Y \qquad (8.28)$$

$$C = XY \qquad (8.29)$$

From these expressions we can produce a logic diagram, showing the build-up of the gating elements required to perform the addition procedure. This is shown in Fig. 8.23(a) with a possible circuit configuration illustrated in Fig. 8.23(b). This is known as a *half adder* circuit.

At this point we can demonstrate the benefits to be obtained from the application of logic to the synthesis procedure. Notice that 8 diodes and 2 transistors are required in the circuit of Fig. 8.23(b). Consider again Equation 8.28 for the sum digit. Application of the rules of the algebra will soon show that this can be factorised to give

$$S = (X + Y)(X' + Y')$$

and since

$$X' + Y' = (XY)' \text{ by De Morgan's Theorem}$$

$$S = (X + Y)(XY)' = (X + Y)C' \qquad (8.30)$$

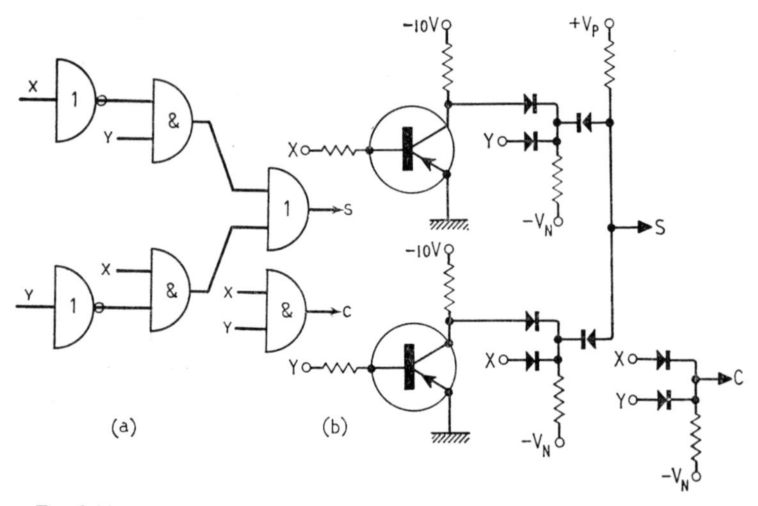

Fig. 8.23. A possible half-adder. (a) Logic diagram; (b) Circuit diagram

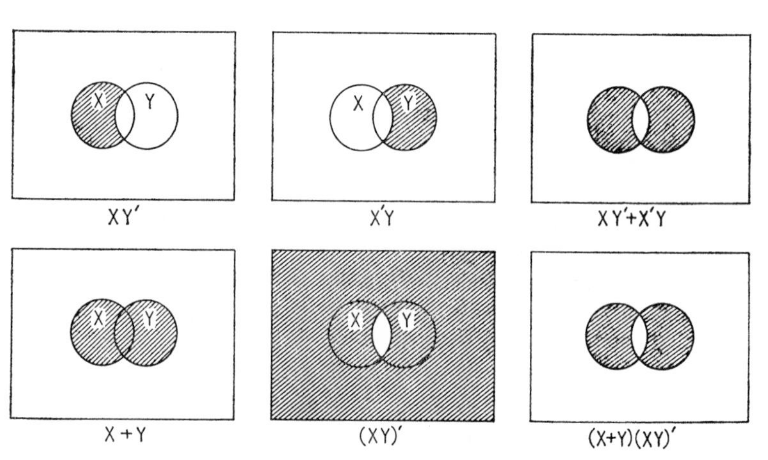

Fig. 8.24. Venn diagrams illustrating the equivalence
$$XY'+X'Y=(X+Y)(XY)'$$

This expression is still correct since it states S is 1 if X or Y is 1 but not if X and Y are 1. A study of the Venn diagram of Fig. 8.24 verifies the equivalence of Equations 8.28 and 8.30.

This is obviously an improvement, since C is now a common factor and only 1 inverter is required. Thus the most economical arrangement for the half-adder is shown in Fig. 8.25(a) and (b). The saving in components is seen to be 2 diodes and 1 transistor.

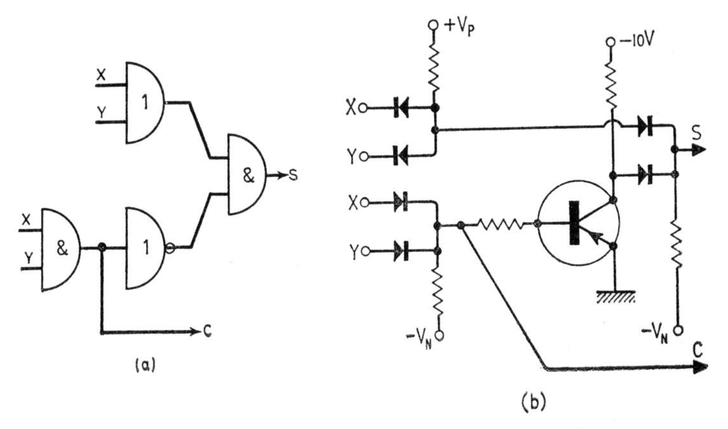

Fig. 8.25. The most economical half adder. (a) Logic diagram; (b) Circuit diagram

The rules which apply to the addition of the carry are exactly the same as above but first it must be shifted to the more significant position. The way in which this is done depends upon the mode of operation, i.e. whether parallel or serial. In both cases we can use another half-adder to effect the actual addition. This half-adder may also produce a carry, but since both half-adders cannot produce a carry at the same time, the 2 carry outputs may be combined in an *OR* gate for transmission to the next higher order. That this is true may be verified from consideration of the rules, for example:

$$C_a = 1 \text{ only if } X = 1 \text{ and } Y = 1$$

But if $X = 1$ and $Y = 1$ then $S_a = 0$

$$C_b = 1 \text{ only if } S_a = 1 \text{ and } C_1 = 1$$

Therefore if C_a is 1, C_b must be 0 and vice versa.

The logic diagram of a 4-digit parallel adder may thus be as shown in Fig. 8.26, where X_1, Y_1 and S_1 are the least significant digits of the numbers X, Y and S respectively.

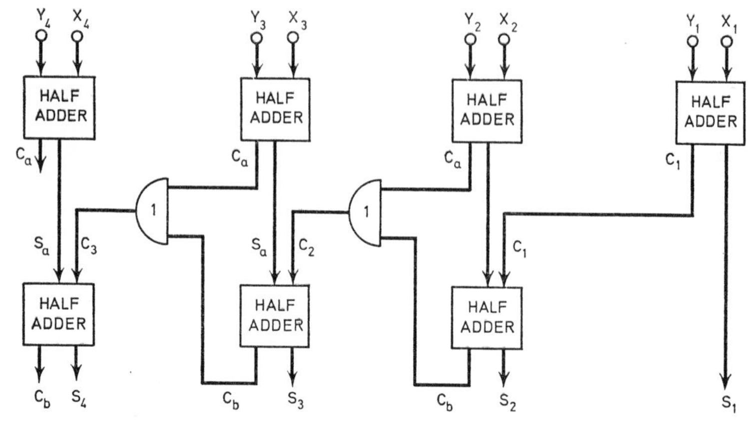

Fig. 8.26. A 4-digit parallel adder

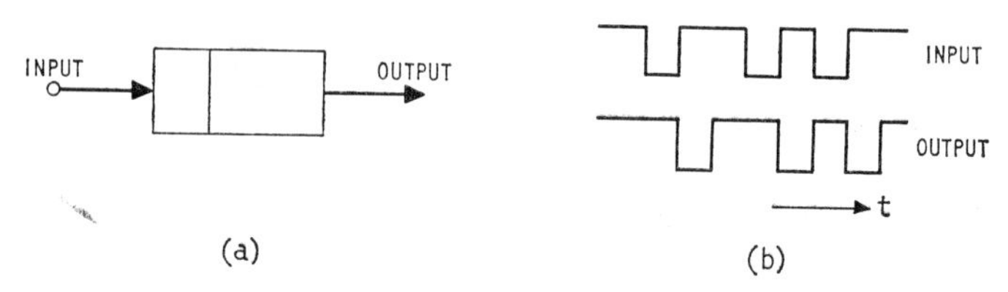

Fig. 8.27. The 1-digit delay unit. (a) Logical symbol; (b) Typical waveforms

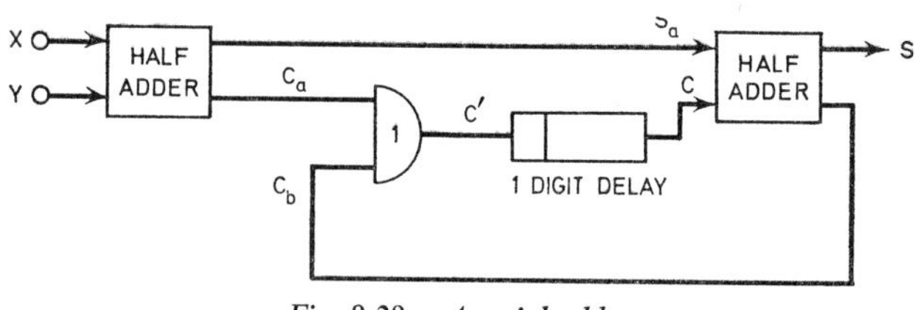

Fig. 8.28. A serial adder

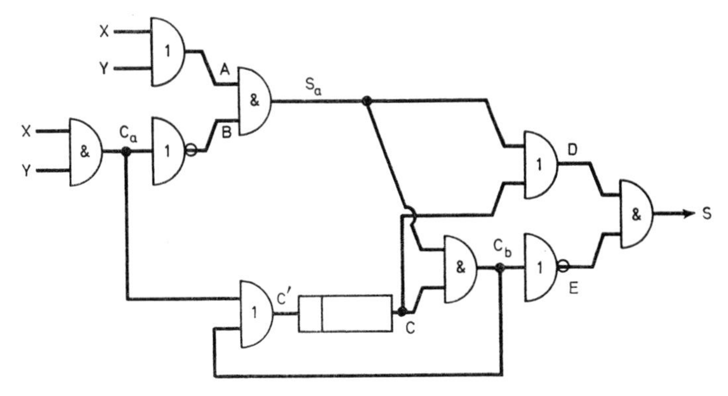

Fig. 8.29. Full logic diagram of the serial adder for use in Example 8.9

In the serial mode of operation, in order to shift a digit to a more significant position it is only necessary to delay it in time by a 1 digit interval. We must thus introduce a new logical element in the form of a device which will produce a prescribed time delay. The logical symbol we shall use with typical waveforms, is shown in Fig. 8.27. The realisation of a circuit to produce the delay we shall consider later.

The logic diagram of a serial adder may thus be as in Fig. 8.28.

Example 8.9

Add 5 to 11 in the 5-digit serial adder shown in Fig. 8.29.

To find the answer consider the waveforms at the points shown in Fig. 8.29 in terms of 0 and 1 and tabulate at digit intervals, least significant digit first, e.g. C is 1 when X and Y are 1, etc. Remembering that the waveform at point C will be identical to that at point C', but delayed by 1 digit interval, and that at the commencement of the addition C is 0, we obtain the result shown in Table 8.4, i.e. 16. In the final digit interval C is 0, i.e. no further carry and so the process is complete.

It may be desirable to realise the half-adder circuit using direct coupled transistors only. To do this it is necessary to reconsider Equations 8.29 and 8.30 as follows

$$S = (X+Y)(XY)' \tag{8.30}$$

$$= X(XY)' + Y(XY)'$$

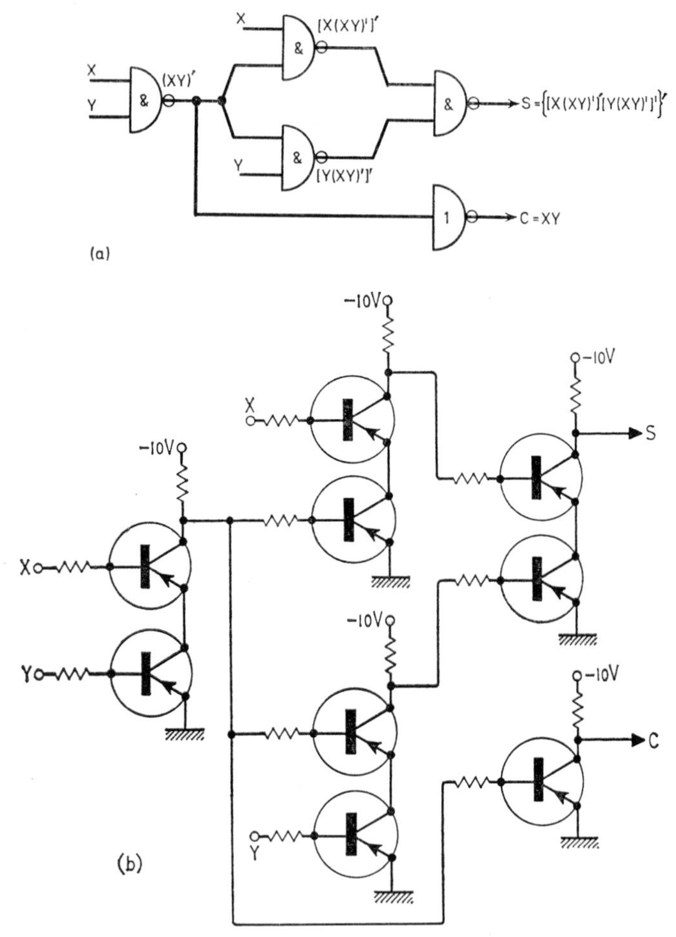

Fig. 8.30. A transistorised-half adder. (a) Logic diagram; (b) Circuit diagram

If we negate this expression twice we obtain

$$S = \{[X(XY)' + Y(XY)']'\}'$$

Now apply De Morgan's Theorem to the expression inside the curly brackets giving

$$S = \{[X(XY)']' [Y(XY)']'\}' \tag{8.31}$$

This function may be realised using *NAND* gates only as shown in Fig. 8.30(a).

Table 8.4

	X	Y	A	C_a	B	S_a	C_b	C'	C	D	E	S
time ↓	1	1	1	1	0	0	0	1	0	0	1	0
	0	1	1	0	1	1	1	1	1	1	0	0
	1	0	1	0	1	1	1	1	1	1	0	0
	0	1	1	0	1	1	1	1	1	1	0	0
	0	0	0	0	1	0	0	0	1	1	1	1

Key:
$X = 00101\ (= 5)$
$Y = 01011\ (= 11)$
$\therefore S = X + Y = 10000\ (= 16)$

Since from Equation 8.29

$$C = XY$$

the carry may be obtained from Equation 8.31 by use of 1 extra inverter.

The transistorised half-adder circuit is shown in Fig. 8.30(b).

8.8.2. Circuits for subtraction

We have seen that there are 2 fundamentally different methods of subtraction available, i.e. either by direct subtraction or by addition of complements.

Direct Subtraction

Consider the problem of subtracting Y from X as in the following example:

	X	0101
	Y	0011
Difference	D	0110
Borrow	B	0010

We have thus produced a difference D and a borrow B, where required, in a manner similar to the function of addition. It is again necessary to shift the borrow digit to a more significant position and complete the subtraction.

From the above table we can deduce

$$D = XY' + X'Y$$

$$B = X'Y \qquad (8.32)$$

Notice that the difference function is the same as the sum in addition. Also the digits X and Y are not now interchangeable.

$$\text{As before} \quad D = XY' + X'Y$$

$$= (X+Y)(XY)' \tag{8.33}$$

and

$$B = X'Y$$

$$= (X'+Y')\,Y \text{ since } YY' = 0$$

$$= Y(XY)' \text{ by De Morgan's Theorem} \tag{8.34}$$

The *half-subtractor* can thus be realised as in Fig. 8.31. The only difference between this circuit and the half-adder of Fig. 8.25 is

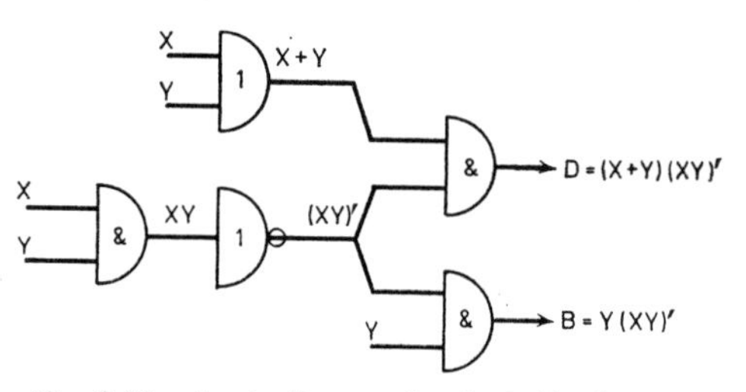

Fig. 8.31. Logic diagram for the half-subtractor

1 *AND* gate. Two of these circuits may be combined in a similar way as for addition to produce full subtraction.

8.8.3. Subtraction by Addition of Complements

In this method of subtraction we must first obtain the binary complement of the number to be subtracted and then feed it to the adder.

A possible method of producing the binary complement results from the following rule for obtaining the complement. ' Examine each digit one at a time, commencing with the least significant. Write each digit up to and including the first 1 exactly as the original number. After that interchange all 1's and 0's, i.e. take the digital complement.' The resulting number is the binary complement of the original number.

A serial circuit which will obtain the binary complement in this way is shown in Fig. 8.32. It uses the bistable multivibrator as a a logical element, which is described in detail in Section 8.9.

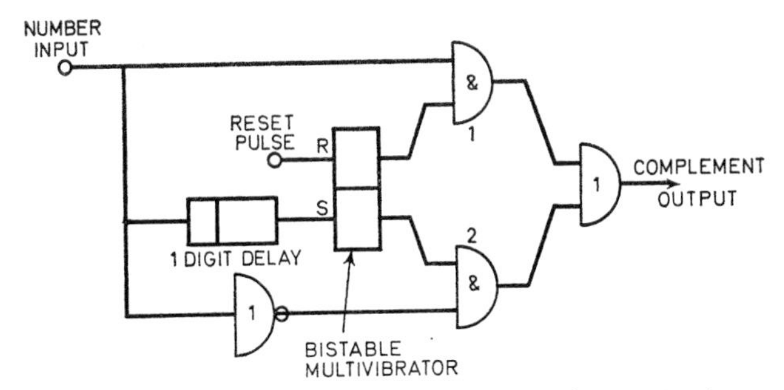

Fig. 8.32. Logic diagram for a serial complement circuit

Initially the bistable circuit is reset, thus opening *AND* gate 1 and closing *AND* gate 2. All digits up to and including the first 1 thus appear at the output unaltered. The first 1 also enters the 1-digit delay unit and changes over the bistable circuit in the next digit interval. This closes *AND* gate 1 and opens *AND* gate 2. For the remainder of the digit train the input digits are passed via the inverter and *AND* gate 2 to the output.

8.8.4. Multiplication and Division

If a machine can add and subtract it can also multiply and divide, since multiplication can be effected by repeated addition and division by repeated subtraction. At the expense of increased circuit complexity it is possible to increase the speed of operation by automatic multiplication and division circuits. To demonstrate this we shall consider a possible multiplier circuit, in which the multiplicand is repeatedly shifted 1 digit more significant and added when called for by a 1 in the multiplier.

Consider the serial circuit of Fig. 8.33 which must be presented with the multiplier in parallel form, least significant digit on the left, and the multiplicand in serial form, least significant digit first. Let the multiplier be decimal 11 (= 1011) and the multiplicand be decimal 10 (= 1010).

The first multiplier digit is 1, therefore the multiplicand is passed through the *AND* gate to the first adder. After passing through the 1-digit delay the multiplicand, shifted 1 digit more significant, is again passed to the first adder, since the second digit of the multiplier is 1. The output of the first adder thus gives the intermediate total. The process continues, the shifted multiplicand being added to the intermediate total when there is a 1 in the multiplier, until the output of the final adder gives the required binary product, i.e. 110 (decimal).

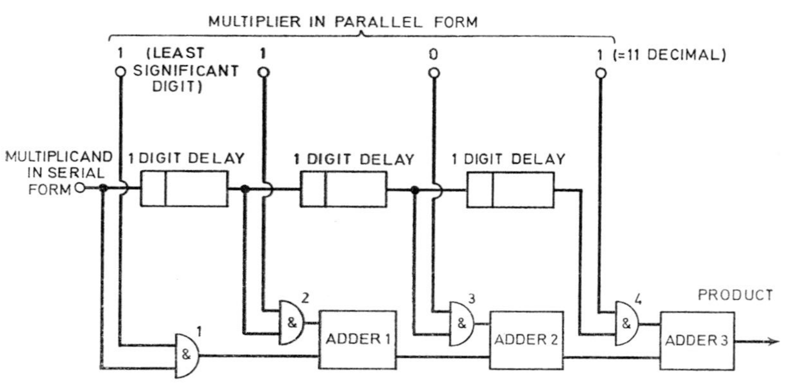

Fig. 8.33. ***Logic diagram for an automatic multiplier***

Table 8.5. WAVEFORMS IN THE AUTOMATIC MULTIPLIER

Output		*Waveforms*								
AND	1	0	0	0	0	1	0	1	0	
AND	2	0	0	0	1	0	1	0	0	
ADDER	1	0	0	0	1	1	1	1	0	
AND	3	0	0	0	0	0	0	0	0	
ADDER	2	0	0	0	1	1	1	1	0	
AND	4	0	1	0	1	0	0	0	0	
ADDER	3	0	1	1	0	1	1	1	0	(= decimal 110)

Waveforms obtained at the outputs of various points in the circuit in terms of 1 and 0 are shown in Table 8.5.

The process has been completed in 8 digit intervals, whereas in repeated addition each of the 10 separate additions would have taken this time. However, as can be seen from Fig. 8.33, this increase in speed requires an appreciable increase in equipment.

8.9. STORAGE DEVICES

We have referred to the requirements for a device to produce time delay in the arithmetic unit and storage of information to be processed. We shall now briefly discuss means of realising these requirements.

8.9.1. The Bistable Multivibrator as a Storage Device

Due to its property of remaining in one of two possible states for an indefinite time the bistable multivibrator is capable of storing information in binary coded form. Referring to the circuit of Fig. 8.34(a), the application of a negative pulse to the *set* input, switches transistor TR1 on, giving 0 (0V) at the A' output, and TR2 off, giving 1 (−10V) at the A output. A negative pulse at the *reset* input gives the opposite condition. Thus there is available from the bistable circuit both A and A' in logical operations which makes it a most useful circuit. The logical symbol for the bistable circuit which we shall use is shown in Fig. 8.34(b).

The logic diagram of a possible storage circuit for a 4-digit serial binary number is shown in Fig. 8.35. Due to the method of operation, to be described, the circuit is referred to as a *shift register*. The

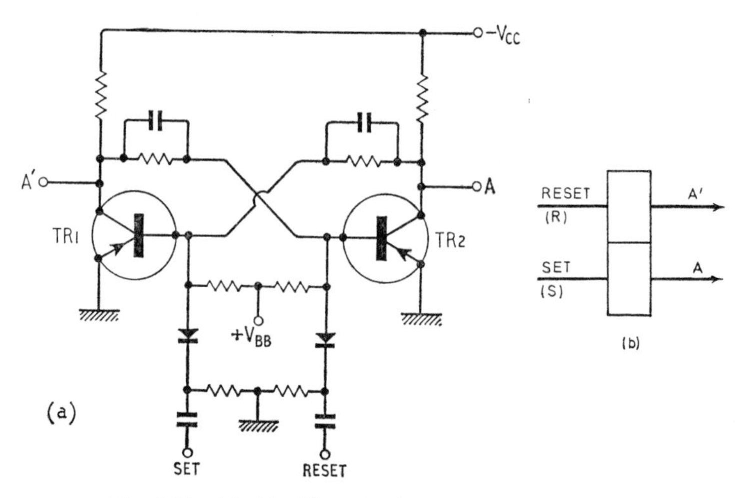

Fig. 8.34. The bistable multivibrator as a storage device. (a) Circuit diagram; (b) Logical symbol

serial information is fed directly to the *AND* gate leading to the *set* input of the first bistable circuit and via an inverter to the *reset* input. Timing is arranged so that at the centre of each digit interval a pulse is applied via the *shift* line to all shift gates. Thus information on the line during digit intervals is written into the first bistable circuit. At the same time the shift pulse writes into any other bistable circuit the information contained in the circuit feeding it. Hence the input information is moved digit by digit along the line of bistable circuits until the first digit to arrive is placed in the last of the chain of circuits. A small amount of time delay must be incorporated in the transfer circuits in order that any given bistable

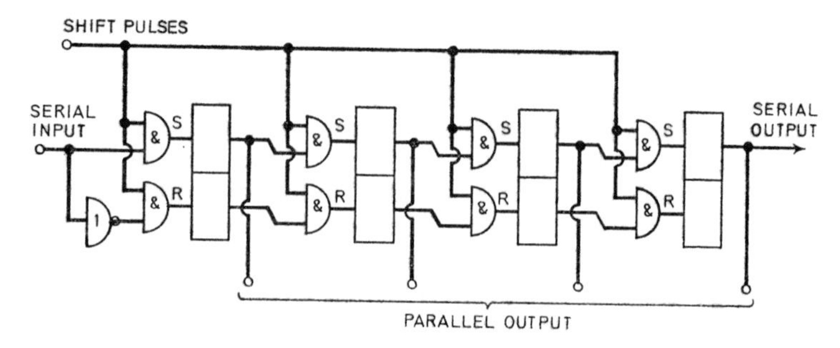

Fig. 8.35. Logic diagram of a shift register

Fig. 8.36. Square-loop magnetising characteristic suitable for binary storage

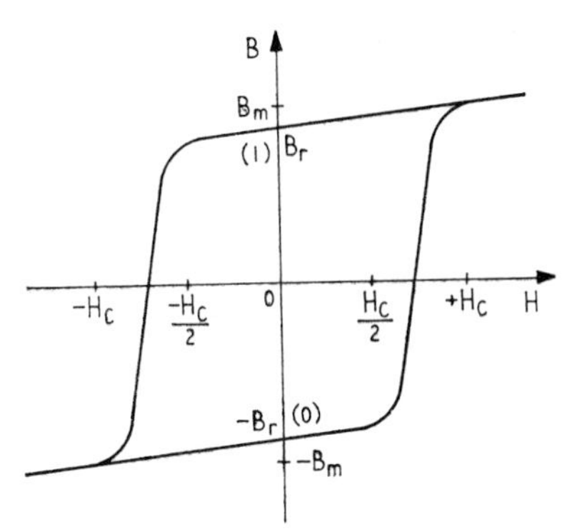

Fig. 8.37. A method of magnetic core storage

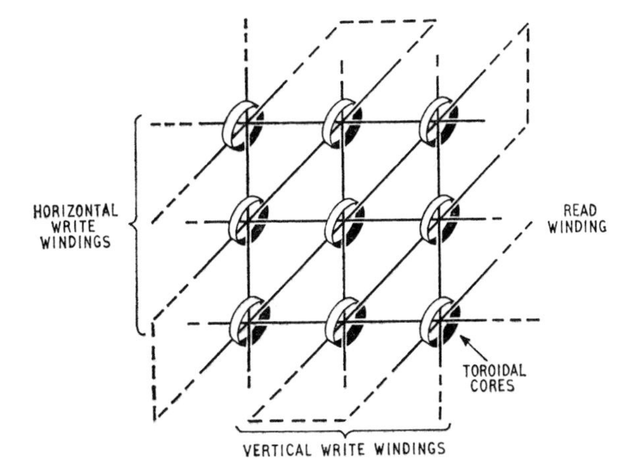

circuit reads the information in the preceding circuit prior to the appearance of the shift pulse and not the new information fed in during the shift period. This may be produced by a *CR* circuit or simply by the changeover time of the bistable circuit. In storing the information the shift register has also acted as a serial to parallel converter. The information can be read out in serial form by applying further shift pulses and taking an output from the right-hand circuit.

8.9.2. Magnetic Core Stores

Very small toroidal magnetic cores are being used in increasing numbers in association with transistor switching circuits to produce means of storage, and a brief description of the method of operation is therefore included.

The possibility of storing information in the magnetic core arises from the property of hysteresis. Magnetic material with a very pronounced hysteresis loop is used as illustrated in Fig. 8.36. If the magnetising force produced by current flowing in a winding on the core is sufficient and of correct polarity to produce the saturation flux density B_m, then when the current is removed the core will remain in a state with a remanent flux density B_r. In this state the core is said to be storing the information '1'. If the current were applied in the opposite direction resulting in flux density $-B_r$, then the core would be storing '0'. Thus the information resides in the direction of the flux in the core.

The determination of whether a 1 or 0 is stored is effected by passing a current in a direction such as to produce state 0. If the core is already in this state there will be no change of flux. If, however, the core were in state 1, then a reversal of flux would occur. A second winding on the core would then produce a voltage pulse in the second case but not the first. Notice that the act of reading the information has destroyed it. Therefore, if continuous storage is required, associated circuits must be provided to rewrite the information.

Blocks of numbers may be stored in an array or *matrix* of cores by a process known as *coincident current selection.* We see in Fig. 8.36 that a magnetising force H_C is sufficient to set the core in either state but $H_C/2$ is not. Therefore if the cores are arranged in a 2-dimensional array as in Fig. 8.37 any one core can be selected by applying a current corresponding to $H_C/2$, to the horizontal and vertical wires coinciding at the desired core. The single wires act as one-turn coils.

The readout winding may be common to all cores, since only 1 core will be read at a time.

CHAPTER 9

MODULATOR AND DEMODULATOR CIRCUITS

9.1. THE MODULATION PROCESS

In electrical communication systems information which usually originates in some other form is converted to a time-varying electrical signal by means of a device called a *transducer*. The electrical signal

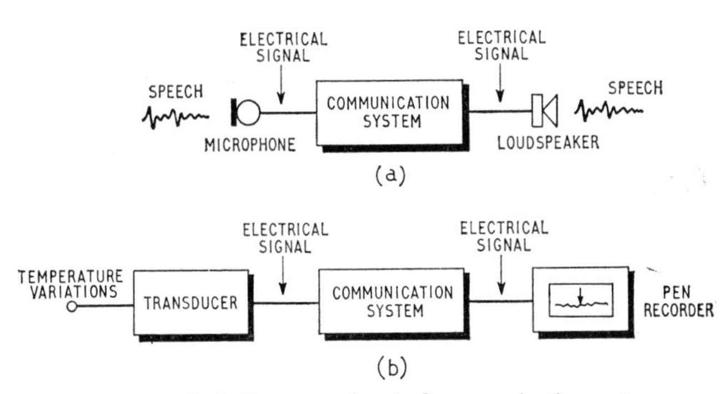

Fig. 9.1. Block diagrams of typical communication systems

is then transmitted over the communication medium and converted to the required information at the receiving end, by another transducer which may be a replica of the transmitter or some other device. For example, speech may be converted by means of a microphone to an electrical signal, passed over the communication system and reproduced in a loudspeaker at the receiver. In another system temperature variations in a furnace may be converted to a corresponding variation in voltage in a transducer, transmitted as an electrical signal and used at the receiver to operate a pen recorder. Fig. 9.1 shows typical block diagrams of such systems.

So far, the communication system has been shown only as a box. If the distance between transmitter and receiver is small, the system may consist simply of a pair of wires. Amplifiers may be included if the distance is such that the received signal level is too low to operate the receiving device. It may be, however, that the communication medium is not suitable for the direct transmission of the electrical signal due to the restriction on operating frequency within the medium. For instance, if the signal is to be transmitted

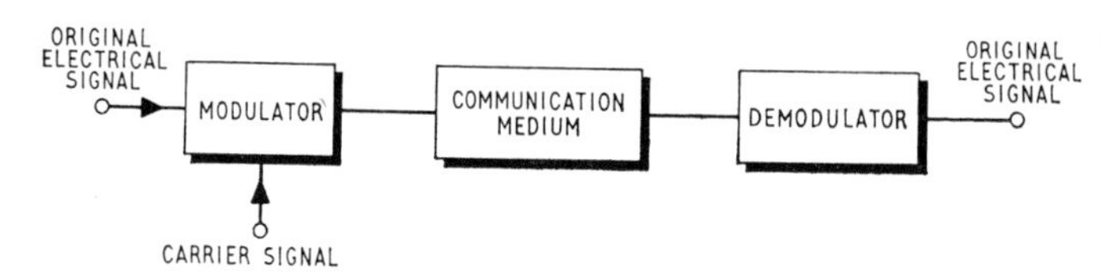

Fig. 9.2. The modulation process

over a radio system then for efficient radiation it should be fairly high in the frequency spectrum, usually above 100kHz. If the signal originates as speech, however, most of the energy will be in the frequency band below 4kHz. It is therefore necessary to translate the signal to a more convenient point in the frequency spectrum prior to transmission.

Alternatively if the signal is to be transmitted over a multichannel telephone system, in which one pair of wires is shared by several different speech circuits, then the signal must either lie in a particular frequency band allocated to the channel in which it is carried or, conversely, only be transmitted for a particular period of time allocated to that channel.

It is necessary in general, therefore, to provide a bearer signal or *carrier* which is acceptable to the communication medium and to superimpose the electrical signal representing the information onto the carrier. This process is referred to as *modulation* and is effected by varying some characteristic of the carrier in accordance with the signal. The signal is recovered at the receiving end by a process of *demodulation*. The block diagram of such a system is shown in Fig. 9.2.

9.1.1. Methods of Modulation

Consider first the case where the carrier is a continuous wave whose instantaneous voltage is given by

$$v_c = A\cos(\omega_c t + \phi) \tag{9.1}$$

where A = peak value of carrier

$\omega_c = 2\pi f_c$ where f_c is the carrier frequency

ϕ = phase angle with reference to some datum.

This steady wave may be modulated by a signal in one of several ways:

(1) *Amplitude modulation*, which results from varying the amplitude of the carrier in accordance with the modulating signal, the angle remaining constant.

(2) *Angle modulation*, which results from varying the angle of the carrier, the amplitude remaining constant. This process may be further subdivided into *frequency modulation* and *phase modulation* as will be discussed later.

It is sometimes more convenient to make the carrier wave discontinuous in the form of a train of rectangular pulses and to vary some characteristic of the pulses in accordance with the modulating signal. This also results in various modulation methods:

(1) *Pulse amplitude modulation*, where the amplitude of a carrier pulse represents the instantaneous amplitude of the modulating signal.

(2) *Pulse duration modulation*, where the time duration of the pulse conveys the modulating signal.

(3) *Pulse code modulation*, where the pulses form a binary coded representation of the instantaneous amplitude of the modulating signal.

There are of course many more ways in which the process of modulation may be carried out but we shall only consider those listed above.

9.2. AMPLITUDE MODULATION

Let the carrier wave be represented by

$$v_c = V_c . \cos \omega_c t \tag{9.2}$$

In this case we assume the reference phase is zero.

Consider the result of amplitude-modulating this wave by a signal represented by

$$v_s = V_s . \cos \omega_s t \tag{9.3}$$

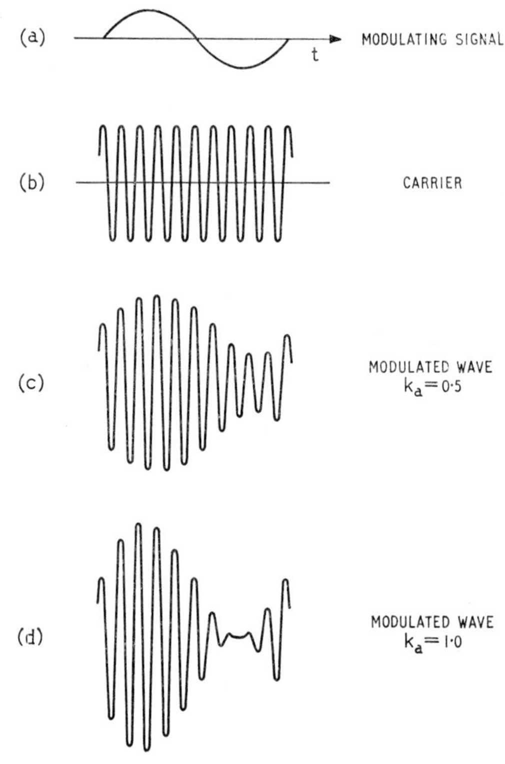

Fig. 9.3. Formation of the a.m. wave

Since the amplitude of the carrier is to be varied in accordance with the signal, then its amplitude at any instant will be

$$(V_c + V_s \cos \omega_s t)$$

Thus the modulated wave may be written as

$$v = (V_c + V_s \cos \omega_s t) \cos \omega_c t$$

$$= V_c(1 + k_a \cos \omega_s t) \cos \omega_c t \qquad (9.4)$$

where $k_a = V_s/V_c$.

k_a is termed the *amplitude modulation index* and $k_a \times 100$ is the *modulation percentage*.

The pictorial representation of the modulation process is shown in Fig. 9.3. Notice that 100% modulation represents the limiting

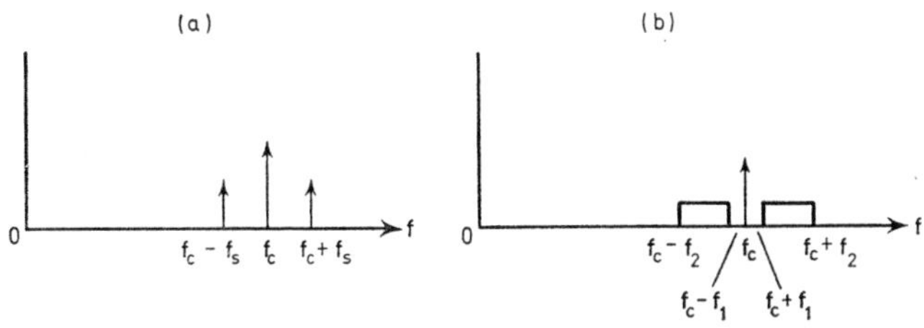

Fig. 9.4. Frequency spectrum of amplitude modulated waves. (a) Single frequency; (b) A band of frequencies

condition above which distortion occurs, i.e. the envelope of the modulated wave no longer follows the modulating signal at all times.

Expanding Equation 9.4 we obtain

$$v = V_c[\cos \omega_c t + k_a \cos \omega_s t . \cos \omega_c t] \tag{9.5}$$

By trigonometrical relations we have

$$\cos(\omega_c - \omega_s)t = \cos \omega_c t \cos \omega_s t + \sin \omega_c t \sin \omega_s t$$

$$\cos(\omega_c + \omega_s)t = \cos \omega_c t \cos \omega_s t - \sin \omega_c t \sin \omega_s t$$

Adding:

$$\cos(\omega_c - \omega_s)t + \cos(\omega_c + \omega_s)t = 2 \cos \omega_c t \cos \omega_s t$$

Substituting this result in Equation 9.5 gives

$$v = V_c\left[\cos \omega_c t + \frac{k_a}{2}\cos(\omega_c - \omega_s)t + \frac{k_a}{2}\cos(\omega_c + \omega_s)t\right] \tag{9.6}$$

The modulated wave thus consists of the original carrier plus 2 new components. The frequencies of these new components are made up of the sum and difference of f_c and f_s and are referred to as the *upper and lower side-frequencies* respectively.

If the modulating signal had contained frequency components from f_1 to f_2 each component would have modulated the carrier, thus producing an *upper side-band* ranging from f_c+f_1 to f_c+f_2 and *lower sideband* from f_c-f_2 to f_c-f_1. These 2 conditions are illustrated in Fig. 9.4.

By means of the modulation process we have thus transferred the original signal to a different part of the frequency spectrum. Notice that each side band contains all the original information while the carrier component contains none. Also the modulated wave covers at least twice the frequency band of the modulating signal.

9.2.1. Power in the Sidebands

The r.m.s. voltage V of the modulated wave is

$$V = V_c\sqrt{\left(\frac{1}{2}+\frac{k_a^2}{8}+\frac{k_a^2}{8}\right)}$$

$$= \frac{V_c}{\sqrt{2}}\sqrt{\left(1+\frac{k_a^2}{2}\right)} \tag{9.7}$$

If this voltage is applied to a resistor R the power developed P is then

$$P = \frac{V_c^2}{2R}\left(1+\frac{k_a^2}{2}\right) \tag{9.8}$$

Notice from this equation that the power in the modulated wave is greater than in the unmodulated carrier. Fig. 9.5 shows both

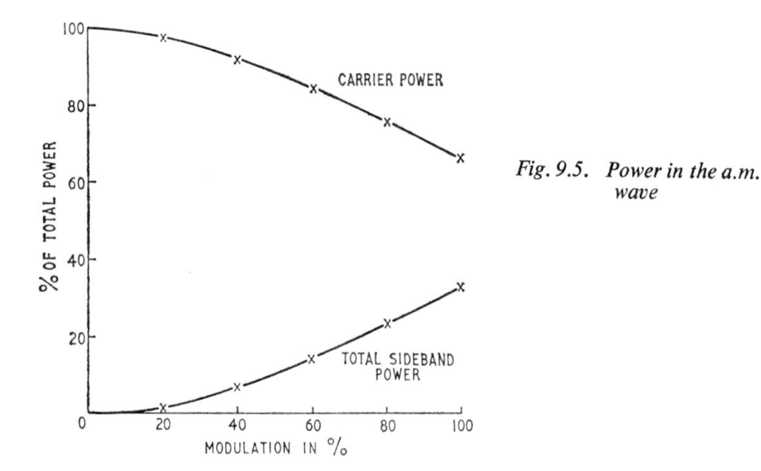

Fig. 9.5. Power in the a.m. wave

carrier power and total sideband power, each as a percentage of total power, plotted as a function of percentage modulation. Even with 100% modulation, only one-third of the total power lies in the sidebands. Since in general the modulation is much less than 100% to prevent distortion due to peaks in the modulating signal, the information constitutes only a small fraction of the total power.

9.2.2. Single Sideband Transmission

The previous discussion has made it clear that transmission of the complete amplitude-modulated wave results in a very inefficient

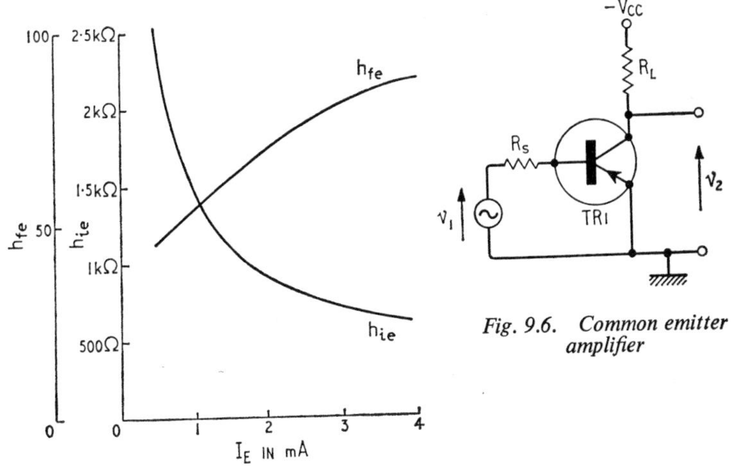

Fig. 9.6. Common emitter amplifier

Fig. 9.7. Variation of h_{fe} and h_{ie} with emitter current

system. Fig. 9.5 shows that even at 100% modulation the majority of the power resides in the carrier. This component itself, however, contains no information. In addition the amplitude-modulated wave requires twice the bandwidth of the original signal to accommodate the 2 sidebands, each of which contains all of the modulating information. It is obvious, therefore, that if we can eliminate the carrier and 1 sideband from the amplitude-modulated wave a much more efficient transmission system will result. Amplifiers will only be required to handle power constituting the actual signal and the bandwidth will be identical to that of the signal.

9.3. AMPLITUDE MODULATION CIRCUITS

9.3.1. Modulated Amplifier

It was shown in Section 3.6 that provided R_L is low the external voltage gain of the common emitter amplifier stage of Fig. 9.6 is given by

$$K = -\frac{h_{fe} \cdot R_L}{h_{ie} + R_s} \tag{9.9}$$

where h_{fe} and h_{ie} are the common emitter h-parameters.

The dependence of these 2 h-parameters on emitter current was discussed in Section 3.8 and is shown in the graphs of Fig. 9.7 for a typical transistor. Fig. 9.8 shows plots of the external voltage gain

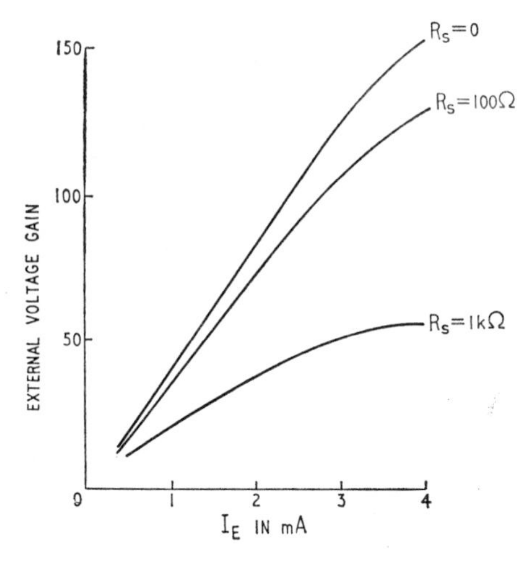

Fig. 9.8. Variation of external voltage gain with emitter current

of this transistor as calculated from Equation 9.9 for $R_L = 1k\Omega$ and 3 different values of R_s.

The curves of Fig. 9.8 show that at low values of emitter current the external voltage gain of the amplifier is almost directly proportional to the emitter current. This provides a method of amplitude modulation, as we shall now show.

Let the effective instantaneous emitter current i_E be given by

$$i_E = I_E + k_1 V_s \cos \omega_s t \tag{9.10}$$

Also let the external voltage gain K be given by

$$\begin{aligned} K &= k_2 . i_E \\ &= k_2(I_E + k_1 V_s \cos \omega_s t) \end{aligned} \tag{9.11}$$

Then if the input voltage is $V_c \cos \omega_c t$ the output voltage will be

$$= K . V_c \cos \omega_c t$$

$$= k_2 V_c \cos \omega_c t (I_E + k_1 V_s \cos \omega_s t)$$

$$= k_2 V_c I_E \cos \omega_c t + k_1 k_2 V_c V_s \cos \omega_s t \cos \omega_c t$$

$$= k_2 V_c I_E [\cos \omega_c t + k_a \cos \omega_s t \cos \omega_c t] \tag{9.12}$$

where $k_a = k_1 V_s / I_E$.

Thus if the emitter current is varied in accordance with the modulating signal the output voltage contains the components of the amplitude modulated wave, which may be selected by a tuned circuit.

A typical circuit is shown in Fig. 9.9(a). Similar results may be obtained if the base current is varied in accordance with the modulating signal, and a typical circuit is shown in Fig. 9.9(b). The value of the capacitor C must be such that it effectively short circuits the biasing resistor R at carrier frequencies but not at modulating signal frequencies.

9.3.2. Balanced Modulators

A modulator circuit which contains no carrier component in the output wave is known as a *balanced modulator*. This circuit is obviously useful in single sideband transmission systems.

9.3.3. The Cowan Modulator

Consider first the circuit of Fig. 9.10(a) which is the well-known *Cowan modulator*. If the carrier level is made large compared to the signal level then the diodes will be either forward or reverse-biased, depending on the polarity of the carrier voltage. When terminal 1 is positive with respect to terminal 2 all 4 diodes will conduct. Assuming ideal diodes the signal path is thus shorted out. Reversal of the polarity of the carrier causes the diodes to be

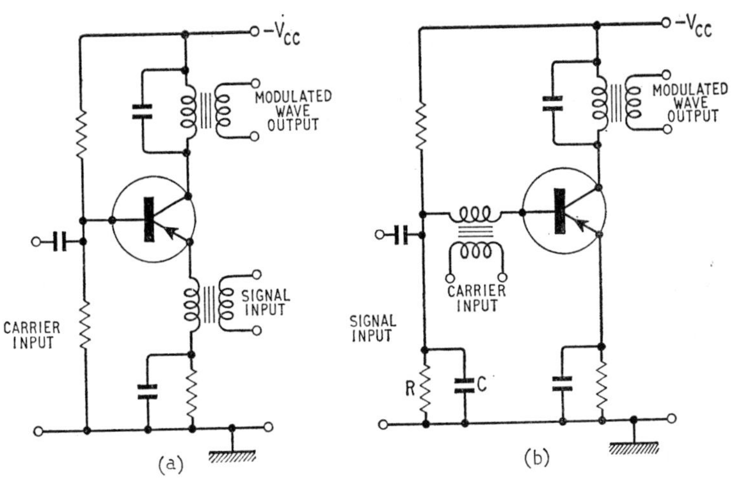

Fig. 9.9. Modulated amplifier. (a) Emitter modulation ; (b) Base modulation

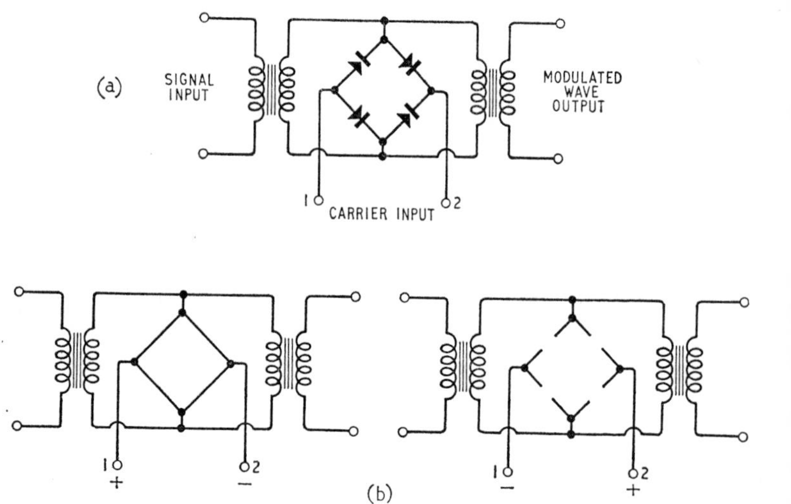

Fig. 9.10. The Cowan modulator. (a) Circuit diagram; (b) Equivalent circuits

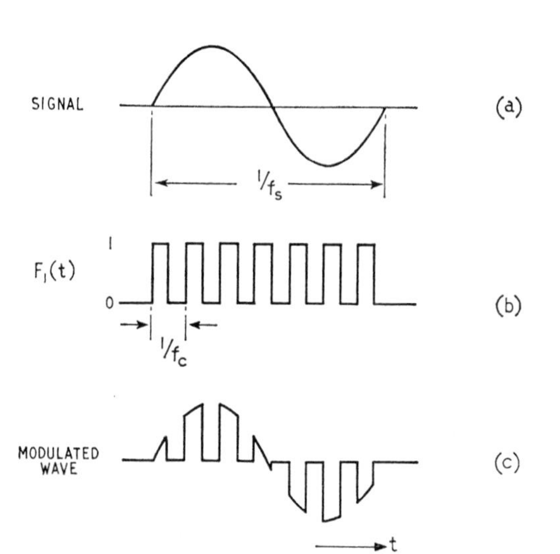

Fig. 9.11. Waveforms in the Cowan modulator

open circuit. Fig. 9.10(b) illustrates these conditions. Thus the signal voltage source is successively short-circuited and open-circuited at a frequency corresponding to that of the carrier. This process produces a waveform varying at carrier frequency with an amplitude that varies at signal frequency as indicated in Fig. 9.11(c) although it contains no carrier component, as can be verified by a simple analysis.

Assuming that the diodes are switched instantaneously from open circuit to short circuit as the carrier changes polarity, we can consider

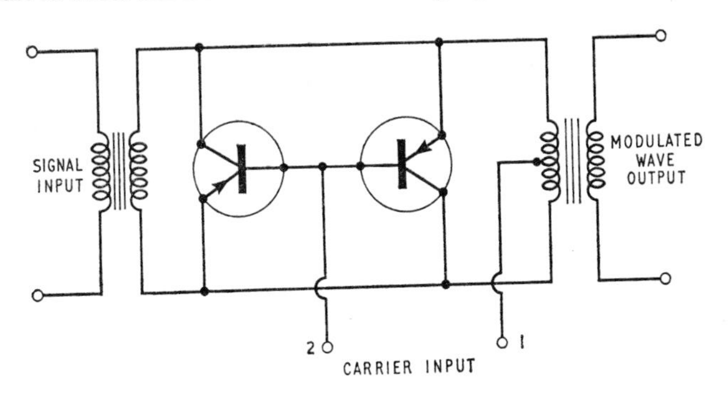

Fig. 9.12. A transistor version of the Cowan modulator

that the signal output is controlled by a switching function $F_1(t)$, having the frequency of the carrier and varying between 0 and 1, as shown in Fig. 9.11(b). The analytic representation of this function according to Fourier's Theorem is

$$F_1(t) = \frac{1}{2} + \frac{2}{\pi}[\cos\omega_c t - \tfrac{1}{3}\cos 3\omega_c t + \ldots] \tag{9.13}$$

If the signal voltage is $V_s \cos\omega_s t$ then the output voltage is given by

$$\begin{aligned} v &= F_1(t)\, V_s \cos\omega_s t \\ &= V_s \cos\omega_s t \left[\frac{1}{2} + \frac{2}{\pi}\cos\omega_c t - \frac{2}{3\pi}\cos 3\omega_c t + \ldots\right] \\ &= \frac{V_s}{2}\cos\omega_s t + \frac{2V_s}{\pi}\cos\omega_s t\cos\omega_c t - \ldots \end{aligned} \tag{9.14}$$

The output voltage thus contains the original signal, upper and lower sidebands and other harmonic components, but no carrier.

A transistor circuit equivalent to the Cowan circuit is shown in Fig. 9.12, where the principle of operation is essentially the same.

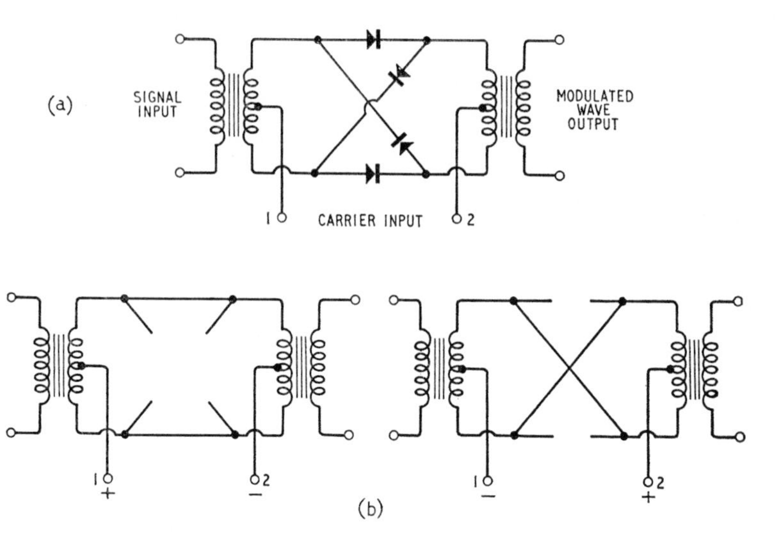

Fig. 9.13. The ring modulator. (a) Circuit diagram; (b) Equivalent circuits

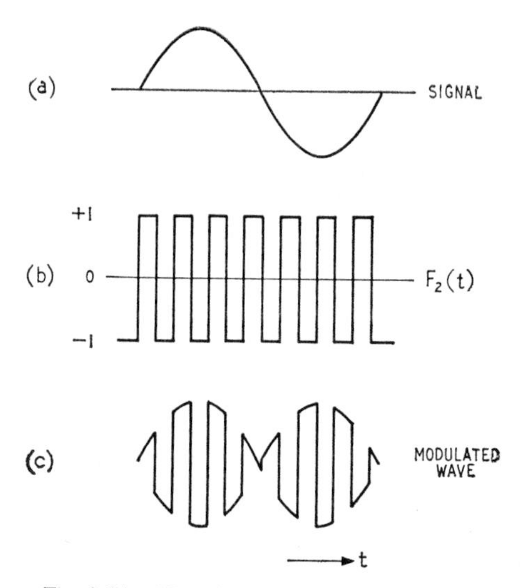

Fig. 9.14. Waveforms in the ring modulator

When the carrier input terminal 2 is positive the transistors are cut off and the modulating signal appears unaltered at the output. Reversal of the carrier signal polarity causes the transistors to saturate and, assuming ideal transistors, the signal path is short circuited. The use of transistors reduces the carrier power required for satisfactory operation of the modulator.

A single symmetrical transistor may be used in place of the two asymmetrical transistors shown. This is a transistor in which the emitter and collector regions are similar, with the result that the device performs equally well in the forward or reverse direction.

9.3.4. The Ring Modulator

Another balanced modulator much used in communications systems is the *ring modulator* shown in Fig. 9.13.

Following the same reasoning as for the Cowan circuit we find that in this case there is a phase reversal of the signal voltage at

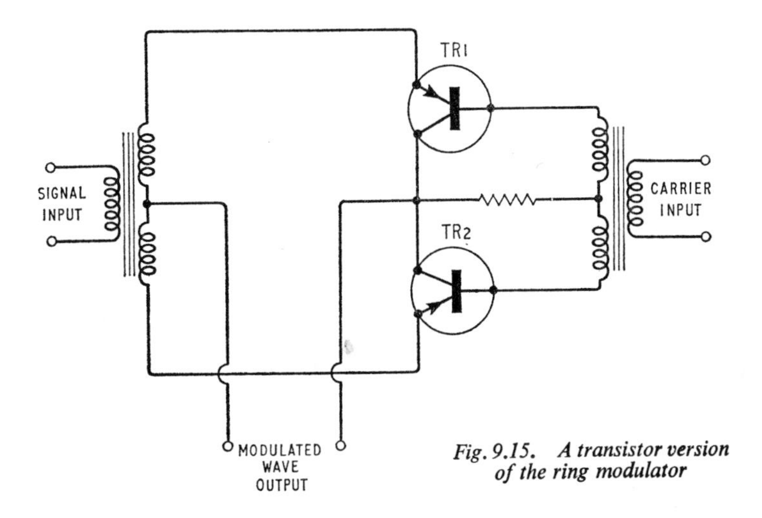

Fig. 9.15. A transistor version of the ring modulator

the output at a rate corresponding to the carrier frequency, as shown in Fig. 9.13.

This time the switching function obviously varies between +1 and −1, giving

$$F_2(t) = \frac{4}{\pi}[\cos \omega_c t - \tfrac{1}{3}\cos 3\omega_c t + \tfrac{1}{5}\cos 5\omega_c t - \dots] \qquad (9.15)$$

Thus the output voltage is

$$\frac{4V_s \cos \omega_s t}{\pi}[\cos \omega_c t - \tfrac{1}{3}\cos 3\omega_c t + \ldots] \tag{9.16}$$

Again the output signal contains the upper and lower sideband components plus harmonics, but no signal or carrier. Relevant waveforms are shown in Fig. 9.14.

An elementary transistor version of the ring modulator is shown in Fig. 9.15.

The carrier signal alternately switches on transistors TR1 and TR2 and the modulating-signal transformer windings are arranged so that the output contains the modulating signal, reversed in phase at a rate corresponding to the carrier frequency. Again the use of transistors reduces the carrier power requirement.

9.3.5. Demodulation

Since the carrier in an amplitude modulated wave is simply acting as a bearer of the modulating signal in the normal case, after transmission over the communication medium it is necessary to separate the 2 components again. This process is known as *demodulation* or *detection.* The simplest method of recovering the signal is by a process of *envelope detection.*

Consider the circuit of Fig. 9.16(a) and assume an ideal diode having the characteristic shown in Fig. 9.16(b). If a sine wave of voltage is applied to this circuit the resulting load voltage will be as in Fig. 9.16(c), i.e. the negative half-cycle has been removed. If now a capacitor is connected across R_L it will charge to the level of the input voltage on the positive half-cycle and it will tend to discharge via R_L during the negative half-cycle. Provided the time constant of the $R_L C$ network is very long compared to the repetition rate of the input wave the effective capacitor voltage will be constant at the peak value of the input signal. In practice there will be a small discharge from the capacitor resulting in a small a.c. ripple superimposed on the d.c. voltage as shown in Fig. 9.16(c).*

Variations of the amplitude of the input voltage will produce a varying output signal which will follow the envelope of the input wave, provided that the speed of variation is sufficiently slow to allow the capacitor voltage to readjust. Thus if the input voltage is the amplitude modulated wave of Fig. 9.3 then the output waveform will be the modulating signal superimposed on a d.c. component.

* Notice that the circuit of Fig. 9.16 is the simple half-wave rectifier referred to in Section 5.6.

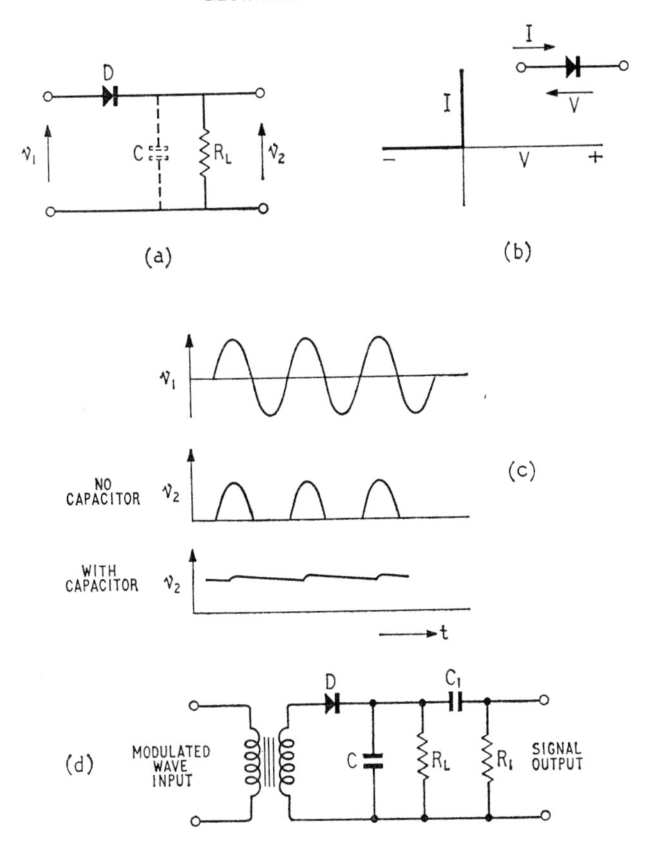

Fig. 9.16. The envelope detector. (a) Basic circuit; (b) Ideal diode characteristic; (c) Waveforms for (a); (d) Practical circuit

This d.c. component may be removed by the $R_1 C_1$ circuit shown in Fig. 9.16(d).

Since there are 2 conflicting requirements on the time constant of the load circuit of the diode, in that it must be long compared to the period of the carrier wave but short compared to the period of the highest modulating wave, this method of detection is only suitable in cases where the carrier frequency is very large compared to the modulating frequency, e.g. radio reception.

The envelope detector circuit may be modified to include a transistor as shown in Fig. 9.17. The base–emitter diode acts as the rectifying element as before but, due to the high input resistance of the common collector stage in the active region, the loading on the source, which will probably be a tuned circuit, is much less. In addition the transistor provides some power gain.

In a single sideband system the carrier must be reintroduced before demodulation can take place and in practice the incoming signal is remodulated with the carrier in a circuit identical to that at the transmitting terminal, e.g. a Cowan or ring or equivalent. In order to prevent distortion it is essential that the frequency of the reintroduced carrier is identical with that at the transmitting

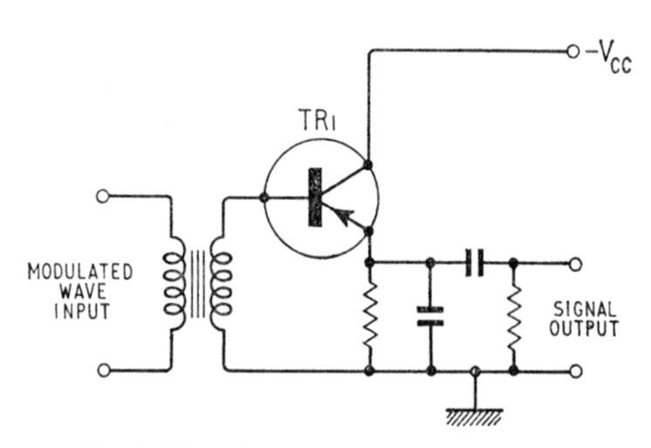

Fig. 9.17. A transistor envelope detector

terminal. This may be ensured by using crystal controlled carrier oscillators or by transmitting a synchronising signal over the medium and using it at the receiver to control the frequency of the receiver carrier oscillators.

9.4. ANGLE MODULATION

Let the carrier wave again be represented by the expression

$$v_c = V_c \cos(\omega_c t + \phi_0) \tag{9.17}$$

We now require to vary the instantaneous angle of this wave in accordance with the modulating signal. The voltage of the modulated wave at any instant will then be given by

$$v = V_c \cos\theta$$

where θ is the angle swept by the voltage vector

i.e. $$\theta = \omega t + \phi_0$$

When the modulation is zero, the angular frequency is ω_c, thus

$$\theta_1 = \omega_c t + \phi_0$$

In the modulated wave let the departure from θ_1 be α

i.e.
$$\theta = \theta_1 + \alpha \tag{9.18}$$

Now α may be varied in one of two ways, either by varying the instantaneous frequency of the modulated wave giving frequency

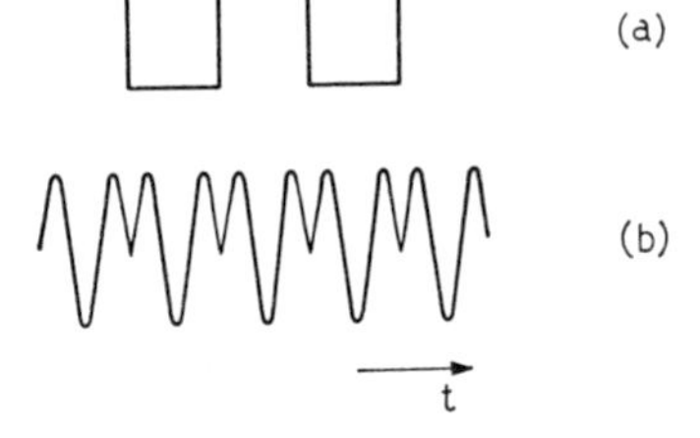

Fig. 9.18. Phase modulation. (a) Modulating signal; (b) Modulated wave

modulation, or by varying the instantaneous phase of the modulated wave giving phase modulation. Consider these two cases separately.

CASE 1

The modulating signal is proportional to α.

Let
$$\alpha = k_1 . V_s \cos \omega_s t \tag{9.19}$$

Then
$$\theta = \theta_1 + \alpha$$
$$= \omega_c t + \phi_0 + k_1 V_s \cos \omega_s t \tag{9.20}$$

Thus the expression for the modulated wave, ignoring the constant phase angle ϕ_0 is
$$v = V_c \cos [\omega_c t + m_p . \cos \omega_s t] \tag{9.21}$$

where
$$m_p = k_1 V_s$$

This expression represents a phase modulated wave and m_p is the *phase modulation index*. Typical waveforms for square wave modulation are shown in Fig. 9.18 where the change in level of the signal causes phase reversal of the carrier.

CASE 2

The modulating signal is proportional to the rate of change of α. The instantaneous frequency of the modulated voltage vector is proportional to the rate of change of the total angle θ.

Thus
$$\omega = \frac{d\theta}{dt} = \frac{d\alpha}{dt} + \frac{d\theta_1}{dt} \tag{9.22}$$

Let

$$\frac{d\alpha}{dt} = k_2 V_s \cos \omega_s t \tag{9.23}$$

Then

$$\omega = k_2 V_s \cos \omega_s t + \omega_c$$

i.e.

$$f = \frac{k_2 V_s}{2\pi} \cos \omega_s t + f_c \tag{9.24}$$

Thus the frequency of the modulated wave is varying in accordance with the modulating signal giving frequency modulation.

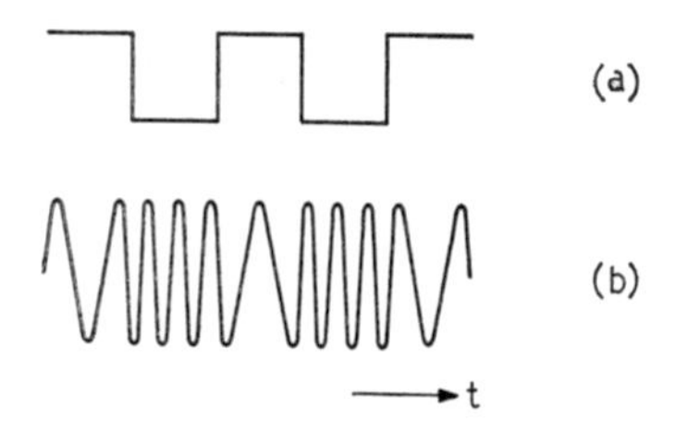

Fig. 9.19. Frequency modulation

The *maximum frequency deviation* f_d from the unmodulated value is

$$f_d = \frac{k_2 V_s}{2\pi} = \frac{\omega_d}{2\pi}$$

From Equation 9.22

$$\theta = \int \omega \, dt + \phi_0$$

$$= \int (\omega_d \cos \omega_s t + \omega_c) \, dt + \phi_0$$

$$= \frac{\omega_d}{\omega_s} . \sin \omega_s t + \omega_c t + \phi_0 \tag{9.25}$$

Again neglecting the constant phase angle ϕ_0 we have

$$v = V_c \cos [\omega_c t + m_f \sin \omega_s t] \tag{9.26}$$

where

$$m_f = \frac{\omega_d}{\omega_s}$$

This expression represents a frequency modulated wave and m_f is the *frequency modulation index*. Typical waveforms for square wave modulation are shown in Fig. 9.19.

Notice that frequency modulation and phase modulation, both being forms of angle modulation, differ only in the modulation indices, m_f being inversely proportional to the modulating signal f_s and m_p being constant, plus a phase shift of 90°. It is thus relatively easy to convert one form to the other, so let us look more closely at the frequency modulation case.

The expression for the frequency modulated wave given in Equation 9.26 cannot be broken down into a few simple terms as was the case for amplitude modulation, but it can be shown,[1] using Bessel Functions, that the wave contains the carrier frequency f_c plus an infinite number of side frequencies of $f_c \pm f_s, f_c \pm 2f_s, f_c \pm 3f_s$, etc. whose amplitudes depend on the degree of modulation.

9.4.1. Power in the Frequency Modulated Wave

It is obvious from a consideration of Equation 9.26 that the total power content of the frequency modulated wave is constant, irrespective of the degree of modulation. The distribution of power between carrier and sidebands depends on the modulation index and in fact certain values of modulation index cause the carrier to disappear completely. Fig. 9.20 shows sketches of the relative amplitudes of

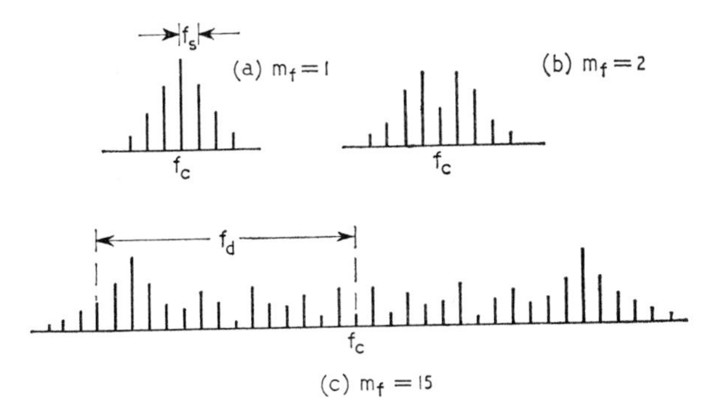

Fig. 9.20. Frequency spectrum of f.m. waves

various frequency components for 3 different values of modulation index. Notice also that for a given frequency deviation, low modulating frequencies, which result in a high modulation index, will produce more sidebands of appreciable amplitude than will high modulating frequencies.

[1] TIBBS, C. E., and JOHNSTONE, G. G., *Frequency Modulation Engineering*, Chapman & Hall Ltd (1947).

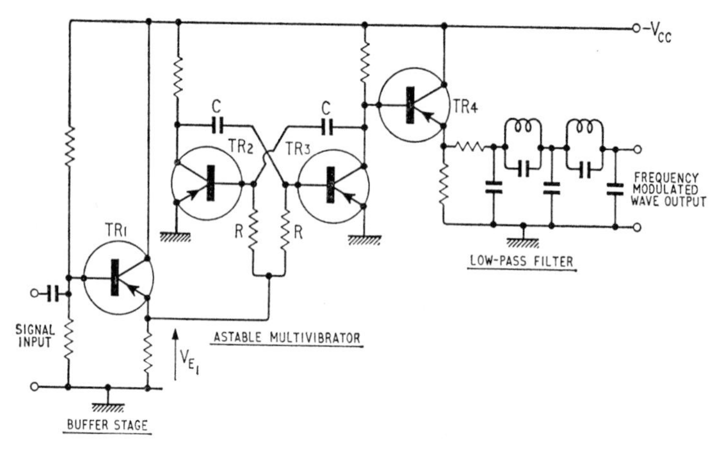

Fig. 9.21. A frequency modulator circuit

9.5. FREQUENCY MODULATION CIRCUITS

The requirement is for a circuit which will convert the variation in amplitude of a modulating signal to a variation in frequency of a carrier wave. A method of producing such a frequency modulated wave is to control the frequency of oscillation of a carrier oscillator

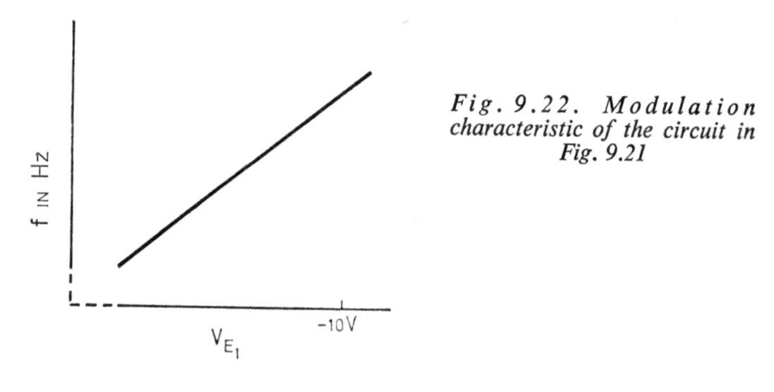

Fig. 9.22. Modulation characteristic of the circuit in Fig. 9.21

in accordance with the modulating signal amplitude. A convenient arrangement for doing this makes use of the astable multivibrator described in Chapter 7 with the base return voltage dependent upon the modulating signal voltage as shown in Fig. 9.21. With no input to transistor TR1, which acts as a buffer stage, the multivibrator frequency will depend on the quiescent emitter–ground voltage V_{E1} of TR1 and the time constant CR, according to Equation 7.11.

That is
$$f=\frac{1}{2CR\log_e\left(1+\frac{V_{CC}}{V_{E1}}\right)}\text{ Hz} \tag{9.27}$$

Application of a modulating signal voltage to the base of TR1 will cause its emitter voltage to vary about the quiescent value. Provided the multivibrator is working on the linear portion of the characteristic of Fig. 9.22 then satisfactory frequency modulation will result. Since the multivibrator generates square waves it is necessary to include the low pass filter at the output to attenuate the unwanted harmonics.

9.5.1. Demodulation

One of the great advantages of frequency modulation systems is that the modulating information is not contained in any amplitude variation. Thus any variation in amplitude of the received signal due to

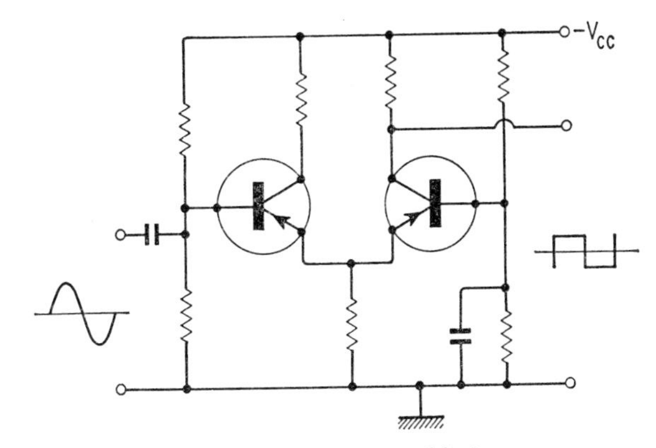

Fig. 9.23. A symmetrical limiter

noise pick-up which would obviously constitute distortion in an amplitude modulation system can easily be removed by clipping or *limiting* the received signal. A suitable circuit for producing the required symmetrical limiting is the emitter coupled pair of Fig. 9.23, which was discussed in Section 6.4. It is then necessary to convert the variation of frequency in the modulated wave to a variation in amplitude, ideally identical to that of the original modulating signal.

An elementary way in which this may be carried out is as follows. Consider the circuit of Fig. 9.24 where transistor TR1 may be part

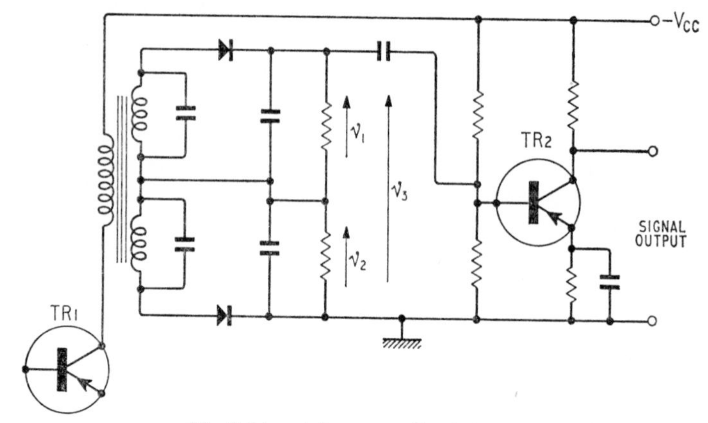

Fig. 9.24. A frequency discriminator

of the limiter circuit and the 2 secondary windings of the transformer are tuned to frequencies just above $f_c+f_{d\,\max}$ and just below $f_c-f_{d\,\max}$ respectively, where f_d is the frequency deviation. The response of the circuit at various points will then be as in Fig. 9.25. Notice that the signal applied to the amplifier stage is quite linearly

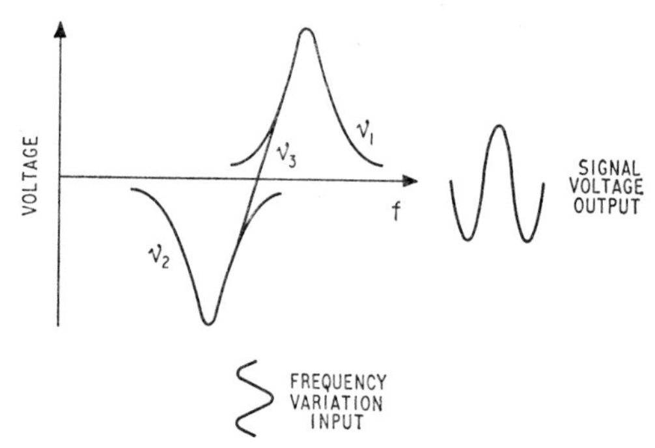

Fig. 9.25. Demodulating action of the frequency discriminator

dependent on the frequency of the input wave (provided the frequency deviation is very small compared to the carrier frequency). Thus the original modulating information is recovered.[1]

A method of demodulation using switching circuits is shown in block diagram form in Fig. 9.26(a). Waveforms obtainable at

[1] A more efficient circuit is the Foster–Seeley Discriminator discussed in TIBBS, C. E. and JOHNSTONE, G. G., *Frequency Modulation Engineering*, Chapman & Hall Ltd. (1947).

various points in the system are shown in Fig. 9.26(b). Briefly the circuit operation is as follows. The square wave output from the limiter is passed to 2 differentiating and clamping circuits, one giving positive pulses at the leading edge of the square wave, the other giving negative pulses at the trailing edge. The negative

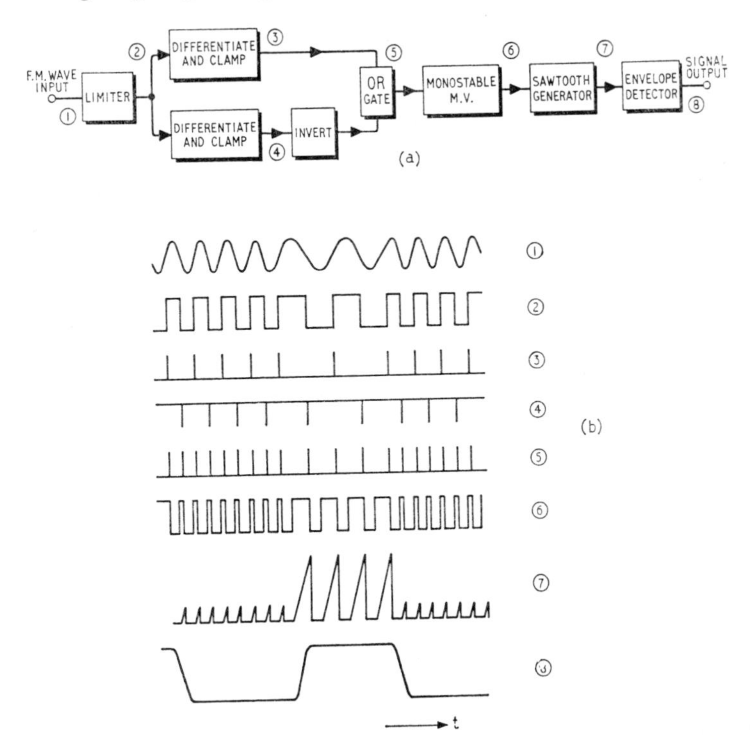

Fig. 9.26. The zero crossing detector. (a) Block diagram; (b) Waveforms in the zero crossing detector

pulses are then inverted and combined with the original positive pulses in an *OR* gate. The output of the *OR* gate is a train of positive trigger pulses occurring at each *zero crossing* of the input wave. These pulses are then used to control the time of occurrence of constant-width pulses generated by the monostable multivibrator circuit. Following the monostable circuit is a linear sawtooth generator whose output is $-V_{CC}$ for the duration of the input pulse, rising linearly between pulses, then again reducing to $-V_{CC}$ at the occurrence of the next pulse. Thus the amplitude of the sawtooth wave is determined by the spacing between the input pulses, i.e. by the frequency of the received modulated wave.

A linear sawtooth voltage waveform may be produced by applying a constant current drive to a capacitor, since the voltage v across the capacitor in Fig. 9.27(a) is given by

$$v = \frac{1}{C}\int i.\mathrm{d}t$$

$$= \frac{I.t}{C} \tag{9.28}$$

A transistor operated in the common base mode acts as a very efficient constant current generator and the circuit of Fig. 9.27(b) will produce the desired sawtooth waveform as follows, with ideal transistors.

Assume that transistor TR1 is initially in saturation, resulting in zero voltage across the capacitor. The collector voltage of TR2 is therefore equal to $-V_{CC}$. If TR1 is switched off at time $t = 0$ the capacitor will begin to charge from the constant current source provided by the common base stage TR2. Thus at any instant

$$v_c = -V_{CC} + \frac{1}{C}\int i_c.\mathrm{d}t$$

$$= -V_{CC} + \frac{I_C t}{C}$$

But

$$I_C = \alpha_B . I_E$$

$$= \frac{\alpha_B V_{BB}}{R_e}$$

Therefore

$$v_c = -V_{CC} + \frac{\alpha_B V_{BB}}{CR_e}.t \tag{9.29}$$

If TR1 is again switched on at time $t = t_0$ the resulting waveform of v_c will then be as shown in Fig. 9.27(c). Envelope detection of this train of sawtooth waves will then yield the original modulating signal.

9.6. PULSE MODULATION

So far we have discussed methods of translating signal frequencies to different parts of the frequency spectrum by modulating a

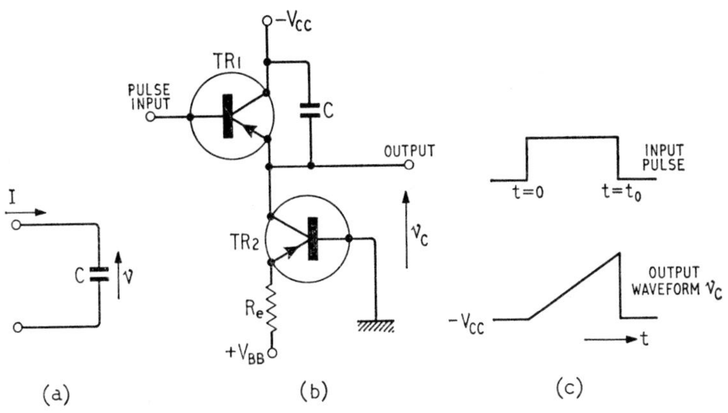

Fig. 9.27. A simple sawtooth generator. (a) Basic circuit; (b) Sawtooth generator circuit; (c) Relevant waveforms

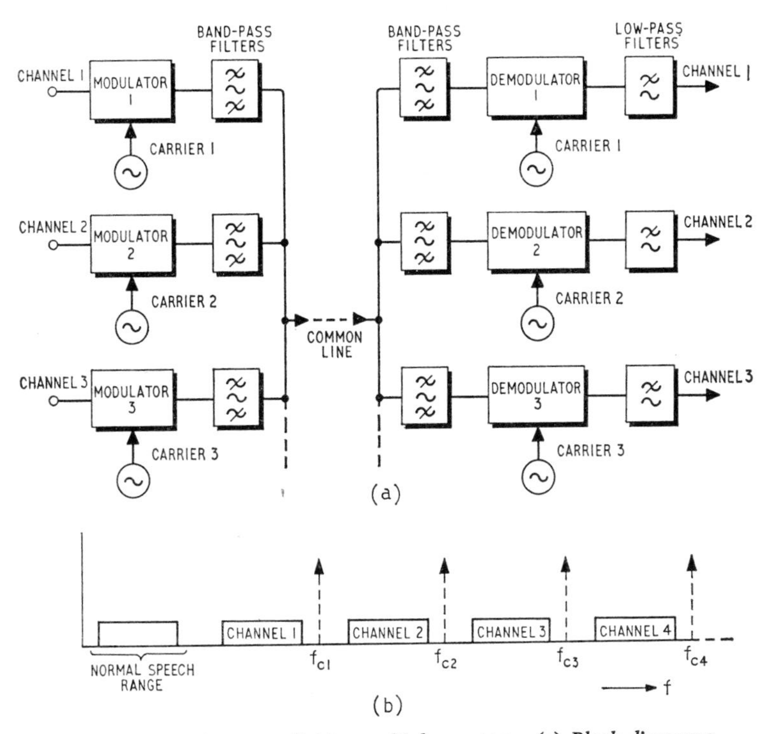

Fig. 9.28. A frequency-division multiplex system. (a) Block diagram; (b) Frequency spectrum of the frequency-division multiplex system

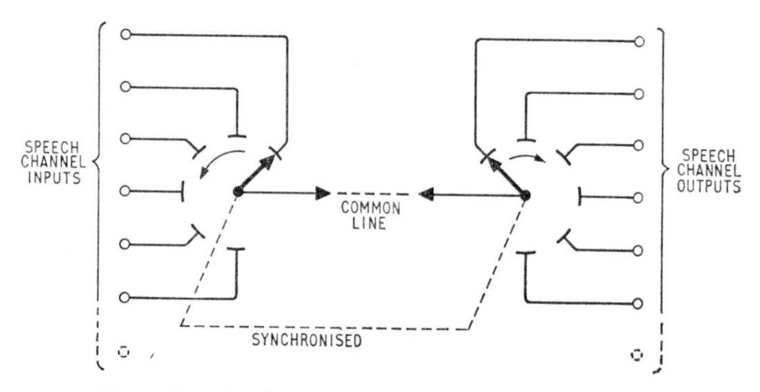

Fig. 9.29. An elementary time-division multiplex system

continuous carrier wave. This frequency translation is used extensively in present-day communication systems for transmitting large numbers of information sources on the same medium. A typical basic block schematic of such a single sideband *frequency-division multiplex system* is shown in Fig. 9.28(a) with the associated frequency spectrum in Fig. 9.28(b). Thus the transmission medium is shared on a frequency basis by several channels, all of which may be using the medium at the same time. Notice from Fig. 9.28(a) that frequency-selective filter networks are used for channel separation and different carrier frequencies are required for each channel.

A simpler system results from the sharing of the medium on a time basis rather than on a frequency basis, resulting in a *time-division multiplex system.* In this method of operation the modulating signal is not continuously transmitted, but samples of the signal are taken at discrete intervals of time and applied to the communication medium. It can be shown that these samples contain all the information in the original modulating signal, provided the samples are taken at a frequency slightly higher than twice the highest significant frequency component of the message wave.[1] During the time interval between the application of these pulses to the medium it is possible to transmit similar samples of other message channels. Thus the transmission medium is made available to each channel in turn for a specific interval of time, during which interval the whole system bandwidth is available to that channel.

In effect we now have a carrier wave which is no longer continuous, but it occurs as a sequence of pulses, some characteristic of which is varied in accordance with the modulating signal.

[1] BLACK, H. S., *Modulation Theory* Van Nostrand (D.) Co. Ltd. (1953).

An elementary system is shown in Fig. 9.29. The 2 commutator switches at the ends of the system rotate in synchronism and connect each message channel in turn to the line. The speed of rotation must be related to the channel frequency in the manner stated above. Thus for a time-division multiplex system transmitting speech for

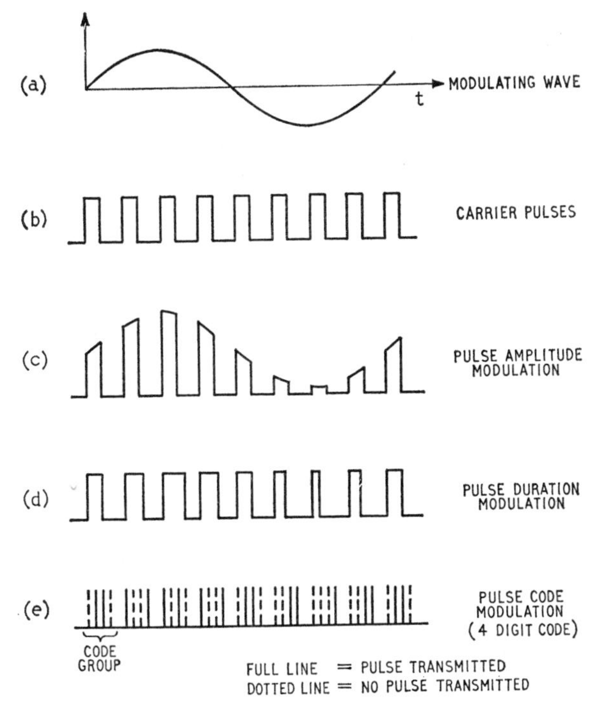

Fig. 9.30. Methods of pulse modulation

which the standard upper frequency limit is 3.4kHz the switches must rotate at a rate exceeding 6800 times per second.

Some ways in which the carrier pulses may be varied in accordance with the modulating signal, as discussed briefly below, are illustrated in Fig. 9.30.

9.6.1. Pulse Amplitude Modulation

This is the simplest form of pulse modulation in which the amplitude of the carrier pulse is varied in accordance with the modulating

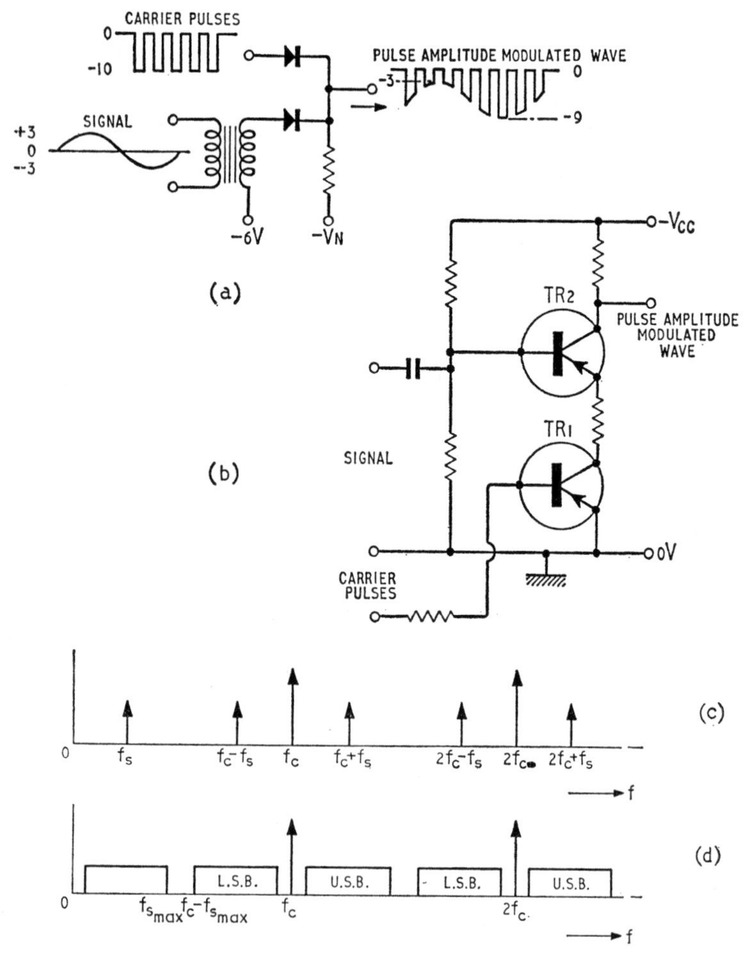

Fig. 9.31. Producing the pulse amplitude modulated wave. (a) A simple pulse amplitude modulator; (b) A transistor modulator; (c) Frequency spectrum for single frequency signal; (d) Frequency spectrum for a complex modulating wave

signal. Fig. 9.31(a) illustrates a basic pulse amplitude modulator in which the train of carrier pulses is applied to one input of an *AND* gate and the modulating signal is applied in series with a d.c. bias to the second input. The gate output will only be negative when both inputs are negative and will then ideally be equal to the least negative value. If the instantaneous sum of the modulating signal voltage and the d.c. bias lies between 0V and the carrier pulse amplitude, then pulse amplitude modulation will result.

An elementary transistor modulator is shown in Fig. 9.31(b). The output signal from this circuit will vary in accordance with the modulating signal when the transistor TR1 is switched on by the negative carrier pulses, thus producing the same modulated wave as the diode circuit, except for inversion. The output waveform will thus consist of carrier pulses with amplitudes varying in accordance with the instantaneous modulating signal. Alternatively we may consider the output to be a succession of samples of the modulating signal occurring at a rate equal to the carrier frequency. An analysis of the frequency spectrum of this output wave shows that, in general, if the modulating wave is a sine wave frequency f_s, and the repetition rate of the carrier pulses is f_r, there will be components in the output of

$$f_s, f_r, f_r \pm f_s, 2f_r, 2f_r \pm f_s, \text{ etc.}$$ [1]

This spectrum is represented by the line spectra of Fig. 9.31(c). If the modulating signal were a complex wave containing significant frequency components up to $f_{s\,max}$ then the spectrum would be modified to include sidebands about the carrier and its harmonics as in Fig. 9.31(d).

Since the modulated wave contains the original signal, then in order to demodulate it is only necessary to pass the modulated wave of Fig. 9.30(c) through a low pass filter. The attenuation/frequency characteristic of the filter is such that the maximum modulating frequency $f_{s\,max}$ is passed unaltered, while the component $f_r - f_{s\,max}$ and all other frequencies are attenuated. This requirement sets a practical minimum limit on the sampling rate, i.e. on the pulse carrier frequency, since the frequency must be high enough to allow a practical filter to change from pass-band to stop-band between $f_{s\,max}$ and $f_r - f_{s\,max}$. Practical systems have been built using an 8kHz pulse-repetition frequency for a channel with 3·4kHz maximum speech frequency.

[1] BLACK, H. S., *Modulation Theory*, Van Nostrand (D.) Co. Ltd. (1953).

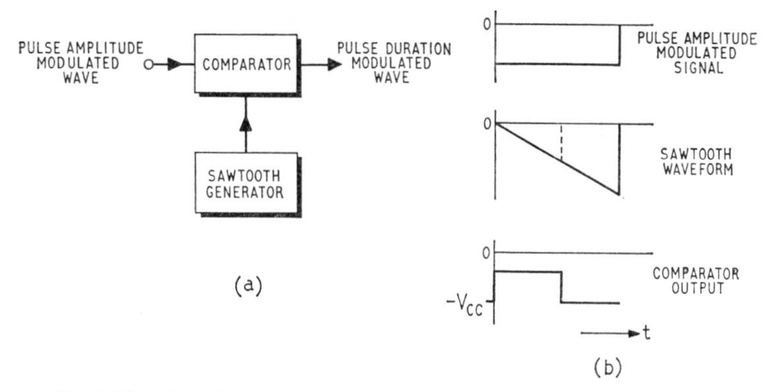

Fig. 9.32. A method of pulse duration modulation. (a) Block diagram; (b) Relevant waveforms

9.6.2. Pulse Duration Modulation

The pulse amplitude modulation system though simple suffers from the same drawback as continuous wave amplitude modulation in that variation of amplitude, due for instance to the introduction of noise in the transmission medium, will cause distortion. If, however, we

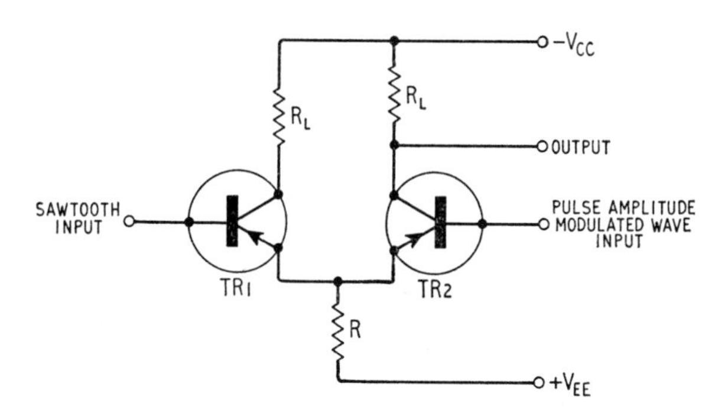

Fig. 9.33. A comparator circuit

vary the duration of the carrier pulses in accordance with the modulating signal the system becomes much less sensitive to noise. A possible way in which this may be carried out is indicated in Fig. 9.32. The principle involved is to convert the amplitude of samples of the modulating signal obtained by pulse amplitude modulation to a time interval. The pulse amplitude modulated signal is applied to one input terminal of a comparator circuit such as shown in Fig. 9.33

and the output of a sawtooth generator is applied to the other terminal. The sawtooth wave commences to rise from zero at the beginning of the sampling period, and when its instantaneous amplitude exceeds the signal amplitude the comparator shows a change in voltage level at its output. The output of this comparator is thus

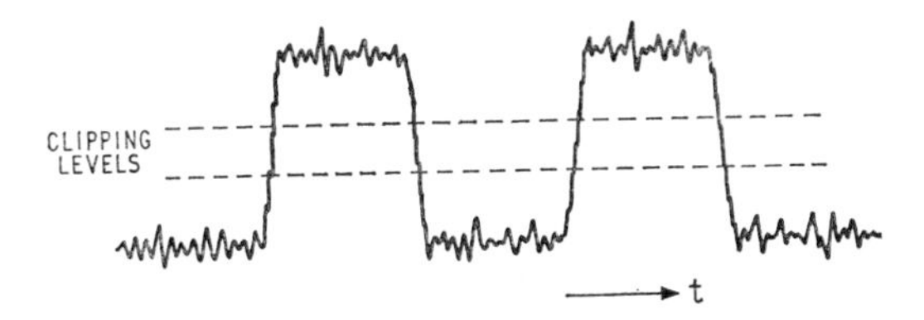

Fig. 9.34. Reduction of noise by clipping

a pulse of voltage commencing at the beginning of the sawtooth wave and ending when the signal amplitude is attained. Therefore provided the sawtooth is linear, the signal amplitude is correctly converted to a pulse duration as shown in Fig. 9.32(b).

The advantage of this method of modulation lies in the fact that at the receiving end of the system the signal may be clipped symmetrically, thus removing most of the added noise without losing the information content, which is in the pulse duration as illustrated in Fig. 9.34.

9.6.3. Pulse Code Modulation

In a pulse-duration modulation system it is still necessary to be able to determine precisely the duration of the pulses in order to obtain distortionless demodulation. In general, edges of the received pulses are not vertical due to the effect of the system bandwidth limitation on the pulse rise-time. Thus noise will not be completely eliminated by clipping and will cause jitter of the edges of the clipped pulse, Fig. 9.35. If the modulating information is converted to a digital code, however, it is only necessary at the receiver to determine the presence or absence of a pulse at a given time instant to obtain the correct result. This is the principle of *pulse code modulation* as illustrated in Fig. 9.30(e), and results in the most efficient method of modulation, enabling communication to take place over extremely noisy systems. Pulse code modulation may be obtained from pulse-duration modulation as illustrated in Fig. 9.36. The pulse duration modulated signal is applied to one input of a 2-input *AND* gate while the other input is fed with a continuous train of clock pulses.

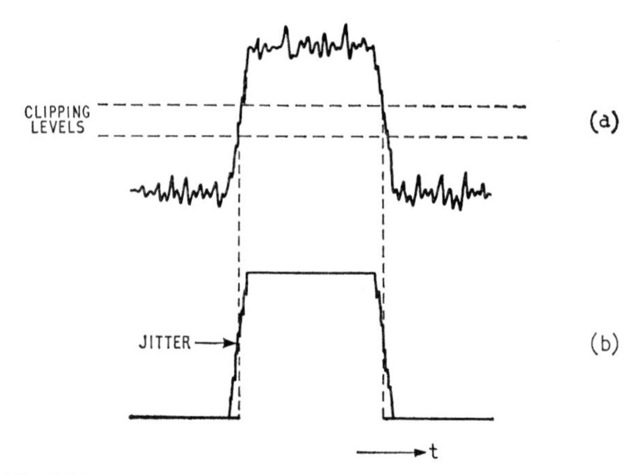

Fig. 9.35. *Jitter of pulse edges caused by noise.* (*a*) *Input signal;* (*b*) *Clipped and amplified signal*

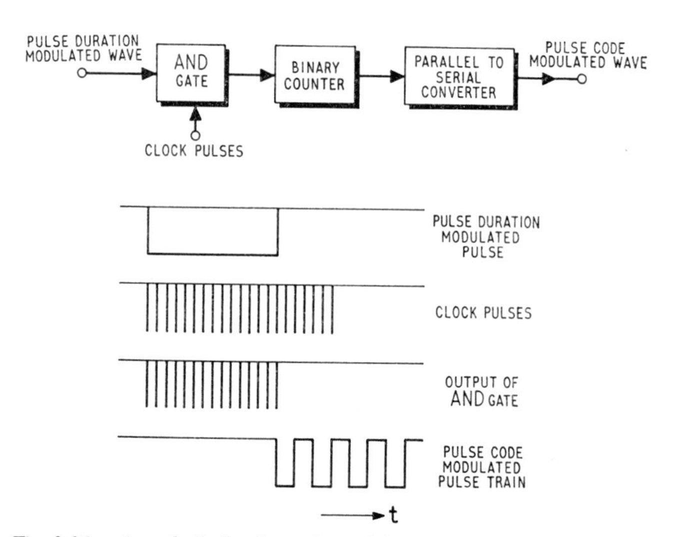

Fig. 9.36. *A method of pulse code modulation, showing block diagram and relevant waveforms*

The output of the *AND* gate is applied to a binary counter. The *AND* gate is open only for the duration of the pulse in the pulse-duration modulated signal. Hence the number of pulses passed to the counter is proportional to the pulse duration and therefore proportional to the signal amplitude. The binary counter gives a digital representation of the signal amplitude in parallel form. Conversion to serial form as a train of pulses may be carried out

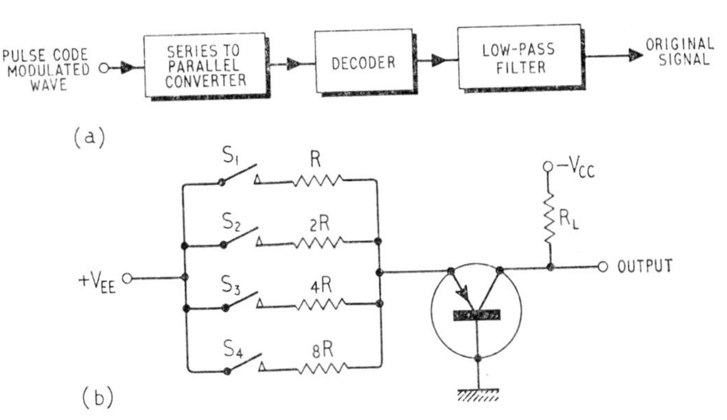

Fig. 9.37. Demodulating the pulse code modulated wave. (a) Block diagram; (b) An elementary decoder

using a shift register in the manner discussed in Section 8.9. Thus after each sampling period a sequence of pulses forming a digital representation of the amplitude of the modulating signal is transmitted over the system. The result of coding the signal is that, in the time interval allocated to one pulse for straightforward pulse amplitude modulation, it is now necessary to transmit n pulses, where n is the number of digits in the binary code. The operating speed is thus increased n times with the result that the bandwidth requirement of the communication system has been increased n times, a penalty which one must always pay for the increased immunity from noise.

After transmission over the communication system it is necessary to decode the signal from the digital representation in order to effect demodulation. A possible means of demodulation is shown in Fig. 9.37.

It is first necessary to convert the digital representation from serial to parallel form. Again a shift register, using bistable multivibrators, may be used. The decoder shown in principle for a 4-digit code in Fig. 9.37(b) may then be used. The switches 1 to 4 correspond

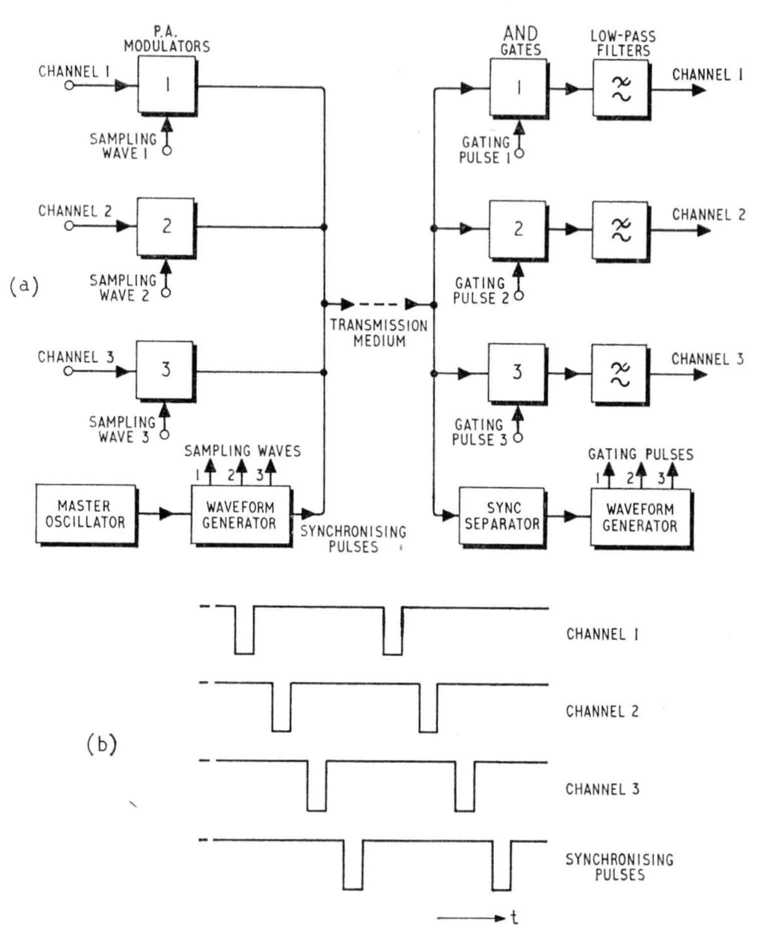

Fig. 9.38. A 3-channel pulse amplitude modulation system. (a) Block diagram; (b) Sampling wave sequence

to the 4 digits of the code and are open for 0, closed for 1. Switch 1 corresponds to the most significant digit and switch 4 to the least significant. The resistors connected to the emitter are weighted in order of significance of digit so that, assuming the transistor input resistance is negligible compared to R, the emitter current is dependent on the switches closed. The voltage developed across

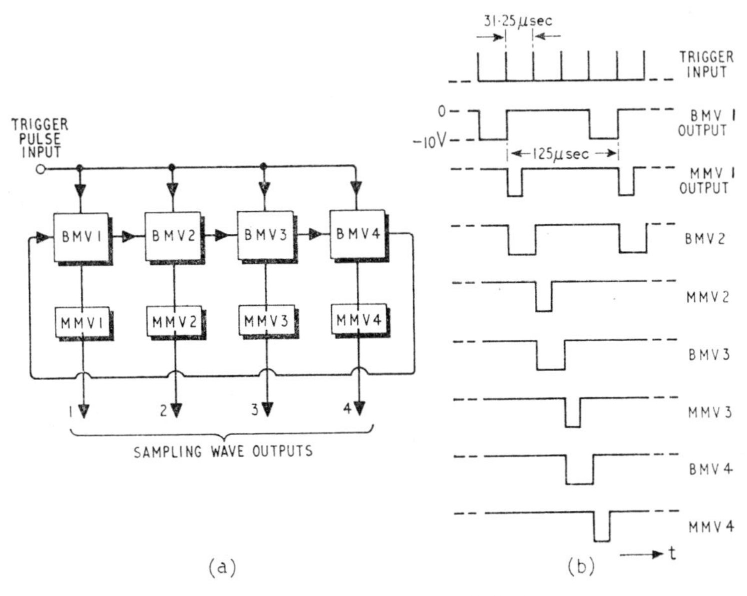

Fig. 9.39. A ring counter. (a) Block diagram; (b) Waveforms in the ring counter

the collector load resistance can then assume any one of 16 different levels, from which pulse amplitude modulated samples may again be taken. It is then only necessary to pass these samples through a low pass filter to recover the modulating signal.

Since the decoded signal can only assume one of a finite number of levels the quality of the system obviously depends on the number of digits used in the code group.

9.6.4. A Typical Pulse Amplitude Modulation 3-Channel System

Fig. 9.38(a) shows the basic block schematic of a typical 3-channel pulse amplitude modulated speech system. Since the channels must be sampled in turn in a regular time sequence, i.e. sample channel 1, then channel 2, etc., then back to channel 1 again, as was the case for the elementary switch, it is necessary to produce individual carrier pulse trains identical in form but with a time occurrence as in Fig. 9.38(b). In order to correctly separate the multiplexed signal into the proper channels it is necessary to keep the receiver in

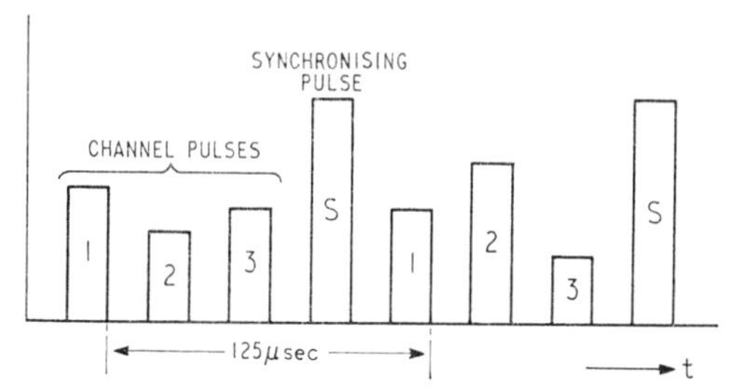

Fig. 9.40. Time spectrum of the time-division multiplex system

synchronism with the transmitter at all times. One set of pulses is thus transmitted purely as synchronising pulses, with the result that the system is effectively a 4-channel system.

A possible method for obtaining the pulse trains in the correct time sequence is indicated in Fig. 9.39(a), where use is made of a ring of 4 bistable multivibrators. Assume that the sampling frequency is to be 8kHz, i.e. the time between sampling pulses is 125μsec. The repetition rate of the triggering pulses will then be 31·25μsec. The multivibrators are connected so that only one is *ON*, represented by −10V at its output, at a given time and the other three are *OFF*, represented by 0V at their outputs. The trigger pulses are applied so as to turn an *ON* multivibrator *OFF* and the stage-to-stage connection is such that an *ON* multivibrator turning *OFF* overrides the trigger and turns the succeeding stage *ON*. At the same time it triggers its associated monostable multivibrator, thus producing an output pulse. The result of this arrangement is that the *ON* state is stepped round the ring by one position for each input trigger pulse resulting in the waveforms of Fig. 9.39(b). The modulated pulses of channels 1, 2 and 3 are transmitted with the synchronising pulses whose amplitudes are appreciably larger than the maximum modulated pulse amplitude. The time spectrum of the line pulses will then be as in Fig. 9.40.

At the receiving terminal, the signal is first applied to a circuit which extracts the synchronising pulse train. This may be a Schmitt trigger circuit which responds only to the large amplitude synchronising pulses. The synchronising pulse train is then used to control a trigger pulse generator and ring counter similar to the equipment at the transmitter, and gating pulses are applied to the channel *AND* gates in the correct time sequence. Thus if the incoming signal is applied to channel 1 *AND* gate coincidentally with channel 1 gating pulses only the samples of the original modulating signal on channel 1 will be passed to the channel 1 demodulator. The demodulator will be a low pass filter with an attenuation/frequency characteristic which passes frequencies up to 3·4kHz and attenuates the $f_c - f_s$ component (i.e. $8 - 3{\cdot}4 = 4{\cdot}6$kHz) and all frequencies above.

CHAPTER 10

INTEGRATED CIRCUITS

10.1 INTRODUCTION

Since the early days of electronics there have been continual efforts to miniaturise electronic equipment. The development of the transistor and the semiconductor diode has stimulated the development of miniature passive components, resistors and capacitors, because transistor circuits could be operated at very low voltage and power levels with corresponding low heat dissipation. Assemblies of transistors with miniature passive components on small printed circuit boards resulted in significant reductions in size and weight of equipment. The technique was still conventional however, since it used assemblies of discrete components.

Further requirements for size reduction led to the development of *microelectronics.* This is the general name given to extremely small electronic components and circuit assemblies made by thin film, thick film or semiconductor technology. An *integrated circuit* is a special case of microelectronics and refers to a circuit fabricated as an inseparable assembly of electronic elements in a single structure that cannot be divided without destroying its intended function. We shall concern ourselves with *monolithic integrated circuits* in which all circuit elements both active and passive are simultaneously formed in a single small wafer of silicon by the diffused planar technique. The elements are interconnected to form the required electronic circuit by metallic strips deposited on the oxidised surface of the silicon wafer using evaporation techniques.

Before discussing the techniques of integrated circuits in further detail it is necessary to consider certain fundamental processes.

10.1.1. The Planar Process

The basic process used for forming monolithic integrated circuits is the planar process, so called since all operations are carried out on the top surface of a slice of semiconductor material, usually silicon. An important property of silicon is that a layer of silicon oxide on the surface of the slice prevents the diffusion of certain elements, including boron and phosphorous, into the silicon. In addition, silicon oxide can be easily removed from the surface by etching with a hydrofluoric acid solution without etching the silicon. Thus, if a slice of silicon is oxidised by heating it in a flow of oxygen to form a layer of silicon oxide on the surface and then the oxide is removed from selected regions by etching, impurities can be diffused into these selected regions only. This process of selective diffusion is the basis of all silicon monolithic integrated circuit fabrication, allowing the simultaneous formation of a number of separate components in a single slice of silicon.

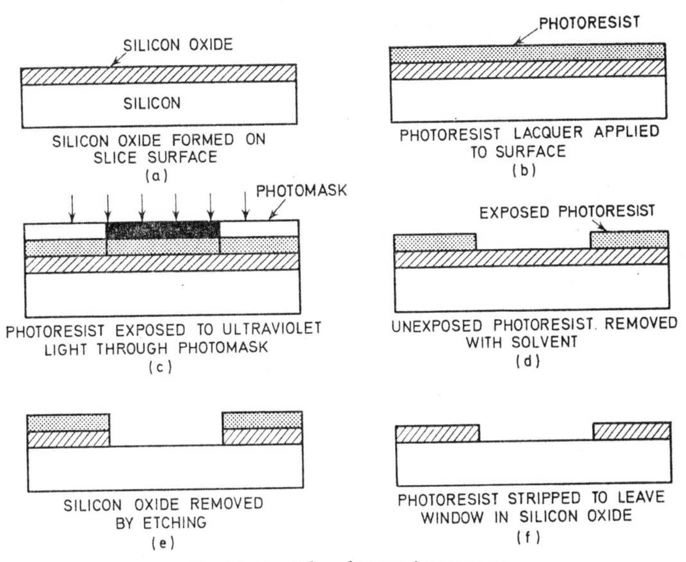

Fig. 10.1. The photoresist process

The selective removal of silicon dioxide is carried out by a photolithographic process using photoresist material. The steps in the process are illustrated in Fig. 10.1. After oxidation the oxidised surface of the slice is coated with a thin layer of photoresist laquer. This is an organic substance which polymerises when exposed to ultraviolet light and then resists attack by acids and solvents.

A photographic mask is placed over the slice and illuminated with ultraviolet light. The photoresist under the opaque regions of the photomask is unaffected by the light and can be removed with a solvent, the exposed photoresist remaining in the other regions. The slice is baked to harden the photoresist and then immersed in a hydrofluoric acid solution to etch away the silicon oxide unprotected by polymerised photoresist. Finally, the photoresist is removed from the surface and the slice is thoroughly washed. It is now ready for diffusion, which only occurs through the openings (called windows) in the oxide. This complete photoresist process must be repeated each time the silicon oxide is selectively removed.

10.1.2. The Discrete npn Planar Transistor

Using the photoresist process just described, an n-type silicon slice is oxidised and windows for the base diffusion are opened in the oxide. Referring to Fig. 10.2 the p-type impurity for the base diffusion,

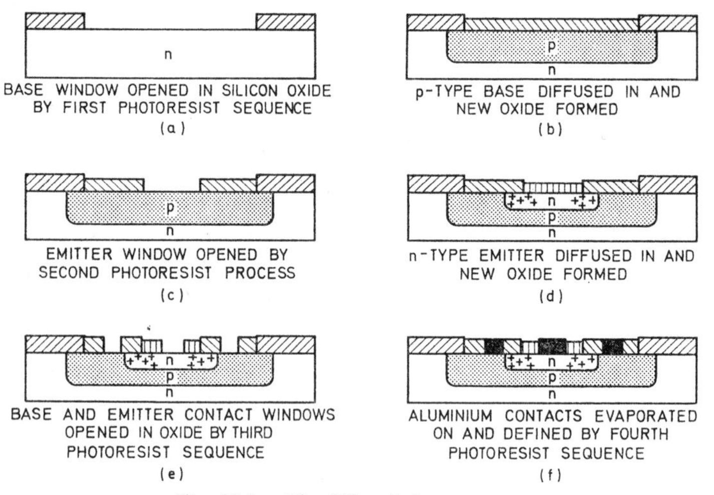

Fig. 10.2. The diffused planar process

usually boron, is deposited on the surface of the silicon and diffused in to form the pn junction at the required depth, typically 0·1 mil. During the latter part of this diffusion, steam is mixed with the nitrogen used in the diffusion process to form a new layer of silicon oxide on the surface of the diffused region. Sideways diffusion of the p-type impurity causes the pn junction to form under the oxide, giving protection against surface contamination.

The slice is now prepared for the emitter diffusion by etching windows in the new oxide grown over the base region, using the same photoresist process as before. Phosphorous is diffused in to a typical depth of about 0·06 mil, to form the n-type emitter region and steam is again used to form silicon oxide on the surface. The resulting base width is of the order of 0·04 mil.

The next process is the forming of metallised contacts to the base and emitter regions. Once again the photoresist process is used and contact windows are opened in the silicon oxide. Aluminium is now evaporated on to the whole surface of the slice and a fourth photoresist sequence carried out with a contact photomask to remove the aluminium from everywhere except the contact windows. The aluminium remaining in the contact windows is then alloyed to the silicon to form a low resistance contact.

10.1.3. The Integrated Circuit Transistor

The technique for fabricating the transistor for integrated circuits is the same as described above, but certain special points must be observed. First, since all the circuits elements are formed in close proximity in a wafer of silicon which is electrically conducting, they must each be electrically isolated from the substrate. The only connection between the elements is then the metallised pattern on the surface.

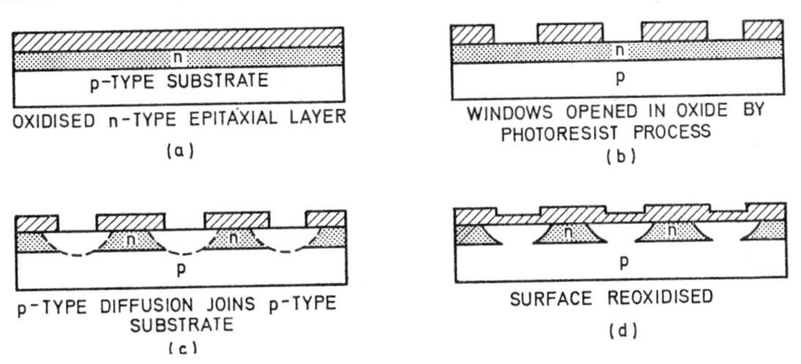

Fig. 10.3. Junction isolation

There are various methods for effecting this isolation and one method commonly used is that of diode isolation, which uses the high resistance of a reverse-biased pn junction. Referring to Fig. 10.3 an n-type epitaxial or surface layer is first grown on a p-type substrate. The surface of the epitaxial layer is oxidised and the oxide selectively removed from everywhere but the regions in which the elements will be formed. A p-type diffusion then forms p-type regions extending down through the epitaxial layer to the p-type substrate. This leaves the n-type regions each separated from the substrate by a pn junction. When the final integrated circuit is operated, the pn junctions are all biased in the reverse direction by connecting the p-type substrate to a potential more negative than any part of the circuit. Each junction then presents a very high resistance which isolates the elements formed in its n-type region.

The second important difference between integrated and discrete planar transistors is that in the integrated circuit device the collector contact is made at the top surface beside the base and emitter contacts. This introduces additional series collector resistance compared with the discrete transistor, in which the collector contact is at the bottom surface. To minimise the series resistance, a low resistance high concentration n-type region is selectively diffused into the substrate slice before the epitaxial growth of the n-type layer as shown in Fig. 10.4, resulting in a lower resistance path from

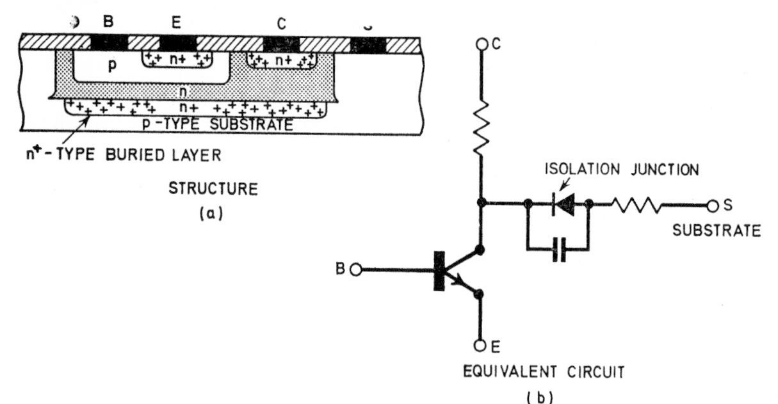

Fig. 10.4. The integrated circuit transistor

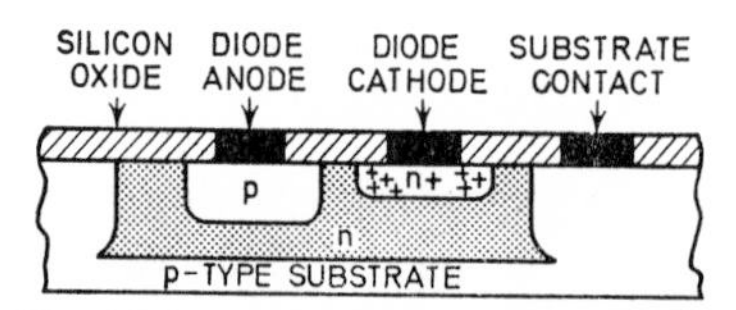

Fig. 10.5. An integrated circuit diode

the collector contact to the junction. The equivalent circuit of the transistor including the isolation junction is shown.

Integrated circuit diodes are prepared by forming pn-junctions at the same time as one of the transistor junctions. Fig. 10.5 shows a diode in which the cathode is the original n-type region. The p-type anode is formed during the transistor base diffusion.

10.1.4. Diffused Resistors and Capacitors

Silicon is a resistive material and its resistivity is dependent on the concentration of current carriers. A resistor can be formed in a silicon wafer by diffusing a suitable impurity into a defined region, the value of the resistor depending on the concentration of the impurity, the dimensions of the region at the surface and the depth of diffusion. Most resistors in integrated circuits are formed at the same time as the p-type base region. Fig. 10.6 shows the cross section of such a diffused resistor, with diode isolation.

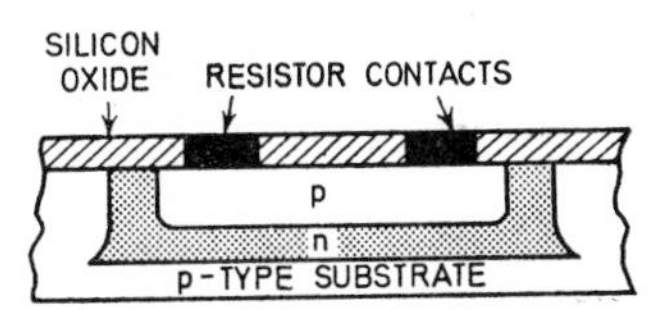

Fig. 10.6. Diffused p-type resistor

A junction capacitor uses the capacitance of the reverse biased pn junction formed at the same time as the emitter junction or the collector junction of the transistor. The value of capacitance attainable is limited to about 100 pF and, since the capacitance of a pn junction is dependent on the value of reverse voltage, it is necessary to maintain the correct voltage bias in the circuit. Such capacitors have the advantage that they can be formed at the same time as the other elements with no additional processes. The arrangement of a capacitor with a junction formed at the same time as the transistor emitter junction is shown in Fig. 10.7.

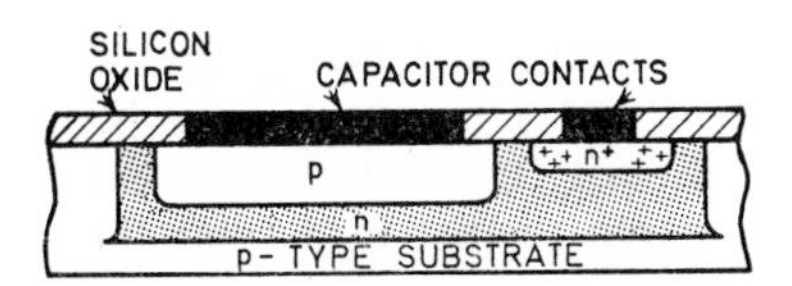

Fig. 10.7. A junction capacitor

10.1.5. Field Effect Transistors (FET)

The junction transistors discussed so far are *bipolar* transistors, i.e. two types of carrier, free electrons and positive holes are involved in their operation. A device developed in recent years and having considerable application in the realms of integrated circuits is the field effect transistor. In particular the surface, or metal-oxide-semiconductor (MOS) type of FET is of considerable importance.

A field effect transistor is essentially a semiconducting current path whose conductance is controlled by applying, perpendicular to the current, an electric field resulting from reverse-biasing a pn junction.

10.1.6. The Unipolar FET

Consider a semiconductor bar with ohmic contacts made at each end. The conductance between the ohmic contacts depends upon the dimensions of the bar and the conductivity. We have seen that a pure semiconductor such as silicon or germanium is a poor conductor, since few mobile carriers (holes and electrons) are available to conduct current. If, however, a large concentration of impurity atoms is introduced into the crystal, then effectively the current carriers available are all of one type, e.g. if the impurity atoms are all donors (pentavalent) then only electrons are available to carry current and the bar is n-type. The conductance of the bar is then proportional to the total number of carriers present.

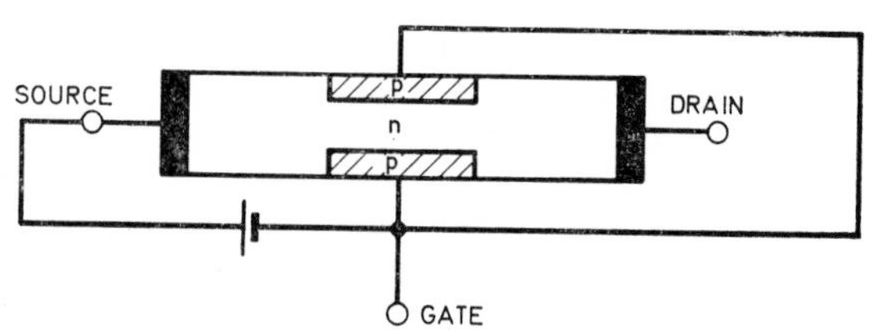

Fig. 10.8. Elementary unipolar FET

Now consider the original n-type bar with p-type impurities introduced into opposite sides as shown in Fig. 10.8, thus forming two pn junctions with the semiconductor bar. The two regions are electrically connected and a reverse voltage is applied to the two junctions. If the impurity concentration is made very high in the p-type regions compared to that in the bar then the space charge layer due to the contact potential and the external bias will extend almost entirely into the region of the bar between the junctions. Since there are virtually no carriers in the space charge layer, the conductance between the so-called source and drain terminals is almost entirely determined by the region between the pn junctions not depleted of free carriers by the reverse junction voltage. Thus the applied gate voltage directly controls the conductance of the bar.

Fig. 10.8 represents an elementary unipolar FET. The terminals drain, gate and source are analagous to the collector, base and emitter respectively, of the bipolar transistor. The part of the bar directly between the pn junctions is referred to as the *channel* and is the active part of the FET, the rest of the bar being effectively bulk resistance. Notice that the drain and source terminals are interchangeable.

10.1.7. The MOS Field Effect Transistor

A variation on the basic FET structure which is particularly useful in integrated circuit applications is the surface, or metal-oxide-semiconductor (MOS), device, in which a conducting channel is induced between two very closely spaced electrode regions by increasing the electric field at the surface of the semiconductor between the electrodes. The general construction of the device is shown in Fig. 10.9.

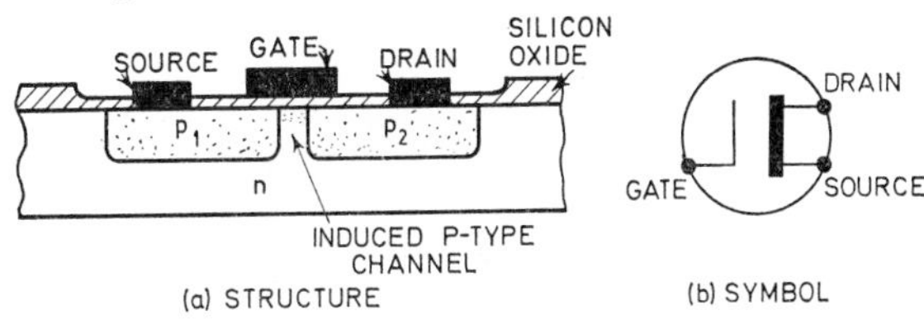

Fig. 10.9. The p-channel MOS transistor

The source and drain electrodes are formed by a p-type diffusion into an n-type silicon wafer, resulting in two pn junctions back to back between the electrodes. The gate is a metal plate separated from the channel by an insulating dielectric. With a voltage applied to the source positive with respect to the drain, the np junction is reverse biased and no current flows from source to drain. If the gate electrode over the space between the source and drain is now made sufficiently negative with respect to the source holes are attracted to the surface of the n-type region and cause it to change to p-type. The result is a p-type channel between the two p-type electrodes and current is able to flow.

The fabrication steps for the MOS transistors are very similar to those described for the diffused planar transistor, except for the thin layer of very pure oxide grown over the channel area to isolate the gate electrode. Because of its construction, the MOS transistor is self isolating. Both the source and the drain are isolated by their own pn junctions, the gate is isolated by the layer of silicon oxide and the channel formed under the gate is also isolated by the pn junction which forms with it. Thus the packing density is much greater than with bipolar transistors. Because of the isolation between gate and channel in the MOS device, the gate leakage current is considerably less than in the unipolar device, resulting in a much higher input impedance. Further the MOS transistor can be operated with a single polarity of supply voltage. A disadvantage is that the MOS device has a considerably lower cut-off frequency and switching speed than the bipolar transistor.

10.1.8. MOS Resistors and Capacitors

The channel between the source and drain of an MOS transistor can be used as a resistor whose value depends on the gate potential and the transconductance of the structure. A relatively wide source-to-drain spacing is used with the gate and drain connected, thus biasing the device to the *ON* state. Such resistors can be made in a much smaller area than that required for diffused resistors, allowing a still further increase in packing density. The structure of an MOS capacitor is shown in Fig. 10.10.

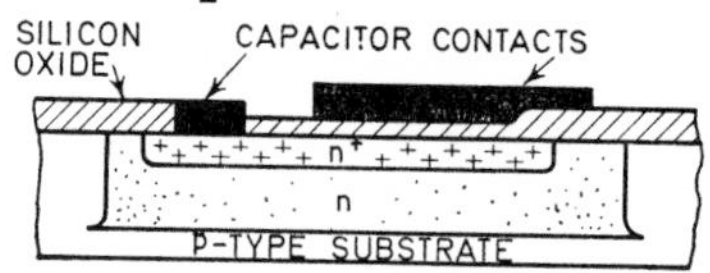

Fig. 10.10. A MOS capacitor

A high concentration n-type region is diffused into the silicon at the same time as the transistor emitter diffusion to form the bottom electrode of the capacitor and a controlled thickness of silicon oxide is formed on the surface of this region to produce the dielectric. The top electrode consists of a layer of metal deposited at the same time as the interconnection pattern. The value of capacitance is still limited to a few hundred picofarads though being somewhat higher than for a comparable junction capacitor.

10.1.9. Circuit Formation

All integrated circuit elements such as indicated in Fig. 10.11 are formed simultaneously by the same sequence of oxidation, selective oxide removal, diffusion and metalisation. The result of such a sequence is illustrated in Fig. 10.12 for a specific circuit. Although the elements are shown formed in a line they can in practice be in any position in the wafer.

An illustration of the smaller area required for an MOS circuit compared to the equivalent bipolar circuit is shown in Fig. 10.13 for the case of a simple inverter. Fig. 10.13(a) shows the arrangement of a bipolar transistor and diffused resistor whereas Fig. 10.13(b) shows the corresponding MOS transistor and MOS resistor drawn to the same scale. The MOS circuit gives an area saving up to 5:1 but, as stated previously, at the expense of poorer switching speed and high frequency performance.

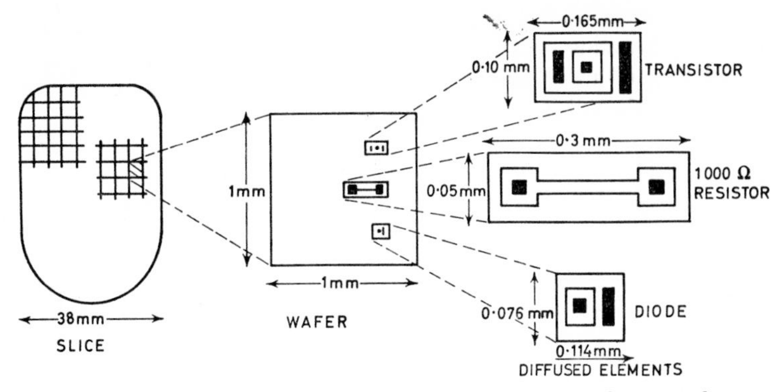

Fig. 10.11. Typical slice, wafer and element relationship. A slice can contain up to 500 wafers and a wafer can contain up to 50 diffused elements

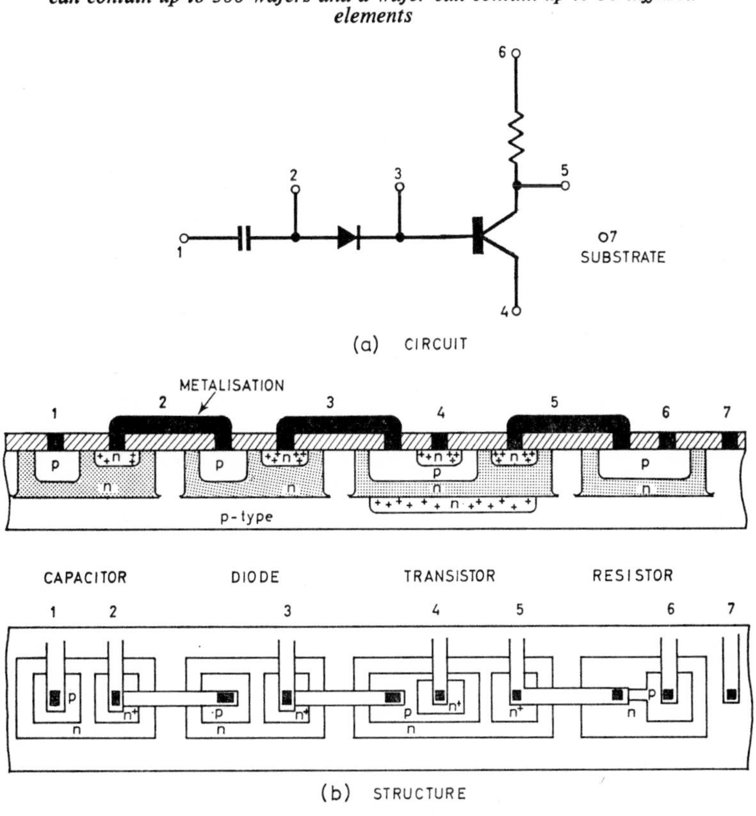

Fig. 10.12. Formation of a four-element integrated circuit

In the development of integrated circuits the MOS device is finding increasing application, both as an alternative to the bipolar transistor and in combined circuits such as the arrangement shown in Fig. 10.14 resulting in higher input impedance and high gain.

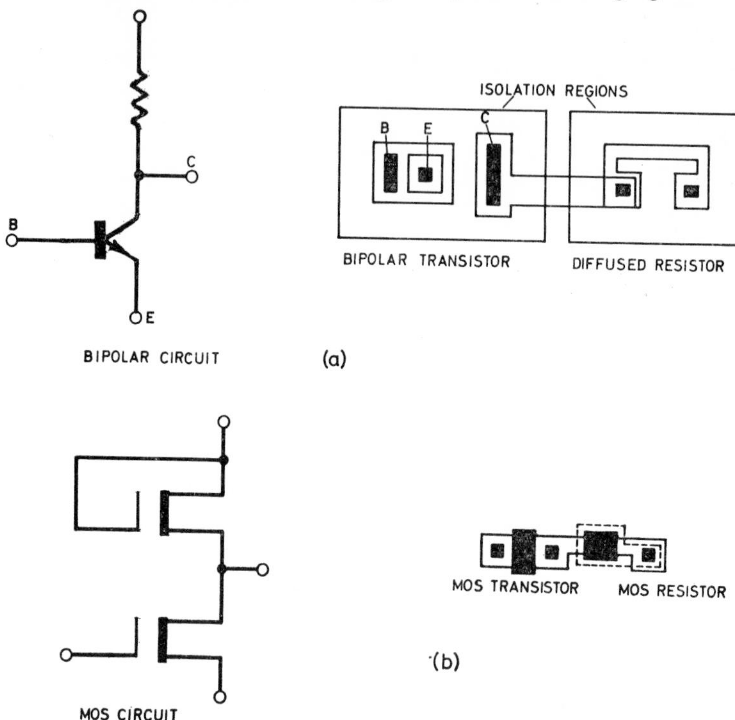

Fig. 10.13. Scaled comparison of equivalent bipolar and MOS inverters

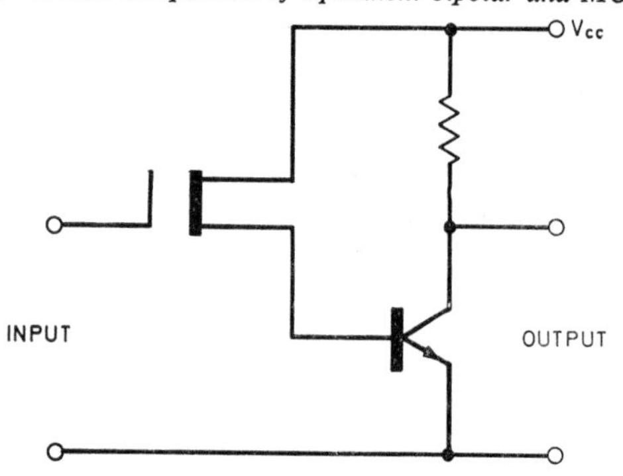

Fig. 10.14. MOS-bipolar combination circuit

10.2 LINEAR INTEGRATED CIRCUITS

10.2.1. Basic Linear Integrated Circuit Design Philosophy

The design of complete circuits such as high gain amplifiers which are to be realised as integrated circuits rather than made up from separate components demands a new concept in the basic approach. This is due not only to certain limitations inherent in integrated circuit manufacturing techniques but also because of new design freedoms that it allows.

In a discrete component circuit cost is approximately proportional to the number of transistors, capacitors and resistors used in descending order of importance. Pnp or npn transistors may be used at will but matching of separate transistors for gain, V_{BE}, etc. incurs considerable cost penalties while the necessity for specifying certain resistor values to close tolerances does not have the same significance.

With integrated circuits it is possible to diffuse in both npn and pnp devices but, for ease of manufacture it is generally preferred to keep to npn exclusively in the more simple circuits. There is more freedom in the number of active three-layer or two-layer junctions which may be used, since doubling the number of such junctions does not increase the cost proportionally, provided no extra processes are involved. Though it is impractical with the techniques used at present to guarantee absolute resistor values to closer than 20% of nominal it is possible to achieve tolerances of better than 5% in the ratio between two resistors of the same value. This is effected by ensuring that they are placed in close physical proximity and have identical layouts. Compared with high stability resistors the temperature coefficient of semiconductor resistors is high.

The fabrication of capacitors and even inductors of very small value is feasible but results in considerable complications in manufacture and is therefore to be avoided in design if possible. Diodes, however are available at small cost by making use of the forward and reverse characteristics of a pn diffused layer where appropriate.

The design philosophy for integrated circuit amplifiers is therefore based on symmetical circuits where possible, with inherent immunity from drift caused by absolute resistor value changes. D.C. coupling between stages is used to eliminate capacitors. Circuit balance is achieved by use of a differential stage such as the emitter-coupled pair.

10.2.2. Biasing Circuits

A basic problem encountered in integrated circuits is bias stabilisation of a common emitter amplifier. Conventional methods usually require substantial d.c. feedback and bypass capacitors to reduce the negative feedback at the operating frequencies (see Section 2.2). With integrated circuits the required bypass capacitors are too large to be practical, but the close matching of components and thermal coupling permit the use of much more radical solutions as shown for example in Fig. 10.15. Assuming that TR_1 and TR_2 are identical

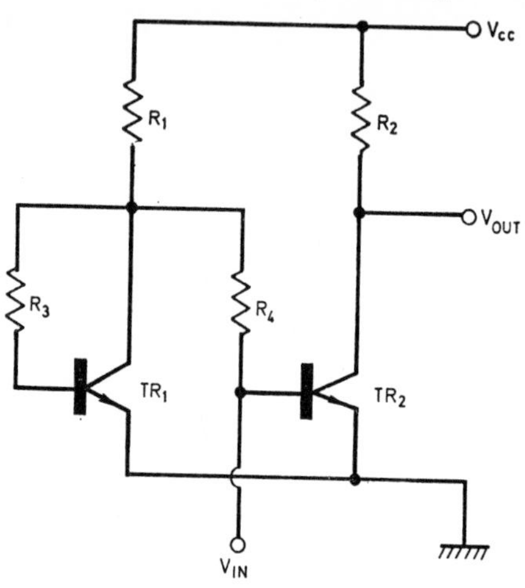

Fig. 10.15. A biasing technique applicable to integrated circuits

transistors and R_3 and R_4 are identical resistors, the collector currents of the two transistors will be equal since their bases are driven from a common voltage point through equal resistances. The collector current of $TR1$ will be

$$I_{c1}=\frac{V_{cc}-V_{BE}}{R_1}-\left(2+\frac{R_3}{R_1}\right)I_B$$

If

$$V_{cc}\gg V_{BE} \quad \text{and} \quad I_{c1}\gg\left(2+\frac{R_3}{R_1}\right)I_B$$

then

$$I_{c1}=I_{c2}\approx\frac{V_{cc}}{R_1} \tag{10.1}$$

Further if

$$R_2=\frac{R_1}{2}, \quad V_0=\frac{V_{cc}}{2} \tag{10.2}$$

and the amplifier will be biased at its optimum operating point, at half the supply voltage, independent of the supply voltages and temperatures. The closeness with which this ideal situation is approximated depends on the degree of match within the integrated circuit.

10.2.3. Constant Current Sources

In the discussion of the emitter-coupled pair in Section 4.1.2 it was shown that for good rejection of the in-phase component (or common-mode rejection) the emitter resistor R should be large, a requirement not compatible with the realisation of semiconductor resistances. A considerable improvement is effected if R is replaced by a constant current source such as discussed in Section 9.5.1. A variation having application in integrated circuits, due to the close matching obtainable is shown in Fig. 10.16.

With such a circuit it is possible to realise a current source with very low collector currents using resistors of only a few kilohms. It makes use of the predictable difference of the emitter-base voltage of two

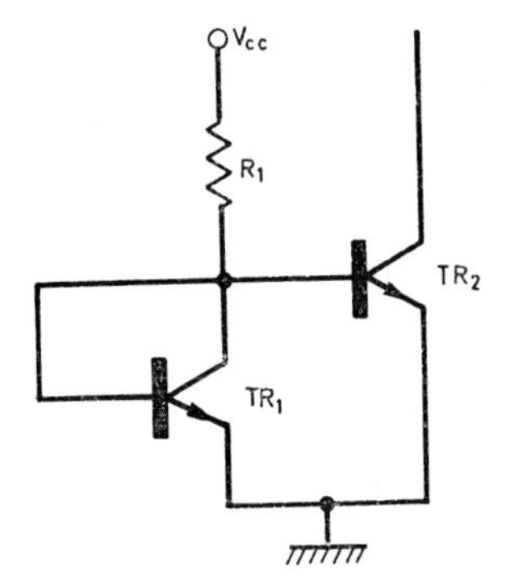

Fig. 10.16. Constant current source

identical transistors operating at different collector currents. Since the collector currents will be equal the operating current can be determined by R_1 and V_{cc}.

10.2.4. A Typical High Gain Amplifier—The Fairchild μA702A

The μA702A is a high-gain amplifier whose operating characteristics are mainly determined by the use of external feedback elements. It is useful as a general purpose d.c. or a.c. amplifier to frequencies as high as 30MHz.

The circuit has been designed according to the basic philosophies outlined above, hence the input stage is a matched pair of bipolar transistors operating at a low collector current in a differential amplifier circuit, TR_2 and TR_3 in Fig. 10.17. The emitters of the input stage are fed from a constant current source TR_1 to obtain good common-mode rejection. The diode-connected transistor TR_9 compensates for the temperature dependency of V_{BE} of the current source transistor.

The second stage converts the balanced signal from the input differential stage to a single ended output. TR_4 and TR_5 are identical transistors placed close to one another. TR_4 functions as a unity gain amplifier which inverts the output of TR_2 and combines it with the output of TR_3 at the base of TR_5. A single-ended output is obtained at the collector of TR_5.

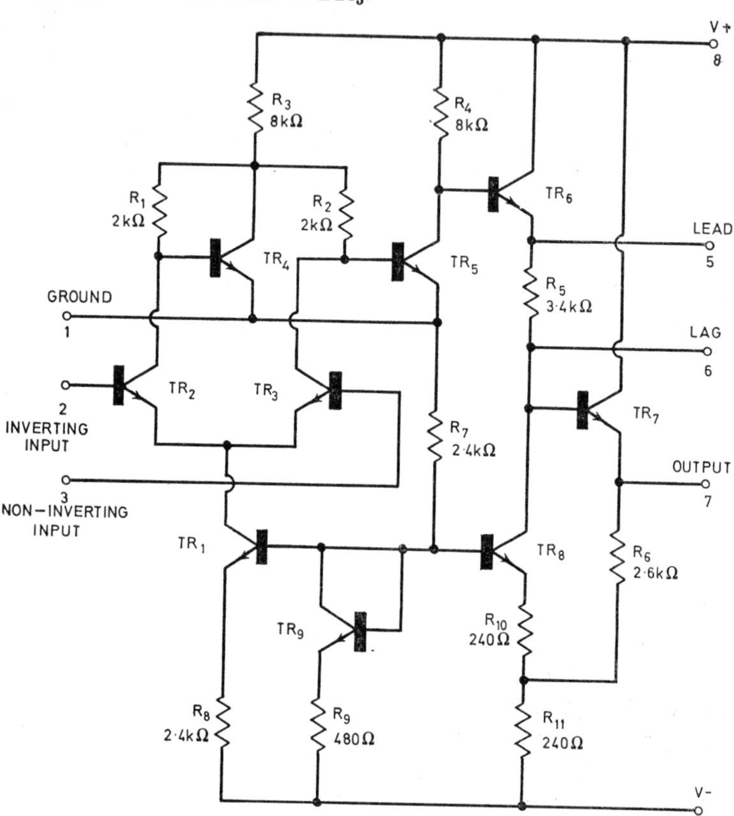

Fig. 10.17. Circuit diagram of Fairchild μA702A high gain amplifier

The first two stages have provided nearly enough voltage gain for the amplifier. Since both positive and negative output swings are desired, some form of d.c. level shifting must be provided as the output of the second stage cannot swing negative. This is effected using only npn transistors as follows. The output of the second stage is buffered by a common collector TR_6 and a current source TR_8 provides a voltage drop across R_5 which gives the basic level shifting. An additional common collector TR_7 is used to give a low output impedance.

Typical performance of the μA702A Amplifier is summarised in Table 10.1, for 25°C temperatures

Table 10.1

	V+ 12V V− = −6V	V+ = 6V V− = −3V
Input resistance	40kΩ	67kΩ
Common-mode rejection ratio	95dB	95dB
Voltage gain	3600	900
Output resistance	200Ω	300Ω
Output voltage swing	±5·3V	±2·7V
Power consumption	80mW	17mW

As has been shown in Section 5.1, application of negative feedback to a high gain amplifier gives a circuit with characteristics dependent almost entirely on the feedback elements. The improvements in gain stability, input impedance, output impedance and linearity are proportional to the amount of feedback. Thus amplification to any degree of accuracy is possible with sufficient feedback. Large amounts of feedback however, require that close attention be given to the amplifier open loop characteristic. Stable circuits must have well defined open loop gain and phase responses to frequencies far above the band of interest. Provision is made therefore in the μA702A amplifier for the external connection of frequency compensating circuits to the terminals marked LEAD and LAG in Fig. 10.17. By suitable choice of these components the amplifier can be made stable under varying conditions of operation.

10.3 BASIC AMPLIFIER CIRCUITS

10.3.1. Inverting Amplifier

With the circuit connected as in Fig. 10.18 the μA702A may be used as a single-ended inverting amplifier with a fixed gain dependent on the values of R_1 and R_2. Assuming an open loop gain approaching infinity for the amplifier then pin 2 may be considered to be a 'virtual earth' since the 'non-inverting' input pin 3 is at earth potential. Then summing the currents at pin 2,

$$\frac{V_{out}}{R_2}+\frac{V_{in}}{R_1}=0$$

$$\frac{V_{out}}{V_{in}}=-\frac{R_2}{R_1} \tag{10.3}$$

R_3 is used to minimise the offset voltage and thermal drift and its optimum value makes the source impedance at both inputs equal, i.e.

$$R_3 = \frac{R_1 R_2}{R_1 + R_2} \tag{10.4}$$

R_4 and C_1 are frequency compensating components used to ensure stability of the amplifier.

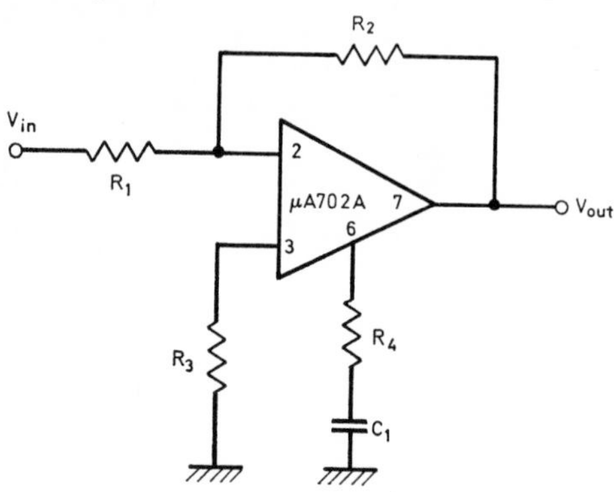

Fig. 10.18. Inverting amplifier

10.3.2. Differential Input Amplifier

With the circuit shown in Fig. 10.19 the μA702A may be used as a differential input, single-ended output amplifier.

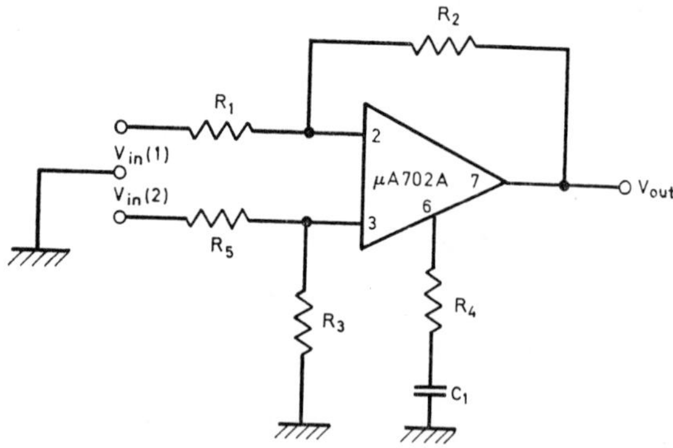

Fig. 10.19. Differential input amplifier

Assuming infinite open loop gain and infinite input impedance then pins 2 and 3 will be at the same common-mode input voltage V_{in}. Since no current flows into pins 2 or 3 then, when $R_1 = R_5$ and $R_2 = R_3$, we have

$$V_{out} = [V_{in}(2) - V_{in}(1)]\frac{R_2}{R_1} \tag{10.5}$$

10.3.3. Low Pass Amplifier

If R_1 and R_2 in Fig. 10.18 are replaced by complex impedances Z_1 and Z_2 respectively then we have

$$\frac{V_{out}}{V_{in}} = -\frac{Z_2}{Z_1}$$

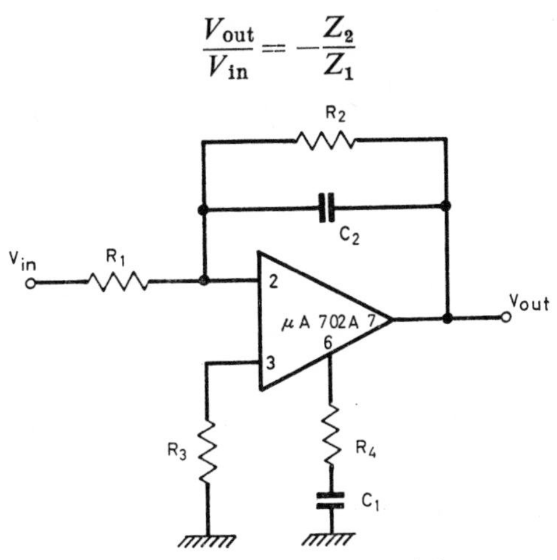

Fig. 10.20. Low-pass amplifier

If Z_1 is a resistor and Z_2 a resistor and capacitor in parallel as in Fig. 10.20 we have

$$Z_1 = R_1$$

$$Z_2 = \frac{R_2}{1 + jwC_2R_2}$$

Then

$$\frac{V_{out}}{V_{in}} = -\frac{R_2}{R_1} \cdot \frac{1}{1 + jwC_2R_2} \tag{10.6}$$

Thus the low frequency gain is $-R_2/R_1$ and the gain is reduced to $1/\sqrt{2}$ of this value at a frequency

$$\frac{1}{2\pi C_2R_2}$$

10.3.4. Sine Wave Oscillator

Fig. 10.21 shows the circuit diagram of a phase shift oscillator of the Wien-bridge type as discussed in Section 5.5.2. Negative feedback is applied to the inverting input of the amplifier through R_3 to stabilize the gain and make it essentially independent of the operational amplifier characteristics. The RC network (consisting of R_1, C_1, R_2 and C_2) applies positive feedback to the non-inverting input of the operational amplifier.

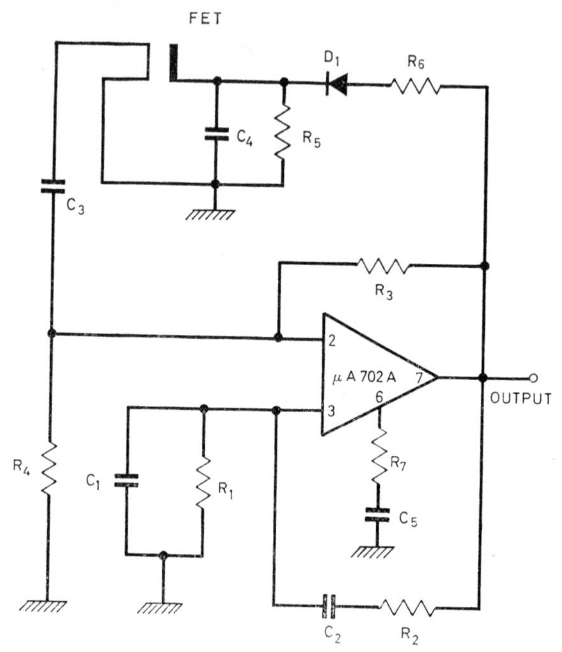

Fig. 10.21. Sine wave oscillator

If $R_1C_1 = R_2C_2$, then, as shown in section 5.5.2, the frequency of oscillation is given by

$$f_0 = \frac{1}{2\pi C_1 R_1} \tag{10.7}$$

The condition for oscillation is that the amplifier must provide a gain equal to or greater than

$$1 + \frac{2R_2}{R_1}$$

The output of the amplifier is rectified by the half wave rectifier circuit (D_1, R_5, C_4) and fed to the gate of the FET. The drain-to-source resistance of the FET is controlled by this voltage to hold the output of the amplifier at a constant level, thus giving automatic gain control.

10.5 DIGITAL INTEGRATED CIRCUITS

In the development of the first digital integrated circuits a discrete component philosophy was used, i.e. an attempt was made to minimise the number of components in the circuit, and in fact an exact duplication of the discrete component circuit was built.

10.5.1. Resistor-Transistor Logic (RTL)

The first circuits developed used the resistor–transistor logic discussed in Section 8.5, based on *NOR* gates such as shown in Fig. 10.22. In this circuit the output is *NOT* 1 if *A OR B OR C* is 1. Major advantages of this system are high speed and low power but the loading capacity is not high and the margin of operation is poor.

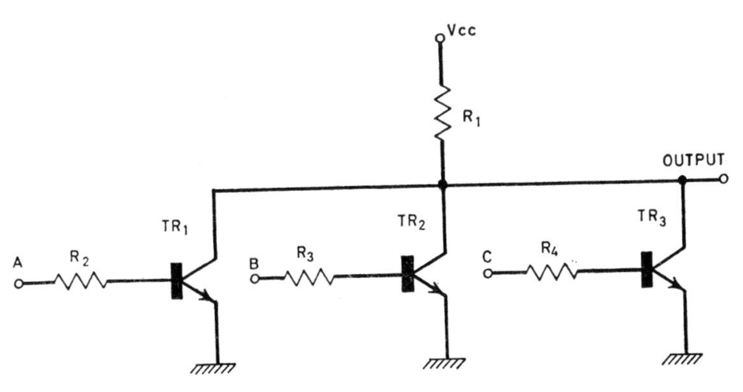

Fig. 10.22. RTL NOR *gate*

10.5.2. Diode-Transistor Logic (DTL)

The next circuits developed used *NAND* logic based pn diode-transistor logic circuits. These circuits such as shown in Fig. 10.23 when based on discrete component circuits are basically diode gates feeding a transistor inverter buffer and the output is *NOT* 1 if *A AND B AND C* are 1. Thus the transistor is turned *OFF* until all the inputs are high. In order to ensure sufficient drive to TR_1, resistor

R_1 must not be too large and R_2 must not be too small. The result is increased loading on previous stages and a necessity for a negative supply at the base of TR_1.

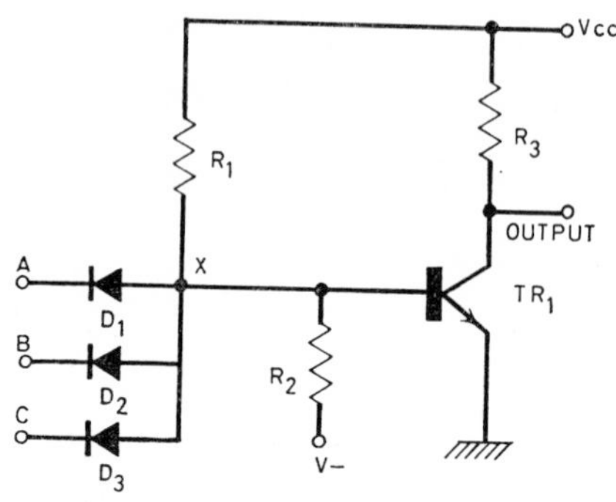

Fig. 10.23. DTL NAND *gate*

The use of integrated circuits allows important variations from the discrete component philosophy with marked improvement in

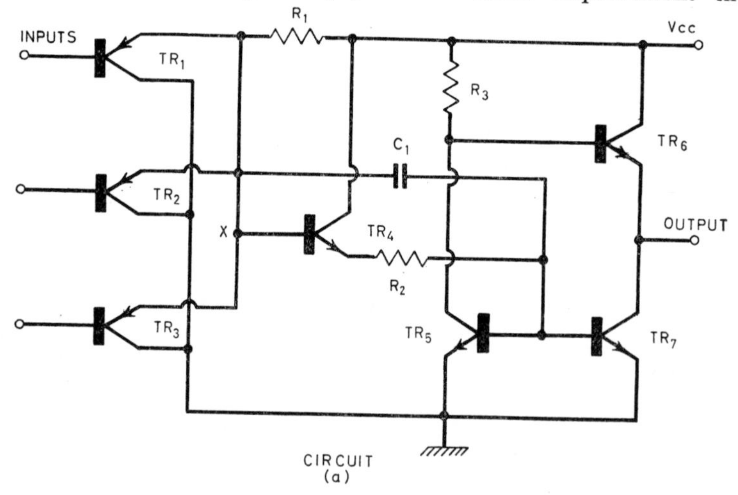

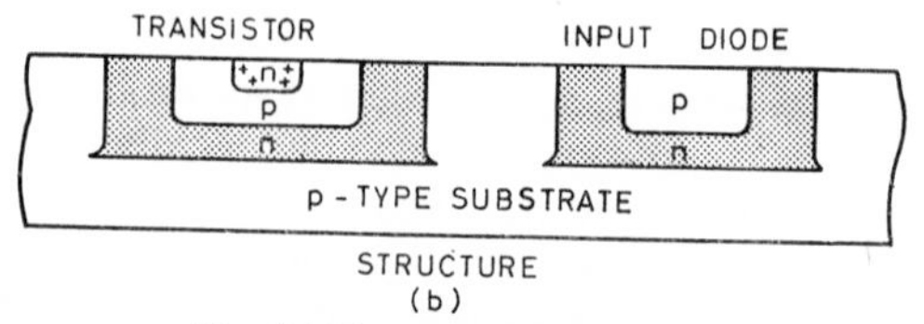

Fig. 10.24. Modified DTL NAND *gate*

performance. First, many more components are used than the minimum required to perform the logic and second, transistors are used wherever possible in place of diodes. Fig. 10.24 shows a modified diode-transistor logic gate. Point X is analagous to the node X in the conventional circuit of Fig. 10.23. Pnp transistors are used in place of the input diodes and an npn transistor is used in place of the series diode. The use of the input pnp transistors is an example of a technical advantage of monolithic integrated circuits which cannot be obtained in discrete component circuits and results from the way in which the circuit is built. Fig. 10.24(b) shows a cross-section of such a circuit in which all components are diffused into a substrate which has been doped with p-type impurity. This p-type substrate is connected to the lowest potential in the circuit (in this case ground) thus ensuring a reverse bias between the substrate and the first diffusion. When an input *diode* is diffused into the substrate, however, the p and n regions of the diode combine with the p-type substrate to form a pnp transistor with common collector. The advantage over the normal diode gate input is that the current which must now be carried by a preceding stage when the input is low is reduced approximately to a value $1/h_{FE}$ of the original input current.

A further advantage from this pnp action occurs in switching circuits when building npn transistors. When the npn transistor is diffused it is again built in a p-type substrate so actually a pnp structure also results. The advantage appears when the npn transistor is driven into saturation, i.e. the base-collector junction is forward biased as discussed in Section 6.3. The disadvantage of driving the transistor into saturation is that it takes longer to remove stored base charge during turn-off, thus increasing switching time. However due to the diffusion process the base-collector junction of the npn transistor is exactly the same as the base emitter junction of the pnp transistor. Therefore as the npn starts into saturation the pnp becomes operative and bypasses the current, which would normally overdrive the npn transistor, into the substrate. As the npn transistor begins to recover from saturation the pnp transistor again becomes inoperative. There is thus a built-in clamp as discussed in the discrete component circuit of Fig. 6.12.

10.5.3. Transistor-Transistor Logic (TTL)

The next integrated circuit system to evolve was transistor–transistor logic circuitry based on *NAND* gates and illustrated by the basic

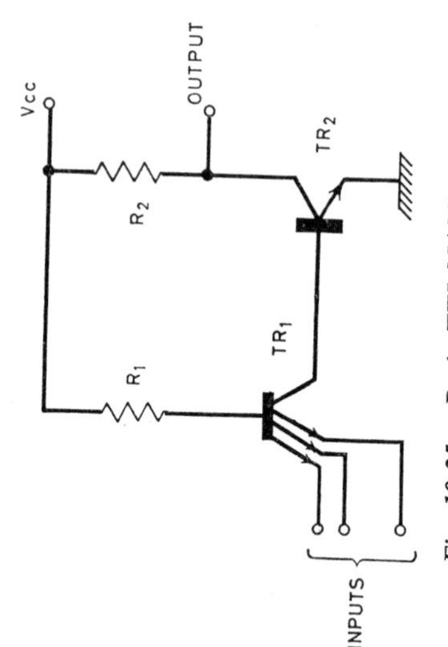

Fig. 10.25. *Basic TTL NAND gate*

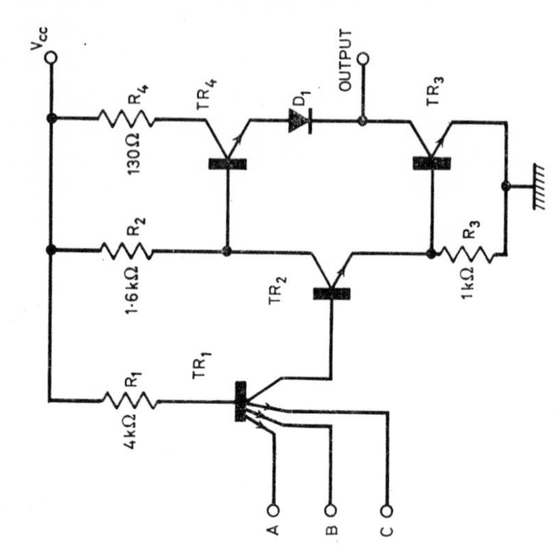

Fig. 10.26. *TTL 3-input NAND gate*

three-input *NAND* gate shown in Fig. 10.25. This system makes use of a component which was non-existent in discrete-component circuitry, i.e. the multiple-emitter transistor. A common collector area is first diffused, a common base is diffused into the collector then several emitters (up to eight is quite feasible) are diffused into the base. By this means more circuitry per unit area is possible, leading to lower cost and an additional advantage of faster speed, since there is lower capacitance associated with the smaller area.

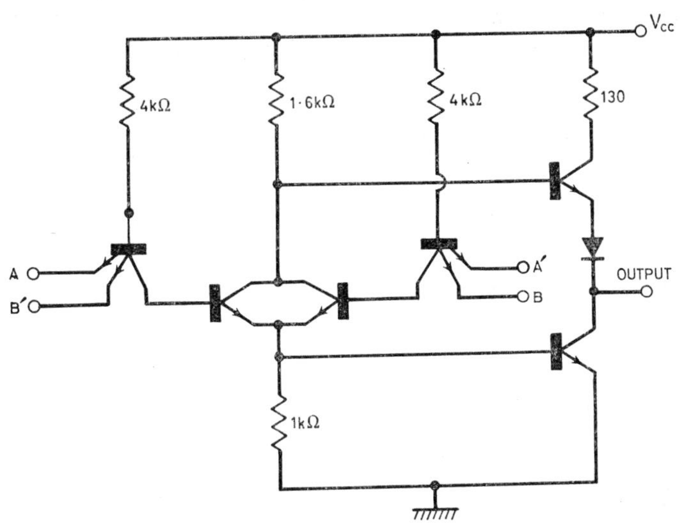

Fig. 10.27. EXCLUSIVE-NOR *gate*

Fig. 10.26 shows the complete schematic of a standard three-input *NAND* gate as used in the Texas SN 7410. TR_1 is the multiple emitter input transistor feeding through a phase splitter TR_2 to the output transistor TR_3. The output circuit arrangement of TR_3, TR_4 and D_1 provides a low output impedance in both the high and low conditions. Diode D_1 ensures that TR_4 is off when TR_3 is on.

The versatility of logic systems built by integrated circuit techniques is demonstrated in Fig. 10.27. By slight modifications to the windows in the diffusion process for producing a two input *NAND* gate an *EXCLUSIVE-NOR* circuit can be obtained, if complemented inputs are available. All that is required is a duplication of the input circuit up to the phase-splitter stage. In this circuit the output is *NOT* 1 if *A AND B′* are 1 *OR* if *A′ AND B* are 1.

10.5.4. Combined Gate Circuits

In the early days of integrated circuit evolution, the simple gates already discussed were used as the basic components and interconnected to form more complex logic systems. The next step was the interconnection of gates on the chips by extending the metallisation pattern, thus producing a self-contained functional circuit on a single silicon chip. Since a complete circuit needs only input, output and supply connection, a more complex circuit is economically possible in one package. Complete logic functions, including up to thirty-five interconnected gates, have become standard types, including such functions as registers, binary counters and adders.

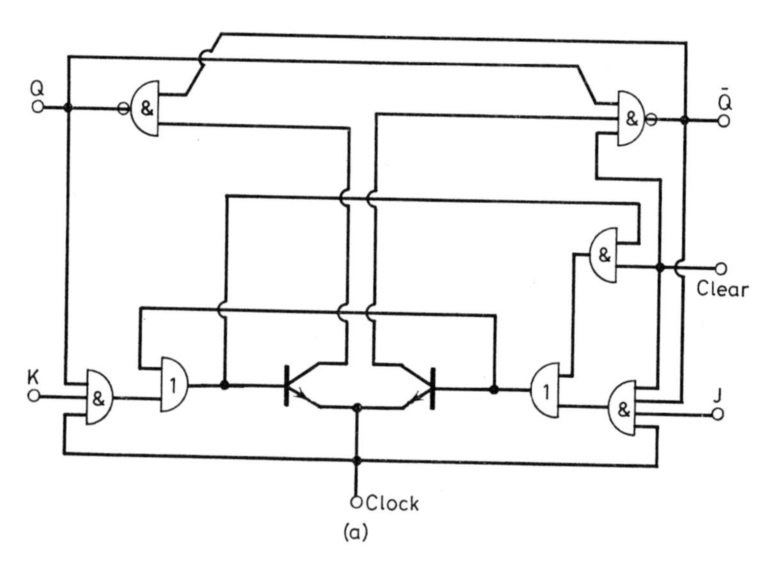

Fig. 10.28 (a). J-K master slave bistable

An example of a widely used bistable circuit with additional logic in the input circuits is the J–K master-slave circuit, shown in Fig. 10.28. In this circuit, inputs to the master section are controlled by the clock pulse. The clock pulse also regulates the state of the coupling transistors which connect the master and slave sections. The operating sequence is such that first the slave is isolated from the master, then the information on the J and K inputs entered into the master. The J and K inputs are then disabled and finally the

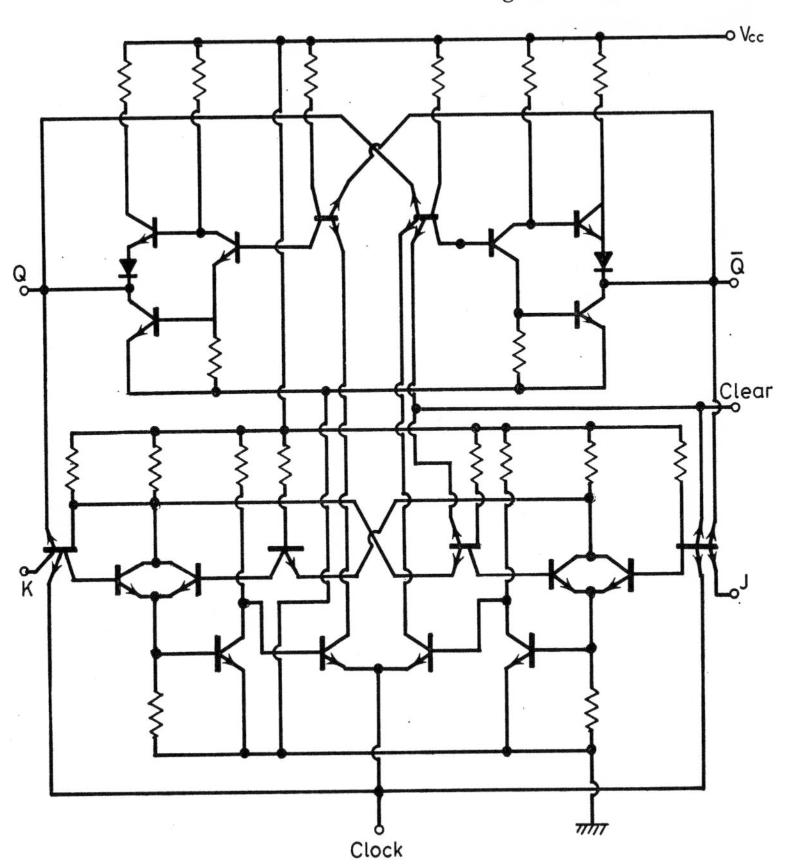

Fig. 10.28 (b). J-K master-slave bistable

information is transferred from the master to the slave. Such bistable circuits as these form the basic logic element of registers and counters.

BIBLIOGRAPHY

1. AMOS, S. W., *Principles of Transistor Circuits*, Second Edition, Iliffe Books Ltd. (1962).
2. *Analysis and Design of Integrated Circuits*, Engineering Staff Motorola Inc. McGraw-Hill Publishing Co. Ltd., New York (1967).
3. BLACK, H. S., *Modulation Theory*, Van Nostrand (D.) Co. Ltd. (1953).
4. CATTERMOLE, K. W., *Transistor Circuits*, Heywood & Co. Ltd. (1959).
5. DEWITT, D., and ROSSOFF, A. L., *Transistor Electronics*, McGraw-Hill Publishing Co. Ltd. (1957).
6. GARTNER, W., *Transistors, Principles, Design and Application*, Van Nostrand (D.) Co. Ltd. (1960).
7. HAYKIN, S. S., *Junction Transistor Circuit Analysis*, Iliffe Books Ltd. (1962).
8. HIBBERD, R. G., *Basic Course in Integrated Circuits*, Orbit Publishing Co. (1969).
9. HOLLINGDALE, S. H., *High Speed Computing; Methods and Applications*, English Universities Press Ltd. (1959).
10. MILLMAN, J., and TAUB, H., *Pulse and Digital Circuits*, McGraw-Hill Publishing Co. Ltd. (1956).
11. NEETESON, P. A., *Junction Transistors in Pulse Circuits*, Philips Technical Library (1959).
12. PETTIT, J. M., *Electronic Switching, Timing and Pulse Circuits*, McGraw-Hill Publishing Co. Ltd. (1959).
13. *Proc. Inst. Elec. Engs.*, 106, Pt. B, Supplements 15–18 (1959).
14. RICHARDS, R. K., *Arithmetic Operations in Digital Computers*, Van Nostrand (D.) Co. Ltd. (1955).
15. RICHARDS, R. K., *Djgital Computer Components and Circuits*, Van Nostrand (D.) Co. Ltd. (1957).
16. SCROGGIE, M. G., *Principles of Semiconductors*, Iliffe Books Ltd. (1961).
17. SHEA, R. E., *Principles of Transistor Circuits*, Wiley (John) & Sons Ltd. (1953).
18. *The Application of Linear Microcircuits*, Engineering Staff, S.G.S.—United Kingdom (1969).
19. TILLMAN, J. R., and ROBERTS, R. F., *Theory and Practice of Transistors*, Pitman (Sir Isaac) & Sons Ltd. (1961).
20. WALTER, D. J. *Integrated Circuit Systems*, Butterworth (1971).
21. WOLFENDALE, E., *The Junction Transistor and its Application*, Heywood & Co. Ltd. (1958).

APPENDIX

A.1. SUPERPOSITION THEOREM

In any network of sources and linear impedances the current flowing at any point is the sum of the currents which would flow if each source were considered separately, with all other sources replaced at the time by impedances equal to their internal impedances.

A.2. THEVENIN'S THEOREM

The current in any impedance Z connected to a pair of terminals $11'$ of a network of sources and linear impedances is the same as if it were connected to a single voltage source whose e.m.f. is the open circuit voltage measured at terminals $11'$ and whose internal impedance is the impedance measured at the open circuited terminals with all sources replaced by impedances equal to their internal impedances.

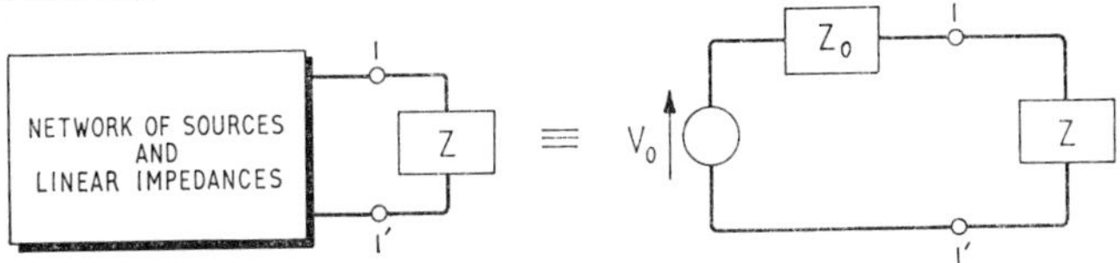

Fig. A.1. Thevenin's theorem in diagrammatic form

V_0 is the voltage measured at terminals $11'$ with Z removed, and Z_0 is the impedance measured at terminals $11'$ with Z removed and all sources replaced by their internal impedances.

A.3. NORTON'S THEOREM

The current in any impedance Z connected to a pair of terminals $11'$ of a network of sources and linear impedances is the same as if Z were connected to a single current generator whose short circuit current is the current flowing between terminals $11'$ when they are short circuited and whose internal impedance is the impedance measured at terminals $11'$ with Z removed and all sources replaced by impedances equal to their internal impedances.

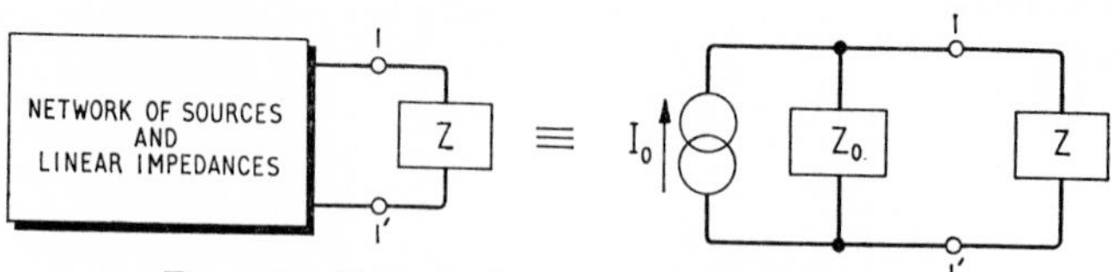

Fig. A.2. Norton's theorem in diagrammatic form

I_0 is the current flowing between terminals 11′ when they are short circuited and Z_0 is the impedance measured at terminals 11′ with Z removed and all sources replaced by their internal impedances.

INDEX